NEAR-FIELD NANO-OPTICS

From Basic Principles to
Nano-Fabrication and
Nano-Photonics

LASERS, PHOTONICS, AND ELECTRO-OPTICS

Series Editor: H. Kogelnik

FUNDAMENTALS OF LASER OPTICS
Kenichi Iga

NEAR-FIELD NANO-OPTICS: From Basic Principles to Nano-Fabrication
and Nano-Photonics
Motoichi Ohtsu, and Hirokazu Hori

OPTICAL–THERMAL RESPONSE OF LASER-IRRADIATED TISSUE
Edited by Ashley J. Welch and Martin J. C. van Gemert

THEORY AND APPLICATION OF LASER CHEMICAL VAPOR
DEPOSITION
Jyoti Mazumder and Aravinda Kar

NEAR-FIELD NANO-OPTICS

From Basic Principles to Nano-Fabrication and Nano-Photonics

MOTOICHI OHTSU
Tokyo Institute of Technology
Yokohama, Japan

and

HIROKAZU HORI
Yamanashi University
Kofu, Japan

SPRINGER SCIENCE+BUSINESS MEDIA, LLC

Library of Congress Cataloging-in-Publication Data

Ohtsu, Motoichi.
    Near-field nano-optics : from basic principles to nani-fabrication
and nano-photonics / Motoichi Ohtsu and Hirokazu Hori.
        p.   cm. -- (Lasers, photonics, and electro-optics)
    Includes bibliographical references and index.
    ISBN 978-1-4613-7192-2   ISBN 978-1-4615-4835-5 (eBook)
    DOI 10.1007/978-1-4615-4835-5
    1. Nanostructure materials.   2. Near-field microscopy.   3. Quantum
optics.   4. Photonics.   I. Hori, Hirokazu.   II. Title.
III. Series.
TA418.9.N35038   1999
621.36--dc21
                                                        99-14419
                                                           CIP

ISBN 978-1-4613-7192-2

© 1999 Springer Science+Business Media New York
Originally published by Kluwer Academic / Plenum Publishers 1999
Softcover reprint of the hardcover 1st edition 19999

10 9 8 7 6 5 4 3 2

A C.I.P. record for this book is available from the Library of Congress

PREFACE

Conventional optical science and technology have been restricted by the diffraction limit from reducing the sizes of optical and photonic devices to nanometric dimensions. Thus, the size of optical integrated circuits has been incompatible with that of their counterpart, integrated electronic circuits, which have much smaller dimensions. This book provides potential ideas and methods to overcome this difficulty.

Near-field optics has developed very rapidly from around the middle 1980s after preliminary trials in the microwave frequency region, as proposed as early as 1928. At the early stages of this development, most technical efforts were devoted to realizing super-high-resolution optical microscopy beyond the diffraction limit. However, the possibility of exploiting the optical near-field, phenomenon of quasistatic electromagnetic interaction at subwavelength distances between nanometric particles has opened new ways to nanometric optical science and technology, and many applications to nanometric fabrication and manipulation have been proposed and implemented.

Building on this historical background, this book describes recent progress in near-field optical science and technology, mainly using research of the author's groups. The title of this book, *Near-Field Nano-Optics—From Basic Principles to Nano-Fabrication and Nano-Photonics,* implies capabilities of the optical near-field not only for imaging/microscopy, but also for fabrication/manipulation/processing on a nanometric scale.

Although a variety of acronyms have been used for the near-field optical microscope/microscopy, e.g., NSOM, SNOM, and PSTM, this book employs the simplest and shortest one, NOM. Chapters 1, 2, 8, and 9 provide detailed theoretical intuitive models based on the concept of interaction-type measurements of short-range electromagnetic correlations by using a spatial/temporal-frequency resonant probe tip. Chapters 3–7 review experimental work.

v

We gratefully thank Drs. T. Saiki, R. Uma Mahewari, S. Mononobe, H. Ito, K. Kurihara, M. Ashino, M. Naya, H. Fukuda, R. Micheletto, J. D. White (Kanagawa Academy of Science and Technology), M. Kourogi (Tokyo Institute of Technology), Drs. I. Banno and T. Inoue (Yamanashi University), and Prof. W. Jhe (Seoul National University) for their collaborations. We also extend a special acknowledgment to Drs. R. Uma Maheswari, J. D. White, Y. Takiguchi (Japan Science and Technology Corporation), S. Lathi, M. Lawrence, and S. Kasapi (Stanford University) for their critical readings and comments on the manuscript. H.H. would like to express his gratitude to Profs. T. Yabuzaki and M. Kitano (Kyoto University) and Prof. K. Kitahara (Tokyo Institute of Technology) for their helpful discussions, and also to Profs. K. Shimoda, T. Sakurai and A. Shimizu for their encouragement. H. H. also thanks T. Matsudo, K. Tsuchiya, and Y. Ohdaira (Yamanashi University) for their assistance and Dr. K. Kobayashi (IBM Japan) for useful comments on the manuscript. Further, we wish to express our special thanks to Dr. L. Nagahara (Phoenix Corporate Research Laboratories, Motorola, Inc.) for his patient reading of the whole manuscript and valuable comments. Finally, we wish to express our gratitude to L. S. Marchand, A. McNamara, and the production staff of Kluwer Academic / Plenum Publishing Corporation for their guidance and suggestions throughout the course of preparation of this book.

Motoichi Ohtsu
(Yokohama, Kanagawa)

Hirokazu Hori
(Kofu, Yamanashi)

CONTENTS

Chapter 1. Introduction

Chapter 2. Principles of Near-Field Optical Microscopy

Chapter 3. Instrumentation

Chapter 4.   Fabrication of Probes

Chapter 5.   Imaging Experiments

Chapter 6.   Diagnostics and Spectroscopy of Photonic Devices and Materials

Chapter 7. Fabrication and Manipulation

Chapter 8. Optical Near-Field Theory

Chapter 9. Theoretical Description of Near-Field Optical Microscope

ONE

INTRODUCTION

1.1. NEAR-FIELD OPTICS AND PHOTONICS

Near-field optics and photonics have been developed extensively in this decade mainly in relation to scanning optical near-field microscopy and related techniques. The underlying physics and potential for applications are, however, spreading into many areas dealing with the interaction of electromagnetic fields with matter. This book is intended to present the basic ideas involved in the rapidly expanding field of near-field optics and photonics to graduate students and researchers, using both experimental and theoretical materials based on original work of the authors. As a general introduction to near-field optics and photonics, we start with a consideration of the physics involved in general optical processes.

The meaning and importance of microscopic electromagnetic interactions depend on the scale which characterizes the apparatus used for their measurement. Although measurements are inevitably concerned with macroscopic physical quantities in their initial and final states, we can deduce from the results that some microscopic events of a certain scale take place during the measurement process if we carefully prepare the experimental apparatus with the help of theoretical considerations.

1.1.1. Optical Processes and Electromagnetic Interactions

An optical system in general consists of a light source, a photodetector, and a light scattering object in between whose optical properties are being investigated (Fig. 1.1a). It is by means of light waves appearing in the initial and final states in the optical system that we communicate with the object of interest. The incident light wave transfers optical energy to the scattering system, and the scattered light wave provides information relevant to the optical properties of the scatterer. For this reason enormous effort has gone into the study of light waves, establishing the discipline of optics. However, we are interested not only in the manner of communication, but also in the physics involved in the optical process, that is, the relevant electromagnetic interactions of material objects.

The light scattering object usually consists of a tremendous number of atoms or molecules on the microscopic level, and each individual atom or molecule

1

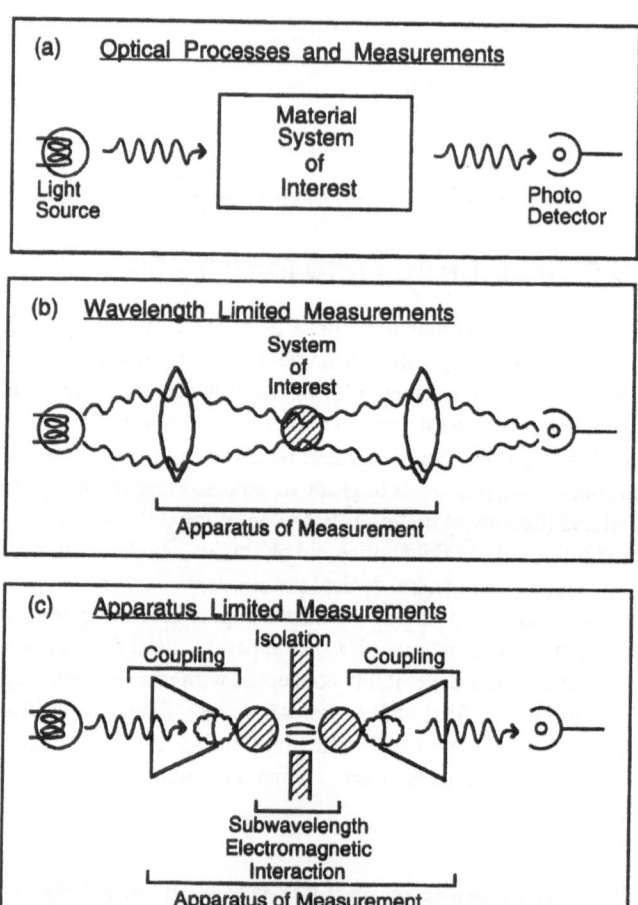

Figure 1.1. Optical processes and measurements. (a) General light source-to-photodetector process. (b) Wavelength-limited measurement using a light wave. (c) Apparatus-limited measurement using a local electromagnetic interaction and an isolation scheme for screening the direct coupling of the source to the photodetector.

participates in the electromagnetic interaction that yields the optical properties of the object. However, not all of the microscopic interaction processes are relevant to the optical properties measured via outgoing light waves. Depending on the way the measurement is made and on the nature of the optical process, one can identify the electromagnetic interactions that are relevant or irrelevant to the detected signal. Irrelevant microscopic interactions are then averaged or traced out over a certain scale of physical parameters, such as spatial coordinates, observation time, energy and momentum spectrum, polarization, and so on.

In an ordinary optical process, the criterion for relevance is set by the properties of the incident and scattered light waves, that is, the characteristics of the light source and the photodetector including the optical elements in between (Fig. 1.1b). For instance, when one measures the refractive index of a dielectric object by means of a direct scattering measurement of a monochromatic light wave, the relevant electromagnetic interaction is the optical response of the microscopic polarization averaged over a volume corresponding to the half optical wavelength cubed. These are *wavelength-limited* measurements of electromagnetic interactions.

The criterion for relevance is completely different when one fabricates a subwavelength structure in a measurement apparatus which enables a microscopic electromagnetic interaction to couple with the incident and outgoing light waves. In this case even an electromagnetic interaction taking place on the subwavelength scale is relevant to the optical properties measured (Fig. 1.1c). The interpretation of the results may require some signal processing and theoretical considerations as well. These are *apparatus-limited* measurements of the electromagnetic interaction. Near-field optics and photonics are classified as this type of optical process.

Here we note that the important idea underlying optical near-field problems is that not only does the light scattering or interaction process at the subwavelength scale matter, but so does the measurement of the result of such a process in a subwavelength vicinity of the event. That is, in near-field problems both the *microscopic scattering* process and its *near-field detection* are important. In order to realize such a near-field measurement of subwavelength interactions, one has to implement the following:

1. A probe tip sensitive to the local field of a subwavelength-size object
2. A screening of longer range interactions, or isolation of the direct coupling of the source field to the photodetector
3. A coupling of the near-field event to a light source and photodetector placed in the far-field region

With respect to these conditions for near-field measurements, one has to deal with at least the following experimental issues:

1. What kind of short-range electromagnetic correlation is utilized for object–probe interaction?
2. How are long-range interaction channels screened?
3. How is the near-field interaction between object and probe tip connected to the light source and photodetector?

These considerations suggest that a near-field measurement should function like a spatial frequency filter to extract microscopic electromagnetic processes from the entire optical process.

In order to make these issues clear, in the following sections we survey the development of scanning near-field optical microscopy and related techniques.

1.2. ULTRA-HIGH-RESOLUTION NEAR-FIELD OPTICAL MICROSCOPY (NOM)

1.2.1. From Interference- to Interaction-Type Optical Microscopy

Optical microscopes have long been popular and powerful tools in a wide area of scientific research, including biology, physics, chemistry, material science, and so on. The principle underlying imaging is the *interference of light waves* coming from a light source, passing through different portions of a macroscopic specimen, and being collimated by a set of well-designed optics to make an interference image showing an enlarged shape and structure to the observer's eyes [1]. The basic principle of interference-type microscopes (or wave-type microscopes) is no different from that of astronomical telescopes.

It is also possible to make interference-type microscopes using not only light waves, but also material waves, such as electron waves [2] or neutron waves [3]. Electron-wave microscopes in particular are very important in modern scientific research and engineering. Furthermore, recent developments in ultracold atom techniques are likely to lead to atomic de Broglie-wave microscopes [4].

One might ask why we should need to proceed from optical to material waves to increase the resolution of microscopy techniques. The answer is, of course, the diffraction limit imposed on any type of interference experiment or wavelength-limited process. Optics textbooks tell us that one cannot identify the interference pattern of light waves coming through a double slit with quarter optical wavelength spacing between the slits [1]. This is in general true for the interference of any type of wave. The smaller the specimen we want to diagnose, the shorter is the wavelength required to produce an interference image.

Here the following questions arise: "Is the interference of waves the only way to construct a microscope?" "Does the diffraction of waves provide the fundamental limit on the diagnosis of microscopic objects?" The answers came when physicists in Zurich obtained an atomic-scale image of a metallic silicon surface by means of *interaction-type* microscopy, named scanning tunneling microscopy (STM) [5–11]. The electronic interaction of atoms on the topmost layer of a metallic surface with an atomic-sized probe tip has an extremely short-range nature. If one can control the position of the probe tip with the precision of the atomic dimension, one can map the strength of the electronic interaction between the specimen and the probe tip to obtain an image of the surface with resolution corresponding to single-atom size. The name *tunneling microscope* comes from the fact that the local electron interaction is understood in terms of electron tunneling across the specimen–probe gap or corresponding vacuum potential barrier [12]. Several types of interactions of short-range nature are available for such scanning-probe interaction-type microscopy, for instance, atomic force microscopy (AFM) [10].

Inspired by this discovery, it was natural to proceed from electrons to light or photons in order to construct an interaction-type microscope [13]. The result is a *scanning near-field optical microscope*, in which the strength of local electromagnetic interaction between a specimen and a scanned optical probe tip is mapped to produce a nanometer-scale image which lies far beyond the diffraction limit [14–19]. In analogy to the electron case, the local electromagnetic interaction is described by photon tunneling, meaning that a photon forming a coupled mode with material excitation tunnels through a vacuum gap between the specimen and probe tip [20–22]. Due to this, the term photon scanning tunneling microscopy is also used for several experimental setups of scanning near-field optical microscopy. The term *optical near-field* is the key to understanding the underlying physics of this novel field of optical science, which covers not only the propagating light waves, as in the wavelength-limited optical process, but also the nonpropagating local electromagnetic interaction or optical near-field in a very narrow vicinity of the illuminated material. The latter belongs to the apparatus-limited optical processes.

For interaction-type microscopy, in general, signal processing techniques and appropriate theoretical background are indispensable for the interpretation of the obtained images. This is because one has to derive topographic information about the specimen via near-field interaction with the scanned probe tip, which involves a *destructive measurement* of the optical near-field associated with the illuminated specimen. The topographic features do not necessarily correspond directly to the strength of the sample–probe interaction. We are reminded here of the case of atomic resolution electron STM, in which the interpretation of the image requires an elaborate computation of the band structures of the substrate and local interaction with the probe tip regardless of how clear the obtained image shows the structure on the atomic scale [23]. Also, in the case of scanning optical near-field microscopes, one has to know a great deal about the optical characteristics of the specimen and the probe tip as well as the means of light illumination, detection, and signal processing for the appropriate interpretation of the obtained images. Fortunately, there exists a *near-field criterion* in the optical case which one can use to take advantage of the long-wavelength limit and make things much simpler. That is, the near-field optical system is characterized by a dimensionless small parameter $Ka = 2\pi a/\lambda$, where a is the size of the object of interest and λ is the optical wavelength. If the system of observation satisfies a certain near-field criterion, we can avoid confusion coming from several long-range effects, such as the interference of optical waves. This point is discussed in detail later.

A large portion of this book is devoted to recent developments in near-field optical microscopes and related optical near-field techniques as well as studies of their physical background; this provides a foundation for the interpretation of images as well as for extensive development of near-field optics and photonics and related quantum electronics and their applications in the near future.

Before proceeding with detailed discussions on general problems in near-field optics, we briefly review the history of near-field optical microscopy and related techniques.

1.2.2. Development of Near-Field Optical Microscopy and Related Techniques

1.2.2.1. First Proposals of Ultra-High-Resolution Microscopy

Recalling the golden age of modern physics in the early 20th century, it is not surprising that the first proposal of optical near-field ultra-high-resolution microscopy appeared in a paper by Synge in 1928 [24]. He proposed the use of a scanned microscopic aperture to construct an ultra-high-resolution microscope with resolution far beyond the diffraction limit, and extensively discussed technical requirements and difficulties. The proposal was of course far beyond the technical capability of those days, and experimental demonstrations in the optical region needed to wait for modern technology to be able to fabricate a subwavelength-size probe tip, control it with nanometer-range reproducibility, and image pictures by means of computer-aided signal processing. It is interesting that the study of a microaperture microscope was reported by O'Keefe in 1956 [25] at the same time great developments were being made in quantum electrodynamics. The near-field microscope was studied by Ash and Nicholls in 1972 [26].

Near-field optical microscopy is supported by modern nanofabrication technology as well as the precision control of scanning probe tips and techniques of image processing transferred from STM and AFM techniques developed in the 1980s.

1.2.2.2. Demonstration of Near-Field Microscopy in the Microwave Region

An important step in realizing the near-field microscope was the study of a microwave version [27]. Taking advantage of longer wavelengths, microwave experiments allow us to proceed far beyond the diffraction limit with even ordinary-size waveguides and field detectors as a subwavelength probe. The results of microwave experiments are a clear demonstration of the resolution limit being dependent not on the wavelength of illumination, but on the size of the probe tip. Several interesting microwave experiments investigated the scattering process of surface waves [28, 29] and provided information on how to proceed with optical experiments on a nanometric scale.

1.2.2.3. Demonstration of Near-Field Microscopy in the Optical Region

Experiments in the optical region were started in the 1980s by several groups [13]. Since then many types of probes and systems have been developed, such as small apertures [17], squeezed glass pipettes [18], and sharpened optical fibers [30]. A sharpened single-mode optical fiber prepared by chemical etching provided the first nanometer-resolution picture of a biological specimen [31]. An important recent advance was the demonstration of operation in water [32], which is essential for an *in vivo* biological measurement. A wide variety of near-field optical probes with nanometer-size probe tips is now available for both dielectric and metal-cladded aperture types as well as combined types [33–35].

1.2.2.4. Different Types of Apparatus and Operations in Near-Field Microscopy

In the course of experimental development, several techniques have been introduced and named depending on such aspects as the nature of the probes, means of illuminating the specimen, and signal detection: for instance, scanning near-field optical microscopy (SNOM), near-field scanning optical microscopy (NSOM), photon scanning tunneling microscopy (PSTM), and scanning tunneling optical microscopy (STOM) in the narrow sense. For their classification, see the proceedings of the international conference on near-field optics and related techniques held at Besancon in 1992 [13].

Although it might be convenient to use different names for near-field optical microscopes and their relatives for the purpose of technical classification, we intend to use the general name *near-field optical microscopy* (NOM) in referring to all techniques of optical near-field microscopy throughout this book. This is due to our objective to describe general and unified ideas underlying optical near-field problems using a general framework of interaction-type scanning-probe microscopy including electron STM, AFM, and so on. Therefore, one should consider the acronym NOM used in this book as referring not to a specific microscopy technique, but to near-field optical microscopy in general. It could also be called SNOM, which corresponds to the general usage *scanning probe microscopy* (SPM) with *near-field optical* (NO) probe [10]. However we use the term NOM, since SNOM is already used in relation to a specific kind of near-field optical microscopy.

1.2.2.5. Mesoscopic Nature of NOM Processes

The spatial scale relevant to NOM techniques lies between the atomic scale (0.1 nm) and the scale of optical wavelengths (100 nm). As in general for such mesoscopic phenomena, we can provide various theoretical descriptions and interpretations of the underlying physics. In other words, we have a good theoretical basis for the microscopic limit of atomic interaction with the electromagnetic field and a basis for the macroscopic limit of a material response in the sense of an averaged field, but the approach to mesoscopic phenomena has not been uniquely mapped. In fact, NOM has been studied by various theoretical means, such as via electromagnetic boundary value problems [36–39], self-consistent approaches for multiple scattering problems [40–44], spatial Fourier analysis [45, 46], quantum field theory [47, 48], and so on. Microscopic theory covers atomic scale phenomena in terms of microscopic Maxwell equations coupled to the Schrödinger equation [49, 50].

As is generally true for mesoscopic problems which involve very complicated interactions of a many-body nature, the theoretical description depends on the selection of a model extracted from appropriate experimental results. What is important is not calculating a complicated field distribution or the exact source distribution on the material surface, but *extracting some physical meaning of the local event* taking place in a system consisting of a subwavelength-size specimen and the probe tip. In other words, the fundamental requirement for the description of an interacting many-body system is an empirical model based on experimental results to help us to understand the underlying physics.

1.2.2.6. Extension of NOM Techniques

As a concomitant of its complex mesoscopic nature, the NOM and related optical near-field techniques involve a wide variety of interesting physics and novel fields of application. For instance, surface plasmon near-field microscopy has been developed as an NOM technique which involves a resonance interaction with a metallic surface [51–53]. Since the resonance character strongly modifies the near-field interaction between the observed surface and the probe tip, one can utilize it to obtain specific information about the sample according to the probe tip. This type of probe is generally referred to as a "supertip" and is of interest in studies of probe fabrication. Dye molecule-doped probe tips and semiconductor microtips are examples of supertips under intensive development [54]. With supertips we can make intensive use of spectroscopic measurements using NOM with illumination from tunable laser light. The microscopic limit of single-exciton STM is also under study, in which a molecular structure is used as a probe tip sensitive to single-atom-level excitation in matter [55]. It should be noted, however, that a resonance-type NOM rejects all the signals coming from interactions lying out of resonance.

It might also be very interesting to extend the ideas of tunneling of elementary excitations or quasiparticles representing coupled modes of the electromagnetic field with matter to nonresonant NOM experiments [48]. This way of describing the NOM provides a generalized view of all types of scanning probe microscopes including electron STM, AFM, and so on.

It is also of great interest to diagnose different characteristics of specimens as well as, for instance, the relation between the shape of a sample surface and buried structures as well as its internal state of excitation. There have been reports of measurement of buried structures of a biological sample [56] and observation of an internal excitation in a semiconductor device [57]. Near-field optical studies of mesoscopic electronic devices such as quantum wires or dots and quantum wells are also under way [58, 59].

1.2.2.7. Combined SPM Techniques

It is also of interest to combine the operation of NOM with several different types of scanning probe microscopy techniques, such as STM, AFM, and so on. A combined operation of NOM with electron STM to control the sample–probe distance has been reported in the early stage of an NOM experiment [18]. One popular technique uses shear-force-controlled positioning of the NOM probe tip, utilizing the atomic force between the sample surface and a vibrating thin probe [60, 61]. The amplitude modulation of the intrinsic vibration of a thin NOM probe due to the atomic force exerted by the substrate surface provides a measure of the sample–probe distance. While combined-type SPMs might be very attractive, one must pay careful attention to interference between fundamental processes. One must be aware that SPMs and, in general, interaction-type microscopes employ a process of destructive measurement. When the types of interaction combined are very different in their nature and in their scales of relevant spatial size, time constant, material character, and so on, the combined apparatus can provide a great deal of useful information indeed. However, if the processes compete with each other, the results may be very different from that expected. In fact, in some situations, a seemingly different measurement of a specimen could result in a similar image due to interference. This difficulty is often referred to as the *artifact problem*. Sometimes the interference provides unexpectedly good results, such as signal enhancement. For example, in the case of atomic-resolution electron STM an exchange force associated with electron tunneling provides an enhancement of the tunneling current and therefore contrast of the STM image [23]. This point has not yet been extensively studied in the case of NOM-related techniques. However, several researchers have reported that there exists an interference between atomic-force images and NOM images when the shear-force technique is employed. Near-field optical measurement of atomic forces has been reported [62].

1.2.2.8.   From Near-Field Microscopes to Manipulators

Developments in NOM and related techniques have not only been in terms of diagnosing matter by operating as a microscope, but also extend to various areas involving the control and manipulation of nanometric objects. Control of a localized photochemical reaction and its detection [63–65] represents an engineering development of ultra-high-density optical data recording [66–69]. The mechanical effect of an optical near-field on small particles is also under study [70, 71]. Control of a biological sample on the molecular level would reveal fundamental processes of excitation transfer and result in the control of living organisms [72]. It also has been proposed to manipulate atomic particles and control their state with nanometric resolution [73–81]. These novel applications are extending the research field of near-field optics, photonics, and related quantum electronics.

1.3.   GENERAL FEATURES OF OPTICAL NEAR-FIELD PROBLEMS

1.3.1.   Optical Processes and the Scale of Interest

Understanding the physical background of optical near-field problems is the key to developing near-field optical microscopy and related techniques. Keeping this in mind, let us study optical near-field phenomena in the framework of general optical processes.

In order to make things clear and simple, let us first consider a light scattering process between two atomic objects Ⓐ and Ⓑ. The point to be noted is that when we consider the electromagnetic interaction of two material systems Ⓐ and Ⓑ that are isolated as an electronic system, several different aspects and theoretical descriptions arise according to the specific spatial extent occupied by the electromagnetic field characterizing the nature of the interaction.

An optical process in general measures the response of a sample system placed between a light source and photodetector as described in Section 1.1:

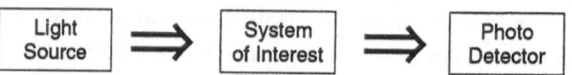

Light source to photodetector system as a whole optical process:

| Light Source | ⟹ | System of Interest | ⟹ | Photo Detector |

There are three typical cases which characterize the optical process for the interacting system via the electromagnetic field according to the spatial distance between Ⓐ and Ⓑ. We refer to these cases as *far*, *near*, and *close* with respect to the electromagnetic interaction of the optical region.

1.3.1.1. Optically "Far" System

In the first case, Ⓐ and Ⓑ are placed a distance far from each other compared to the optical wavelength of the incident light. An interaction taking place in this material system is unidirectional; the incident light wave is scattered first by Ⓐ and then propagates to Ⓑ, and the field rescattered by Ⓑ extends to the photodetector. In this case the electromagnetic field mediating the interaction between Ⓐ and Ⓑ is a *propagating light wave*, or photons in the quantum view. The light waves or photons propagating between Ⓐ and Ⓑ satisfy the usual dispersion relation for light waves or photons, so that the system is separable into two successive light scattering processes. This is the electromagnetic interaction of Ⓐ and Ⓑ in the *far-field* regime.

Far-Field
Material systems Ⓐ and Ⓑ placed far apart compared with the optical wavelength
Picture: Ⓐ to Ⓑ unidirectional action via propagating light wave

Separable system even under electromagnetic interaction

1.3.1.2. Optically "Near" System

The second case is related to optical near-field problems. Ⓐ and Ⓑ are placed apart at a distance shorter than the optical wavelength. However, it is still far with respect to an intrinsic interaction of Ⓐ and Ⓑ. The objects Ⓐ and Ⓑ are also well separated with respect to their electronic state at this distance. When the system is illuminated, however, the electromagnetic interaction between Ⓐ and Ⓑ does not follow a single path. The scattered field reflected back from Ⓑ to Ⓐ has a considerable effect on the process depending on the distance between them. The shorter the distance, the more significant is the near-field effect. Therefore optical properties such as the polarizability of Ⓐ are under the influence of the material system Ⓑ. The electromagnetic interaction is in this case quasistatic and has a mesoscopic nature. The system of Ⓐ plus Ⓑ is therefore inseparable when it is irradiated by an optical field. If one makes a cut between Ⓐ and Ⓑ one sees an electromagnetic field with a curious nature different from a simple propagating light wave or photon. If one expands the electromagnetic field between Ⓐ and Ⓑ in terms of waves, one finds so-called *evanescent waves* with a complex wavenumber, i.e., lying out of the dispersion relation of usual light waves or photons in vacuum. We consider such an electromagnetic interaction process the *near-field* regime.

Near-Field
Material systems Ⓐ and Ⓑ placed closer than the optical wavelength
Picture: Interaction of Ⓐ and Ⓑ via multiple scattering of light waves

Separable under no incident light field; inseparable when the system is illuminated

1.3.1.3. Optically "Close" System

The third case is an electromagnetically tightly coupled system Ⓐ and Ⓑ due to internal interactions existing even when no incident field is applied. The electromagnetic interactions in this case are due to intrinsic fluctuations of vacuum and material polarizations, which give rise to the van der Waals interaction, for instance. In this case it is a good description to consider the system Ⓐ plus Ⓑ as a kind of molecular state when one considers the optical response of the coupled system. If one makes a cut between Ⓐ and Ⓑ and tries to describe the interaction, one needs to resort to quantum electrodynamics and finds virtual photons being exchanged back and forth between Ⓐ and Ⓑ lying out of the dispersion relation of photons in free space.

Internal Close-Field
Material systems Ⓐ and Ⓑ placed extremely close
Picture: Coupled state of Ⓐ and Ⓑ or light scattering by molecular-like state similar to van der Waals molecules

Intrinsically inseparable

Next, let us consider the optical processes involved in a more complicated material system.

1.3.2. Effective Fields and Interacting Subsystems

In an optical process such as a light source-to-photodetector system, one can extract an interacting subsystem which is characterized by its spatial dimensions. Such a subsystem can be described as follows:

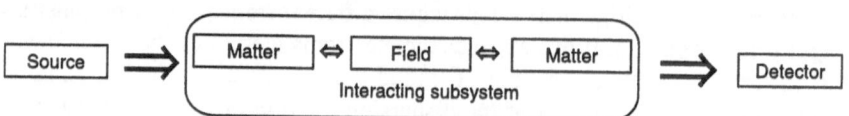

where we have an internal electromagnetic field bounded by matter. The internal field has no direct connection to external fields, which extend to the light source or photodetector. According to the temporal and spatial characteristics of the interaction, the internal field shows a peculiarity which cannot appear in the case of optical waves propagating in free space.

A parameter which characterizes the optical properties of the subsystem is given by $\omega l/c$, where ω is the temporal frequency of the optical field, l is the spatial size of the subsystem, and c is the velocity of light in vacuum. This parameter is equal to the wavenumber $k (= \omega/c)$ times the spatial size l, i.e., kl. According to the magnitude of the characteristic parameter $kl = \omega l/c$, we have three typical cases: macroscopic, mesoscopic, and microscopic.

First, let us consider an interacting subsystem which occupies a spatial volume characterized by a size parameter l larger than the optical wavelength, $\omega l/c > 1$. In this subsystem, one finds propagating waves as internal fields mediating the interaction between the material objects in the subsystem. This corresponds to the *macroscopic* case in which spatial dimensions l of the subsystem are much larger than the optical wavelength $\lambda = 2\pi c/\omega$. In such a macroscopic system the optical field or photonic system can be clearly distinguished from the material system.

On the other hand, when we consider the *microscopic* case, $\omega l/c < 0.01$, we find a material system tightly coupled via electromagnetic interaction, for which it is hard to find an independent meaning of the internal field. That is, the internal field in this case acts as a binding potential or force between the material objects. In other words, the material properties dominate in the microscopic system.

In the intermediate case considered above, $0.01 < \omega l/c < 1$, we encounter an internal field of *mesoscopic* nature, which is related to optical near-field effects. In this case, we can give an independent meaning for the internal field to a good approximation.

1.3.2.1. Effective Field and Scale of Interest

In some cases, the internal field itself is characterized as an effective field which represents an averaged effect of microscopic interaction processes over a volume $V \approx l^3$ as follows:

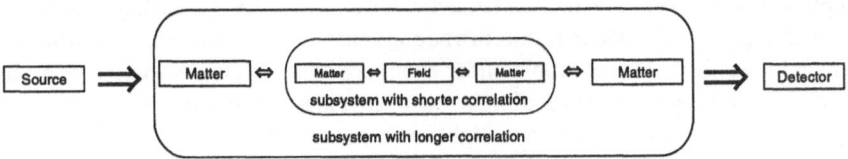

where the size parameter l characterizes the optical process taking place in the subsystem with shorter correlation. The interactions with shorter correlation lengths averaged over the volume l^3 behave as an effective field when observed on a scale much larger than l. In this case, we can consider the internal field as an effective field described by

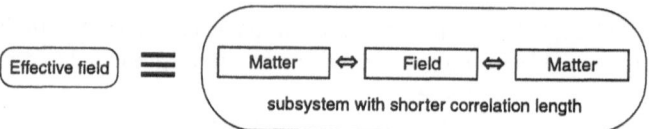

The effective field is considered to be a renormalized field or a coupled mode of the electromagnetic field with matter in the smaller subsystem. As is seen later in this section, these "source-to-detector" processes show peculiarities according to the way the interaction processes are described. That is, according to the macroscopic, mesoscopic, or microscopic nature of the effective field, there arise different aspects of electromagnetic interactions.

1.3.2.2. Effective Field and Space-Time Correlation in Optical Processes

In some cases, a specific internal field exhibits a significant effect on the optical properties of a material system. For example, this occurs when the effective field exhibits a resonance behavior. Here the resonance behavior represents a state in which a specific space-time correlation of internal interaction processes or a cumulative motion of internal degrees of freedom becomes significant. In these cases we can describe the corresponding effective field as a well-established mode. An example of such a case is given by plasma oscillations in metallic media. The remaining part of the internal interactions with shorter correlation lengths gives rise to fluctuations in the optical properties of the system. The process of optical excitation and decay in such a well-defined mode is often described in terms of quantum field theory on the basis of second quantization: creation and annihilation of photons or quasiparticles in the mode. Examples are plasmons and exciton polaritons in the loose sense [82].

Such a well-defined mode of an effective field in matter is very useful in diagnosing the optical response of matter with a specific correlation length. In fact, a scanning plasmon near-field microscope uses such a strong resonance behavior in metallic samples [51, 52]. It is also instructive to note that even for a macroscopic optical system such a resonance behavior arises when the material system has a specific spatial order which results in a strong space-time correlation in electromagnetic interactions. One example under extensive study is the photonic crystal, in which the multiple scattering of light waves in a highly ordered dielectric system gives rise to a band structure in its light propagation characteristics [46, 83, 84].

Besides these cases with strong resonance behaviors, we can construct a system in which we can extract some specific space-time correlations as a superposition of quasimodes to a good approximation. This problem is closely related to general near-field microscopy. Then the problem is to find the appropriate subsystem for describing the important characteristics of the optical process and the circumstances under which it is justified to extract a specific interacting subsystem.

1.3.3. Electromagnetic Interaction in a Dielectric System

1.3.3.1. Effective Field in a Dielectric Medium

In order to study the meaning of effective fields and scales of measurement, let us consider a practical case: the optical response of a dielectric body.

First, let us consider the macroscopic case: light waves propagating in a dielectric medium. A light wave incident onto a dielectric medium induces electric polarizations in the medium, as illustrated in Fig. 1.2. The induced polarizations in the medium are not free, but are influenced by their neighbors. As a result, the incident light wave induces an internal stress field in the medium, so that the incident electromagnetic energy and momentum are distributed to the dielectric medium. The total energy and momentum are transported in the medium in the form of a coupled mode of electromagnetic field with matter. This is why a light wave propagates in a material medium with velocity slower than the velocity of light in

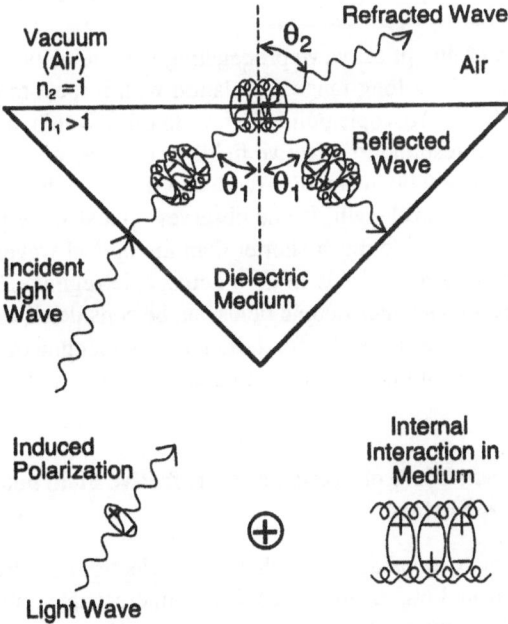

Figure 1.2. Reflection and refraction of a light wave at a dielectric–vacuum (air) boundary and the optical response of matter in the dielectric medium. The optical response of the dielectric medium is determined by electromagnetic interactions of induced polarizations. When the refraction angle θ_2 exceeds $\pi/2$ the refracted wave becomes an evanescent wave showing an exponential decay of field intensity into the vacuum (air) side.

vacuum c. Although the microscopic interactions in the medium are very complicated, it is still possible to describe the macroscopic behavior of coupled modes in terms of waves propagating with velocity $v = c/n$, where the macroscopic index n is known as the refractive index of the medium. This corresponds to an effective field on the macroscopic level, characterized by a correlation length as large as the optical wavelength.

According to the complicated issues involved in the idea of the effective field, there has long been controversy about the field momentum associated with electromagnetic waves propagating in dielectric media: the so-called Abraham–Minkowski controversy [85–88]. Let us briefly survey the essence of the controversy; due to the coupling of electromagnetic waves with matter the purely electromagnetic momentum, as a macroscopic average, should be less than that in vacuum, whereas a wave propagating in a material medium seems to have a wavenumber that is n times larger than that in vacuum. The answer can be seen upon introducing the idea of "pseudo-momentum" or "wave momentum" associated with the coupled mode of the electromagnetic field with matter, which represents a conserved quantity under spatial translation of the macroscopic field-plus-matter system [88]. In these cases purely electromagnetic quantities alone are no longer conserved quantities in the coupled system.

Here the idea of an optical wave propagating with velocity c/n in a dielectric medium turns out to be a long-range correlation with respect to electromagnetic interaction from the macroscopic point of view. In other words, optical waves in a material medium correspond to effective fields in the macroscopic or long-range limit where electromagnetic interactions with shorter correlation lengths are averaged over the optical wavelength. If one observes optical wave propagation in a dielectric medium on a scale much shorter than the optical wavelength, one sees complicated electromagnetic fields in the microscopic regime. From the macroscopic point of view these microscopic fields can be considered as fluctuations.

Similar complications arise also for the angular momentum of electromagnetic waves in a material medium, which is also under extensive study [89, 90].

1.3.3.2. Evanescent Wave of Fresnel as an Effective Field near a Planar Dielectric Surface

It is important to consider the behavior of electromagnetic waves near a dielectric surface from both the macroscopic and microscopic points of view.

Let us consider a steep, planar dielectric–air boundary where the behavior of the material changes abruptly compared with the optical wavelength. Here, the optical wavelength corresponds to the long-range correlation length of electromagnetic interactions. There exist shorter range interactions down to the atomic scale, which usually give rise to no observable effect in the macroscopic regime. However, due to the abrupt termination of the medium, electromagnetic interactions of any correlation length are exposed to the outer half-space around the boundary.

First, let us investigate electromagnetic interactions corresponding to long-range correlations at a dielectric–air boundary illuminated from the inside. Induced polarizations on the boundary surface then give rise to a surface polarization wave and associated electromagnetic fields. That is, the polarization wave on the surface produces both a reflected wave propagating back into the dielectric medium and a refracted wave propagating into the air, provided that the refracted wave satisfies the dispersion relation of light waves in air. The macroscopic relation between surface polarization waves and light waves is usually represented by a set of boundary conditions for electromagnetic fields. A useful result for a planar dielectric boundary is Snell's law; for the geometrical configuration shown in Fig. 1.2, the angle of reflection is equal to the angle of incidence θ_1, and the angle of refraction θ_2 satisfies the phase matching relation

$$n_1 \sin \theta_1 = n_2 \sin \theta_2 \qquad (1.3.1)$$

For a wave incident at an angle within the critical angle θ_c,

$$\theta_1 < \theta_c = \sin^{-1} \frac{n_2}{n_1}, \qquad n_2 < n_1 \qquad (1.3.2)$$

the angle of refraction θ_2 remains real. This implies that the refracted light wave satisfies the dispersion relation in air, keeping coherence with the polarization wave on the surface.

An interesting phenomenon arises when the incident angle θ_2 exceeds the critical angle θ_c for which no propagating light wave in air is able to maintain coherent interaction with the induced polarization wave on the surface. As a result, the incident wave is reflected perfectly into the dielectric medium. Although no propagating wave is excited on the air side, there still exists an electromagnetic field associated with the surface polarization wave, which, however, disappears at a distance far from the boundary surface. This phenomenon is called *total internal reflection*. A refracted wave showing an exponential decay in the direction normal to the boundary surface is called an *evanescent wave* [1]. Evanescent waves exist only as a coupled modes of an electromagnetic field with material excitations on a dielectric surface and carry energy and pseudomomentum along the surface. The range of interaction associated with evanescent waves is called the *penetration depth*, which indicates the distance corresponding to the $1/e$ decay of the field amplitude from the surface. The penetration depth is equivalent to the inverse of the imaginary part of the complex wavenumber of the evanescent wave.

Evanescent waves have two important physical implications in the context of the effective field and its relation to near-field optics. First, evanescent waves provide an example of an effective field showing a finite-range electromagnetic interaction. Second, evanescent waves provide an example of effective fields exposed to the outer half-space of a dielectric medium. These two points are relevant to the foundation of near-field optics and near-field photonics.

Perfect elimination of refracted propagating waves and the appearance of evanescent waves of specific wavenumber arise only for the case of planar dielectric boundaries. In fact, these effects are due to the perfect translational symmetry of the system along its boundary surface. For the case of a nonplanar surface, there arise both refracted propagating waves and evanescent waves with scattered values of the wavevector which are determined by the topography of the surface. Before proceeding to these general cases, let us briefly consider the observation of evanescent waves. This will provide a simple example showing the fundamental process of near-field optical microscopy, i.e., *near-field detection of a near-field event*.

1.3.3.3. Frustration of Total Internal Reflection and Observation of Optical Near-Field

Let us consider an optical plane wave incident on the surface of a prism at the angle of total internal reflection (Fig. 1.3a). There arises an evanescent wave propagating along the prism surface. It is easy to imagine what happens when one puts another prism with the same refractive index in direct contact with the first one (Fig. 1.3b). The propagation of optical waves in the medium is continued perfectly, so that the reflected wave disappears. It is quite interesting to consider the case that the second prism is held at a short distance above the surface of the first prism (Fig. 1.3c). Due to electromagnetic interactions of the two closely facing prism surfaces, a polarization wave is induced on the secondary surface, which in turn excites a propagating wave into the second prism. As a result, part of the optical wave is transmitted into the second prism. Then the process of total internal reflection is said to be *frustrated*. The magnitude of the transmitted intensity through the air gap is dependent on the gap separation. This can be attributed to the exponential decay of the evanescent wave. The effect of frustrated total internal reflection tells us that we can observe evanescent waves by immersing a probe into the evanescent wave and measuring the scattered optical waves by a photodetector.

Such a probe need not necessarily be a planar dielectric surface, but can be anything able to frustrate the total internal reflection and to produce a propagating optical field sensible in the far-field (Fig. 1.3d). We usually describe such a process as follows: an evanescent field is scattered into a propagating wave which extends to the far-field. The evanescent wave of Fresnel was first demonstrated by Raman, who immersed a knife edge into the near-field region in order to produce scattered light into a photodetector.

The idea of frustrated total internal reflection can be extended for near-field optical measurements of a confined field which exists only in the narrow vicinity

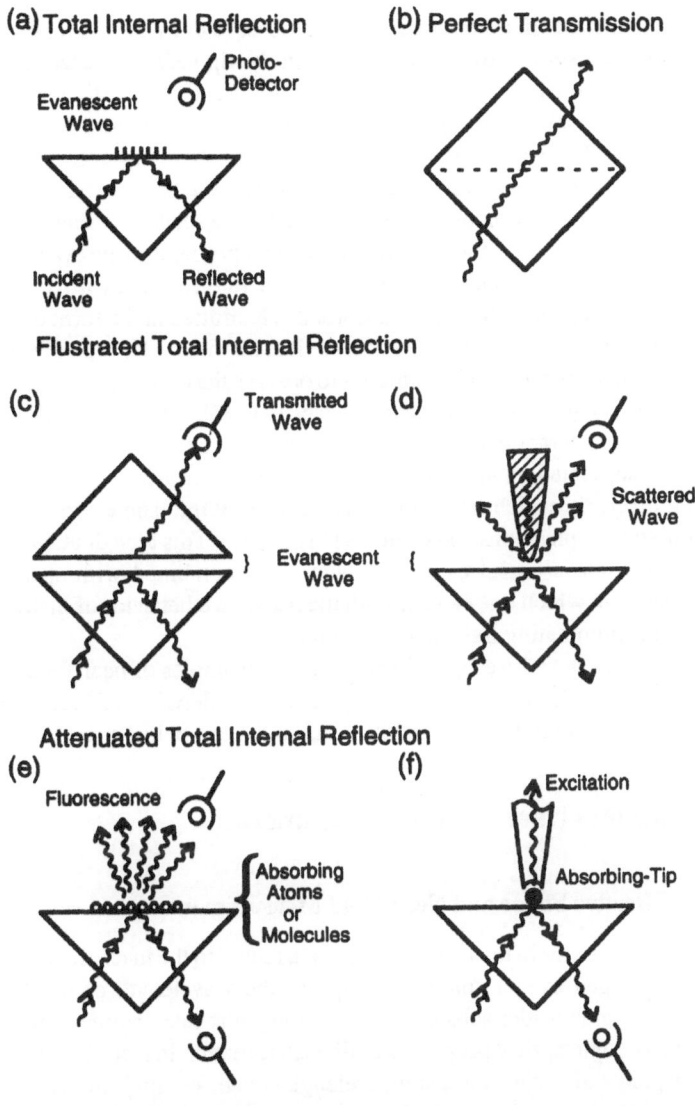

Figure 1.3. Evanescent wave and its measurement. (a) The evanescent wave of Fresnel arises for light incident at the angle of total internal reflection. (b) The evanescent wave disappears when continuity is restored at the boundary surface. (c) Frustrated total internal reflection: part of the evanescent wave is transmitted into a propagating wave in a second prism put at a distance as close as the penetration depth of the evanescent wave. (d) Frustrated total internal reflection by a local probe tip: basic principle of near-field optical microscopy (NOM). (e) Attenuated total internal reflection: light-absorbing matter on the surface absorbs part of an evanescent wave, and as a result the reflected light is attenuated. (f) Attenuated total internal reflection by a local probe tip with light-absorbing characteristics.

of a material surface. In other words, we can observe even a short-range effective field exposed at a surface by putting a small probe tip near the surface as close as the penetration depth of the effective field. In this case, the probe tip makes a direct coupling with a local oscillating polarization induced on the illuminated surface.

There is another way to observe evanescent waves, by using atoms or molecules which can resonantly absorb optical energy from evanescent waves. In this case one can measure the attenuation of a reflected wave as one sends absorbers in the close vicinity of the surface, i.e., one immerses absorbing probes into the evanescent field (Fig. 1.3e). The corresponding phenomenon is referred to as *attenuated total internal reflection*. When the energy absorbed is reemitted in the form of atomic or molecular fluorescence, one can also observe it at far-field. This type of experiment has been one of the most useful techniques to observe the optical properties of atoms and molecules adsorbed on a dielectric surface [91, 92]. It has been utilized also to observe atomic interactions with material surfaces, such as the van der Waals interaction and related quantum optical effects [93]. It is also of interest to use an absorbing probe tip which transforms optical energy into some form of material excitation, such as plasmons and excitons [51, 52, 55]. This type of attenuated total internal reflection is useful especially when it is combined with spectroscopic measurements, in which one can use both the resonance behavior of matter and the coherence of illuminating light sources such as lasers.

Now that we have investigated some simple examples of near-field detection of effective fields, let us return to the general consideration of electromagnetic interactions in a material system and its relation to near-field optics.

1.3.4. Optical Near-Field Measurements

1.3.4.1. Effective Fields and Near-Field Measurements

Optical properties in the usual sense are attributed to the macroscopic response of matter averaged over a volume as large as the wavelength of incident light. However, we can consider substantial electromagnetic interactions as responsible for the macroscopic optical properties of dielectric matter. In fact, if we observe the optical response of matter on a subwavelength scale, we find that shorter range electromagnetic interactions in subsystems constitute the macroscopic optical response. Then we see that the nature of the interaction is dependent on the scale at which we observe the subsystem. In principle, we can trace the interaction processes down to the atomic scale.

Figure 1.4a pictures the relation between macroscopic optical properties and interactions viewed at several different spatial scales: a set of smaller subprocesses constitutes the optical properties viewed at the larger scale. Subsystems at each scale are indicated by circles, each of which has two pairs of incoming and outgoing

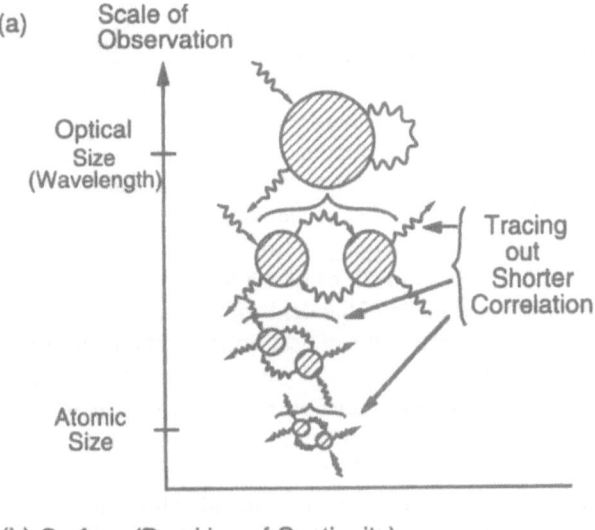

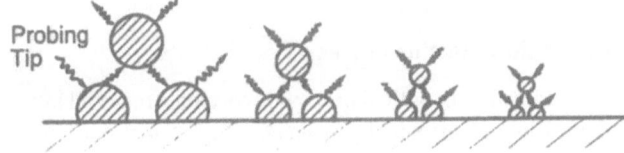

Figure 1.4. (a) Macroscopic optical properties and electromagnetic interactions observed at different spatial dimensions. Shorter range interactions are traced out or averaged to produce larger scale optical responses. (b) Microscopic electromagnetic interactions exposed to the outer half-space near a boundary surface. (c) Detection of local fields by using probe tips capable of making a good correlation with local effective fields.

arrows indicating incident and scattered fields. A set of closed arrows between a pair of subsystems shows the internal interaction between them. The optical response at each scale is pictured by a circle with two pairs of incoming and outgoing arms, and the interaction between subsystems is pictured by a contraction of one incoming arm and one outgoing arm as a pair.

This pictorial representation corresponds to the theoretical procedure of *tracing out* or taking a statistical average of internal interactions between subsystems. The term *tracing out* actually refers to taking the trace of density matrices which represent the material behavior in a statistical framework.

When we trace out a pair of incoming and outgoing arms of a single system we obtain a description of self-interaction which determines the optical response to an incoming field and gives rise to a scattered field. When we trace out a pair of incoming and outgoing arms belonging to two neighboring bodies we obtain the description of the interaction. In both cases the resulting inner lines represent an internal field or the coupled mode of the field with matter.

In general, when we consider the optical response of a dielectric material we can trace out interactions over a volume corresponding to the optical wavelength cubed. This gives the macroscopic dielectric characteristics. In this case the interactions at the subwavelength range are considered as fluctuations.

The situation is very different in the case of surface phenomena. At the material surface all of the significant interactions are exposed to the outer half-space due to the abrupt termination of matter, as shown in Fig. 1.4b. Then it becomes possible to observe microscopic effective fields, using a probe tip capable of making a direct interaction with these effective fields with short penetration depth, as shown in Fig. 1.4c. In this case the probe tip scatters the local effective field into a propagating field which is observable in the far-field regime. Such a near-field detection of effective fields on a material surface is the essence of scanning-probe-type microscopes including STM, AFM, NOM, and so on. Thus the key to these near-field measurements is to understand the role of local probes, which we study in detail in the following.

1.3.4.2. Near-Field Detection of Near-Field Events

Next, let us consider an optical process involving a material system which can be separated into two electronically isolated subsystems; one corresponds to an object and the other to a probe. Suppose that an isolation scheme is implemented between a light source and a photodetector, so that signal transmission from the light source to the photodetector is only via the interaction between the object and the probe. Such an isolation scheme is one of the fundamental requirements for near-field measurements.

A light wave incident on the object drives an effective field both inside and outside of the object. In order to observe the effective field near the object we have to use a probe tip with the following characteristics. First, the probe tip must have dimensions as small as the object, which determines the penetration depth of the effective field, since the effective field corresponds to the electromagnetic correlation produced in the object itself. Second, the probe tip must be put near the object, where the effective field with short penetration depth remains significant. The small object and the probe tip then form an interacting subsystem. As a result, the effective field of the object turns out to be an internal field of the coupled system, which produces a light wave scattered from the probe tip and extended to the photodetector. Such an idea for near-field detection of near-field events is the essence of the optical near-field diagnosis of material systems.

1.3.4.3. Criterion for Near-Field Measurements

The question now arises, "What is the criterion for near-field measurements?" One of the most important purposes of this book is to answer this question. One of the keys is the *existence of a narrow region* in the optical system. We will find in Chapter 8 a dimensionless parameter Ka characterizing light scattering processes, where K is the wavenumber of incident and outgoing light waves and a is the size parameter of the small subsystem, i.e., the object plus probe-tip system, as well as the distance between these. The existence of a narrow region corresponds to Ka being a small parameter characterizing the scattering process in the subsystem. From the theoretical point of view, this enables us to expand the object–probe interaction as a perturbation series in terms of powers of Ka. In fact, as discussed in Chapter 8, light scattering processes in a subwavelength-size dielectric system with size a can be characterized by a surface scattering potential and a bulk potential which have magnitudes characterized respectively by the first and second powers of Ka. This provides us a way to set a criterion for near-field measurement of a near-field event. We describe this relation by

$$1 \gg \beta_S[Ka] \gg \beta_V[(Ka)^2] \tag{1.3.3}$$

where β_S and β_V are coefficients with magnitudes of the order of unity describing, respectively, the surface and bulk terms in the scattering potential. This condition holds when both the size a of scatterers and the distance r of the observation point are small compared with the incident wavelength $2\pi k^{-1}$,

$$ka \leq kr \ll 1 \tag{1.3.4}$$

Near-field effects are then extracted as first-order processes in the perturbation series of light scattering phenomena. Therefore, it is essential to use a scheme which eliminates the zeroth-order process corresponding to the far-field coupling of the light source and the photodetector. In other words, in order to extract a near-field phenomenon from the whole light scattering process, we need to use a probe tip which is sensitive only to high-spatial-frequency components relevant to the local effective field around the object as well as a signal processing scheme which eliminates the signal background. This is why the techniques of fabricating fine near-field probes are of primary importance. It should be noted that the probe tip as a spatial-frequency filter should have a specific bandwidth which corresponds to the localization of the effective field confined around the object. Otherwise, the lateral resolution is lost even though high-spatial-frequency fields are detected. This point is studied in Chapter 9 on the theoretical basis of the angular spectrum representation.

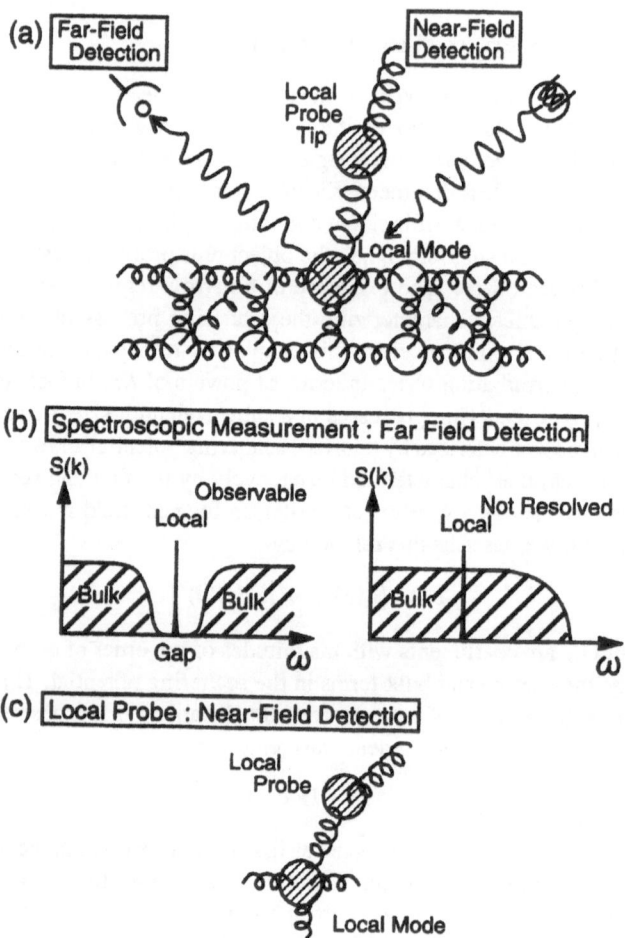

Figure 1.5. Comparison between macroscopic and local probing methods. (a) Spectroscopic and local-probe measurement of surface vibrational modes. (b) Spectroscopic measurement (far-field measurement) of the local mode is possible only when its resonant frequency falls in the band gap of the bulk mode spectra. (c) Local probe making a direct contact with the local mode provides information regardless of its spectral relation to the bulk mode. This provides an example of a near-field measurement of a local event.

Here, it is instructive to consider the meaning of local probes by using an example. Let us consider a crystalline system in which a characteristic vibrational motion determined by the crystal structure and an irregular local oscillation on the surface are excited at the same time (Fig. 1.5a). We often meet such a situation in surface physics. We can observe the behavior of crystalline vibrational modes by a spectroscopic method: measurements of resonance frequencies, spectral width, band gaps, and so on. On the other hand, the irregular part of the vibrational motion is observable by spectroscopy only when the spectrum due to the localized vibration falls in the band gap of the bulk modes. That is, one can observe the spectroscopic character of the local mode only when the corresponding resonance lies in the band gap of the bulk spectra (Fig. 1.5b). This is the limitation imposed on macroscopic techniques of measurement. However, the situation is completely different when we implement a local probe which is able to make a direct interaction with the local vibrational mode on the surface. In this case, the probe tip enables us to make a near-field observation of the local mode and provides a scheme equivalent to a spatial frequency filter which eliminates long-range correlations induced on the surface.

Note that, although the electron STM is one of the most impressive examples of local-probe measurement of local states, the principle is based on the fact that the electronic wavenumber K_e is of the order of the inverse atomic size. This implies that the criterion $K_e a \ll 1$ does not hold for the electron STM. Due to this fact, in order to interpret STM images one needs to account for complicated scattering processes of electron waves in the bulk region, such as band structures of bulk electronic states. Nevertheless, the localization of the probe tip is, of course, on the atomic scale.

In contrast, we can take advantage of the long-wavelength limit in the case of NOM measurements.

1.4. THEORETICAL TREATMENT OF OPTICAL NEAR-FIELD PROBLEMS

1.4.1. Near-Field Optics and Inhomogeneous Waves

In this section, we briefly survey the theoretical aspects of optical near-field problems. One can study a wide variety of interesting problems in relation to near-field optics and photonics by referring to the articles and textbooks cited in the following. The reader interested mainly in near-field microscopy may skip the following sections of this chapter and proceed to Chapter 2.

1.4.1.1. Evanescent Wave of Fresnel and Its Representation

As discussed in Section 1.3, a popular example of optical near-field phenomena appears in the total internal reflection of light waves at a planar dielectric–air boundary. However, even a description of this simple case involves a number of intriguing features relevant to the theoretical treatment of general optical near-field problems. The optical near-field in this case is described in terms of the evanescent wave of Fresnel, which describes the coupled mode of an electromagnetic field with dielectric matter, exposed to the outer half-space of the matter due to the discontinuity at the surface [1].

The evanescent wave propagating along a planar dielectric surface is characterized by its penetration depth into the air, which is determined by both the incident angle of the light waves and the optical response of the matter making up the dielectric. One theoretical description of waves with finite penetration depth uses plane waves with complex wavevectors. The imaginary part of the wavenumber represents the exponential decay of the field intensity in the direction normal to the surface, and the real wavevector lying in the plane parallel to the surface represents a near surface wave in phase with the incident and reflected waves in the dielectric medium.

1.4.1.2. Angular Spectrum Representation of Scattered Field

An important idea involved in treating evanescent waves is as follows. In general, a plane wave with a real wavevector serves as a mode function appropriate for describing the electromagnetic field in a homogeneous system which exhibits a symmetry or invariance under spatial translation (parallel displacement). However, one can apply the plane-wave description even for a system without any spatial translation symmetry. This is possible by extending the wavevector into the complex region. Such a theoretical procedure is well known as *analytic continuation* [94].

Analytic continuation is utilized in a wide variety of theoretical studies in physics and provides much useful information. For instance, one can describe the scattered light field from a material body with arbitrary shape in terms of propagating plane waves plus evanescent waves defined for a half-space separated by an assumed planar boundary. This treatment is called the *angular spectrum representation of scattered waves* [1, 95].

In this theoretical framework, waves with complex wavevector are called *inhomogeneous waves*, whereas real waves are referred to as *homogeneous waves*. One can make good use of analytic continuation when one needs to transform the basis of the field representation for scattering systems with different types of symmetry, such as plane waves, cylindrical waves, spherical waves, and so on. This is discussed in Chapter 9.

The theoretical description on the basis of plane waves with both homogeneous and inhomogeneous components is useful especially in the theoretical description of NOM processes. This is due to the fact that we consider an electromagnetic interaction between a subwavelength-size specimen and probe tip. That is, the spherical basis is appropriate for describing the light scattering process from a small object, whereas the planar basis is convenient for representing the interaction between a specimen and a probe tip in terms of spatial translations. In fact, in the angular spectrum representation of a scattered field from a subwavelength-size object, we can find inhomogeneous waves with very short penetration depth and an associated large wavenumber normal to it, which accounts for ultrahigh resolution in NOM far beyond the diffraction limit. In Chapter 9 we will find a theoretical description of the NOM process based on the angular spectrum representation.

1.4.1.3. Evanescent Fields and Tunneling Problems

There exist two types of inhomogeneous waves in the angular spectrum representation, waves with wavevectors corresponding to exponential decay (Im $\{k_z\} > 0$) and to exponential increase (Im $\{k_z\} < 0$), respectively. In handling a single-body scattering problem, one usually discards the latter because its amplitude goes to infinity in the far-field limit ($z \rightarrow +\infty$). However, when we consider a two-body problem in which a scattered field from one body is rescattered by the other, both increasing and decreasing inhomogeneous waves remain finite between the scatterers. The strength of coupling between the scatterers can be described in terms of an overlap integral of the inhomogeneous waves. Then the two-body light scattering problem corresponds to an optical analogue of the tunneling effect [96, 97]. Actually one can find in quantum mechanics textbooks a description of electron tunneling in terms of evanescent waves [98]. This shows one possible way to describe the sample–probe interaction in the NOM process in the general framework of scanning tunneling microscopy [12]. This point is discussed in detail in Chapter 2.

Based on the similarity between evanescent waves of electrons and light, proposals have been made to simulate the tunneling features of fermions, usually electrons, by experiments using optical waves [96, 97]. Actually, the problem of total internal reflection at a planar dielectric surface is basically one dimensional due to the translational symmetry along the boundary surface. The mathematical formalism is exactly the same as the potential barrier problem for Schrödinger waves in the case of the specific (*TE*) polarization of incident light waves. These studies are of current interest in relation to quantum kinematics, for example, in studying the traversal time of a wave packet through a tunneling gap [96, 97, 99]. This is because a single quantum state can be occupied by a large number of bosons (photons), so that one can expect to observe the behavior of photons by means of a destructive measurement of a small portion of the photons without changing any basic characteristics of the problem. One can foresee a wide variety of intriguing discussions related to such experiments and theoretical treatments.

1.4.2. Field-Theoretic Treatment of Optical Near-Field Problems

1.4.2.1. Diffraction Problems and Optical Near-Field

There is another way to describe optical near-field phenomena in terms of diffraction theory [37]. One of the most important problems is small-aperture diffraction, where a planar incident wave is diffracted by a subwavelength-size aperture bored in a planar conducting plane. The problem was related first to the scattered intensity in the far-field [36]; however, the importance of the optical near-field behavior was recognized later [37, 38]. Recent research has recognized the importance of these studies in relation to near-field microscopy [15, 48, 100]. Many theoretical approaches to handling general scattering problems were developed through the 1940s and 1950s in connection with small-aperture diffraction, such as the differential method [101]. Many of these have provided the basis for further developments in quantum field theory. A so-called "exact" solution in the near-field region was provided by Leviatan [38], which provides the basis for the theoretical description of aperture-type NOM. In fact, several extensions of Leviatan's calculations have appeared, such as for a conducting plane with a finite thickness and array of apertures [39].

1.4.2.2. Scattering of Light by a Small Sphere

The problem of light scattering by a small spherical object has been subject to intensive study. The problem is based on Mie's theory for scattering from metallic spheres, which is concerned with plasmon resonance [1, 102]. The theory describes vector fields in terms of spherical cylindrical functions and spherical vector harmonics.

Microscopic metallic sphere diffraction has both dielectric and conducting aspects. An extensive study of small-sphere polarizability and scattering has been made on the basis of additional boundary conditions accounting for the nonlocal response of matter to an optical field [103–105]. These studies offer a number of alternative ways to describe the electromagnetic interaction of matter.

1.4.2.3. Many-Body Optics and Self-Consistent Approach

Many-body optics accounting for the multiple scattering of light waves in a complicated optical system has been developed on the basis of the interaction propagator description and self-consistent approach to finding optimal mode functions for the system [106–111]. This provides the basis for numerical treatment of NOM processes as well as van der Waals interaction in AFM [112].

In the history of modern physics, scattering theories in optics aided developments in quantum scattering theories. In turn, one can now develop optical theories by utilizing the mathematical techniques extensively developed in quantum field theory [44, 113–118]. One example is the microscopic Maxwell–Schrödinger description of the nonlocal electromagnetic interaction of an electronic system, in which the local density functional method developed in modern solid-state physics is utilized [49, 50]. Scattering theories based on spatial Fourier analysis are also useful in handling systems with resonance behaviors, which involves an extension to fundamental excitations in the second-quantization framework [45, 46].

1.4.2.4. Evanescent Field and Virtual Photon

The evanescent field has been considered as an alternative to the virtual photon description appearing in quantum electrodynamics [119]. Here the virtual photon is the carrier of electromagnetic interaction in the quantum regime, which lies outside the ordinary photon dispersion relation and so cannot be observed outside of the interacting system, but can be observed inside the system by means of a destructive measurement via a direct interaction with a probe.

Here it is instructive to enter into a little more detail about the electromagnetic interaction and the virtual photon picture from the viewpoint of near-field optics. As discussed in previous sections, evanescent waves are distinguished from ordinary light waves by lying outside the dispersion relation of photons, as illustrated in Figs. 1.6b and 1.6f. This is due to the fact that the evanescent field is a coupled mode of the electromagnetic field with matter. In general, light waves or photons serve as the carriers of the electromagnetic interaction of matter as discussed in Section 1.1 [120]. In this sense, the only difference between real and virtual photons or the corresponding light waves is due to the fact that the former lie on the photon dispersion curve and the latter do not [121, 122]. This distinguishes between different types of electromagnetic interaction with matter. As pictured in Figs. 1.6c and 1.6d, an interaction via real photons or homogeneous light waves is reducible to two succeeding interaction processes of photons with each material system. That is, one can find a real photon at any cut of these successive interaction processes. On the other hand, as pictured in Figs. 1.6g and 1.6h the interaction process via virtual photons or inhomogeneous light waves is irreducible, so that if one makes a cut within the interacting system one sees an electromagnetic field showing behavior very different from real photons. Such a separation of a system undergoing an irreducible interaction is relevant to the theoretical description of interaction-type microscopy. That is, we prefer to describe the near-field interaction in terms of a sample object observed by a probe tip, though they are coupled tightly. This is

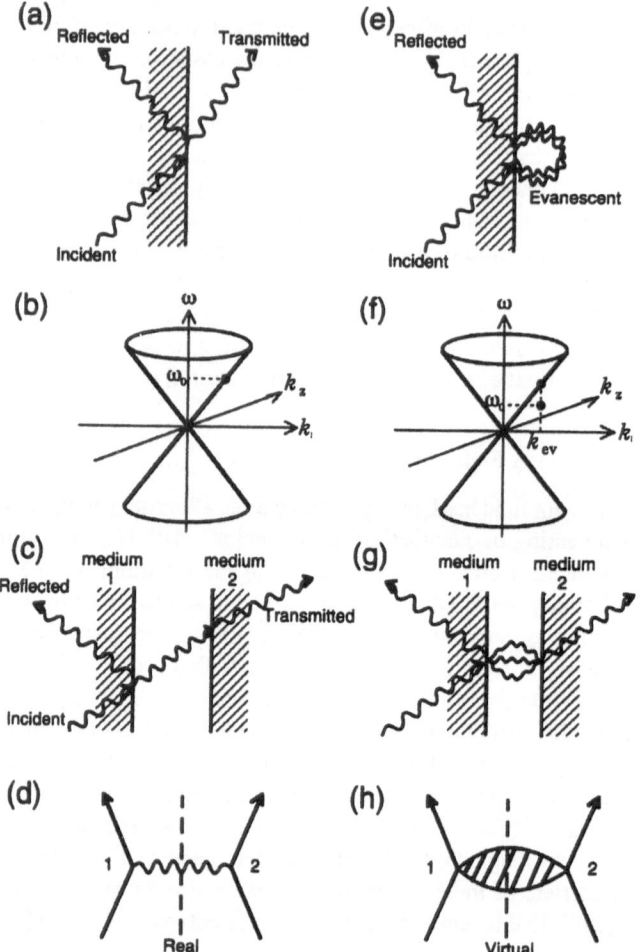

Figure 1.6. Propagating (homogeneous) and evanescent (inhomogeneous) waves and dispersion relation of light waves (photons). (a) Reflection and refraction of homogeneous waves. (b) Homogeneous waves lying on the dispersion relation of light waves or photons. (c) Interaction of two dielectric materials via propagating light wave. (d) Interaction of matter via propagating waves: this process is reducible to two separate parts consisting of the interaction of light waves with matter. One can find an optical wave at any section of these successive interaction processes which lies on the dispersion relation of light waves or photons. (e) Total internal reflection and evanescent waves at a dielectric–air boundary. (f) Inhomogeneous waves lying off the dispersion relation of light waves or photons. (g) Interaction of two dielectric materials via an evanescent wave. (h) Interaction of matter via evanescent waves: this process is irreducible into two separate parts because of its many-body nature. If one cuts this process, one sees an electromagnetic field which is qualitatively different from the light waves lying on the dispersion relation. We can describe such an effective field as evanescent waves or virtual photons lying off the photon dispersion curves.

why we employ evanescent waves or virtual photons in the description of optical near-field processes. We will return to this point after we introduce the NOM process in Chapter 2.

Some virtual-photon-related phenomena in quantum electrodynamics have been explained in terms of evanescent waves. For instance, based on radiation from moving charged particles, an evanescent wave representation of Cerenkov radiation was discussed extensively by di Francia [123] and Tidman [124]. It has also been of interest to study the radiation characteristics of oscillating electric dipoles, such as excited atoms, via coupling with evanescent fields in the close vicinity of a dielectric medium [93]. These types of optical near-field problems are concerned with quantum field theory in the weak-field limit.

Since from the physical point of view an evanescent field is a coupled mode of the electromagnetic field with matter, problems in which an evanescent field interacts with electrons and atoms are considered cavity quantum electrodynamic phenomena in the broad sense [125–133]. A quantum description of the optical near-field is therefore very interesting not only from the theoretical point of view, but also from the experimental point of view for diagnosing and controlling quantum states of electrons and atoms by taking advantage of the remarkable features of the evanescent field as a coupled mode.

1.4.2.5. Quantization of Evanescent Waves

The first attempt to quantize the evanescent field was reported by Carniglia and Mandel in 1971 for the case of evanescent waves at a planar dielectric–air boundary, i.e., total internal reflection [119]. Since the evanescent wave represents a coupled mode of the electromagnetic field with a dielectric surface, it alone cannot be considered as a free field. Coupled with incident and reflected waves in the half-space in a dielectric via Fresnel's relation at the dielectric–air boundary, the evanescent wave is treated as part of an effectively free field appropriate for quantization.

The important idea is the following: when one finds a set of mode functions appropriate for describing the electromagnetic field for a specific material system one can consider these modes to be effectively free in the sense that the interactions are already accounted for or "renormalized." Therefore, there are photons which behave as evanescent waves only in the outer half-space of the dielectric medium and as two homogeneous waves, incident plus reflected, in the dielectric medium. The discussion here recalls several important physical implications of evanescent or inhomogeneous waves as a description of a coupled mode of an electromagnetic field with matter.

1.4.2.6. Evanescent Field and Surface Excitation

It is very informative to compare the quantization of the evanescent wave with theories of elementary excitations in a material medium, such as the idea of exciton polariton first proposed by Hopfield in 1958 [134]. Although the surface polarization wave at a dielectric surface shows no specific resonance behavior in its dispersion relation, such a comparison provides us a way to understand the fundamental processes involved in optical near-field problems.

There is a similarity of these theories, except for the explicit use of the constitutive equation in the polariton case, whereas the material response is hidden behind the boundary conditions or Fresnel's relations for the electromagnetic modes in the case of evanescent modes. Both the quantized evanescent mode and the polariton mode are introduced as the spatial modulation of the coupling process between incident and outgoing waves and the associated self-interaction. In fact, several authors have pointed out that one can consider an optical wave in a dielectric as a kind of polariton [135].

Scanning plasmon near-field microscopy is a well-established technique in one branch of near-field optics. On the other hand, evanescent waves on a prism surface have been utilized both in the excitation and observation of surface plasmons, or plasmon polaritons. In addition, the multiple scattering of optical waves on a random surface has been studied in terms of quasiparticles or so-called "localitons" [47].

1.4.3. Explicit Treatment of Field–Matter Interaction

As seen above, optical near-field problems are a stage where the electromagnetic interaction of matter plays an important role. There remain a number of difficulties in the explicit treatment of the electromagnetic interaction of matter, and a number of theoretical treatments have tried to replace the interactions by using such ideas as the dielectric function of matter and appropriate boundary conditions.

Such problems are based on the quantum mechanical description of the interaction of a charged particle with an electromagnetic field, which has been one of the most intriguing problems in modern physics [136, 137]. Since the direct interaction of a photon with an electron is prohibited due to the momentum conservation rule, the assistance of a material field is an indispensable factor for such direct interaction. This corresponds to the Bremsstrahlung problem. Cerenkov radiation is an example [123, 124]. An intriguing effect of the direct modulation of an electronic wave by an optical field, known as the Schwarz–Hore effect, is under study [138–141]. These problems have a close relation to photon-assisted electron processes, such as photon-assisted tunneling [142–144] and light emission from the electron STM [145–147].

Some of these studies in quantum electrodynamics have a close relation to problems in near-field optics; for example, pseudomomentum transfer between an effective field and atomic particles provides the basis for the application of optical near-field techniques to the manipulation of atomic particles by controlling their kinematic motion as well as their quantum state.

1.5. REMARKS ON NEAR-FIELD OPTICS AND OUTLINE OF THIS BOOK

To conclude this introduction to near-field optics and photonics, let us survey the present status and possible extension of our research. Since it is the main concern of Chapters 2–7 to introduce near-field optical microscopy and related techniques, here we briefly consider the present state and future for near-field optics and related problems.

We take notice of the two scales characterizing modern optical science and photonics. One is the size of material objects, and the other is the spatial extent of the electromagnetic field relevant to specific optical or photonic processes, i.e., the sample–probe distance in optical measurements. Below we give a way to classify these problems.

1.5.1. Near-Field Optics and Related Problems

We have long been interested in the measurement of the optical response of matter to propagating light waves, where the nature of the process is characterized by the size of the object and the optical wavelength. We have developed techniques for near-field measurements to where we can now diagnose the optical response of matter on a scale shorter than the optical wavelength. That is, we can measure the effective field of matter on different scales of localization by means of the near-field detection of near-field events.

We have two different scales in the optical measurement of matter: one is given by the size of the material sample and the other by the spatial extent of the electromagnetic fields relevant to the specific measurement technique. As shown in Section 1.2, there arise three typical regions: *far-field processes* are related to measurements with propagating light waves, *near-field processes* are related to measurements of effective fields of interacting subsystems, and measurements of single material excitations represent a microscopic limit which is left for future research.

It is instructive to map optical near-field phenomena and related problems introduced in the previous sections with respect to these characteristic scales. An example of such a map is shown in Fig. 1.7. It is the task of the reader to complete the map according to his or her understanding and insight for these problems.

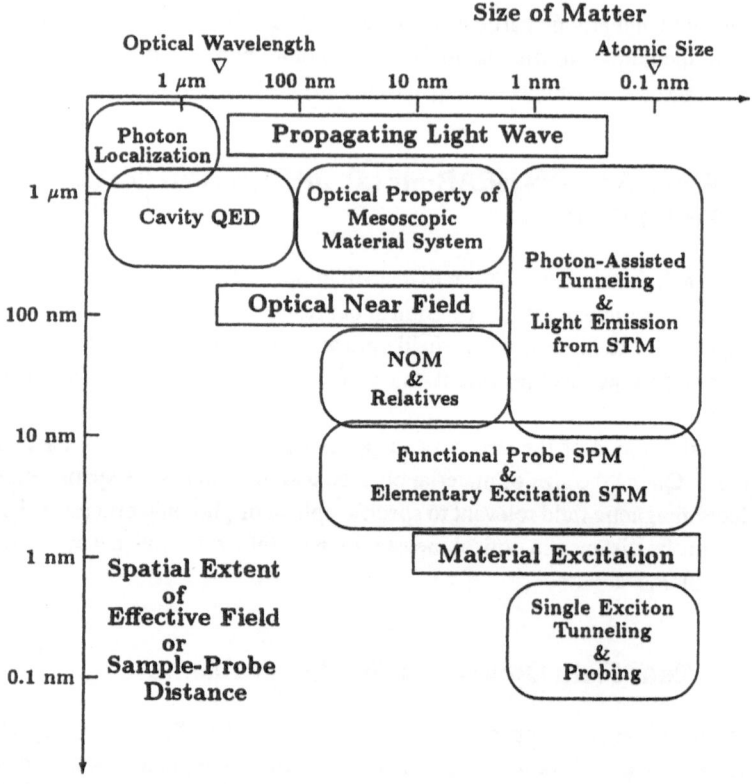

Figure 1.7. Near-field optics and related problems. Optical processes are shown in terms of the scales given by the size of the material object and the spatial extent of the effective field. The latter can be considered as the distance of the observation point from the object. Incidence of optical waves in the visible region is assumed in this chart. According to the spatial extent of the effective field, we can distinguish three characteristic regions of effective fields: the region for propagating light waves, the near-field, and local material excitations. One can shift the scale according to the wavelength used for illumination. The area corresponding to the bottom left is considered as the screened region.

1.5.2. Outline of This Book

This book consists of nine chapters.

A brief introduction to the principles and construction of near-field optical microscopes (NOM) is presented in Chapter 2. Figure 2.2 outlines the relation between the structure of this book and the basic concepts and components involved in general NOM processes.

Chapters 3–7 present experimental aspects of NOM and related techniques. Chapter 3 discusses NOM instrumentation in detail. Chapter 4 describes techniques for NOM-probe fabrication, the most important tool which enables us to go beyond the diffraction limit in optical diagnosis. Examples of NOM imaging experiments are described in Chapter 5, which also offers experimental demonstrations of fundamental processes involved in NOMs and in general near-field optical measurements, such as size-dependent localization of the optical near-field. Interpretations of NOM images are also discussed on the basis of these experimental results with the help of theoretical treatments of the optical near-field discussed in the last two chapters of the book. Chapter 6 presents novel applications of NOM and related techniques for the near-field diagnosis of matter, including electronic devices. Near-field optics is of interest not only for microscopy, but also for the manipulation of microscopic matter, such as atomic particles. This provides a technique for the nanometer-scale fabrication of novel material structures. Chapter 7 discusses these developments in near-field optics with experimental demonstrations of the potential of near-field techniques.

Chapters 8 and 9 present theoretical treatments of near-field optics and fundamental processes involved in NOM. Given the mesoscopic character of near-field optics, there are several approaches to theoretical treatment. Chapter 8 illustrates such aspects of near-field optics with several theoretical treatments of the electromagnetic interaction of matter. It is also important to consider an intuitive model of NOM processes in order to provide a clear understanding of the underlying physics and a theoretical basis for the interpretation of NOM images. These issues are also dealt with in Chapter 8. Chapter 9 describes the theoretical treatment of localized electromagnetic interactions between subwavelength-size material objects. One of the most important tools is the angular spectrum representation of electromagnetic interactions or of light scattering processes. The angular spectrum representation provides a way to evaluate the localization and short-range nature of optical near-field interactions. Throughout the theoretical treatments presented in this book, problems are treated analytically. This is because understanding of the underlying physics is usually related more closely to analytical treatment than to elaborate numerical calculations. Readers interested in the numerical treatment of optical near-field processes should refer to the articles cited in the relevant chapters.

This book is intended to introduce the basic ideas of near-field optics and nanooptics using material based on the original work of the authors. For a broader review of near-field optical research and related techniques, we refer to the book edited by Pohl and Courjon [13] and those written by Paesler and Moyer [148] and by Fillard [149].

1.6. REFERENCES

1. M. Born and E. Wolf, *Principles of Optics*, 3rd ed., Pergamon Press, Oxford, 1965.
2. A. Tonomura, N. Osakabe, T. Matsuda, T. Kawasaki, J. Endo, S. Yano, and H. Yamada, Evidence for Aharonov–Bohm effect with magnetic field completely shielded from electron wave, *Phys. Rev. Lett.* **56**: 792–795 (1986).

3. A. Zeilinger, R. Gähler, C. G. Shull, W. Treimer, and W. Mampe, Single- and double-slit diffraction of neutrons, *Rev. Mod. Phys.* **60**: 1067–1073 (1988).

4. O. Carnal, M. Siegel, T. Sleator, H. Takuma, and J. Mlynek, Imaging and focusing of atoms by a Fresnel zone plate, *Phys. Rev. Lett.* **67**: 3231–3234 (1991).

5. G. Binnig, H. Rohrer, Ch. Gerber, and E. Weibel, Tunneling through a controllable vacuum gap, *Appl. Phys. Lett.* **40**: 178–180 (1982).

6. G. Binnig, H. Rohrer, Ch. Gerber, and E. Weibel, Surface studies by scanning tunneling microscopy, *Phys. Rev. Lett.* **49**: 57–61 (1982).

7. G. Binnig, H. Rohrer, Ch. Gerber, and E. Weibel, 7×7 reconstruction on Si(111) resolved in real space, *Phys. Rev. Lett.* **50**: 120–123 (1983).

8. C. J. Chen, *Introduction to Scanning Tunneling Microscopy*, Oxford University Press, Oxford, 1993.

9. H. J. Güntherodt and R. Wiesendanger, eds., *Scanning Tunneling Microscopy I*, 2nd ed., Springer-Verlag, Berlin, 1994.

10. R. Wiesendanger and H. J. Güntherodt, eds., *Scanning Tunneling Microscopy II*, 2nd ed., Springer-Verlag, Berlin, 1995.

11. R. Wiesendanger and H. J. Güntherodt, eds., *Scanning Tunneling Microscopy III*, 2nd ed., Springer-Verlag, Berlin, 1996.

12. J. Bardeen, Tunneling from a many-particle point of view, *Phys. Rev. Lett.* **6**: 57–59 (1961).

13. D. W. Pohl and D. Courjon, eds., *Near Field Optics*, Kluwer, Dordrecht, 1993.

14. G. Massey, Microscopy and pattern generation with scanned evanescent waves, *Appl. Opt.* **23**: 658–660 (1984).

15. D. W. Pohl, W. Denk, and M. Lanz, Optical stethoscopy: Image recording with resolution $\lambda/20$, *Appl. Phys. Lett.* **44**: 651–653 (1984).

16. U. Ch. Fischer, Optical characteristics of 0.1 μm circular aperture in a metal film as a light sources for scanning ultramicroscopy, *J. Vac. Sci. Technol.* **B3**: 386–390 (1985).

17. U. Dürig, D. W. Pohl, and F. Rohner, Near-field optical-scanning microscopy, *J. Appl. Phys.* **59**: 3318–3327 (1986).

18. E. Betzig, M. Isaacson, and A. Lewis, Collection mode near-field scanning optical microscopy, *Appl. Phys. Lett.* **51**: 2088–2090 (1987).

19. D. Courjon, J.-M. Vigoureux, M. Spajer, K. Sarayeddine, and S. Leblanc, External and internal reflection near field microscopy: Experiments and results, *Appl. Opt.* **29**: 3734–3740 (1990).

20. R. C. Reddick, R. J. Warmack, and T. L. Ferrel, New form of scanning optical microscopy, *Phys. Rev. B* **39**: 767–770 (1989).

21. J. M. Guerra, Photon tunneling microscopy, *Appl. Opt.* **29**: 3741–3752 (1990).

22. S. Jiang, N. Tomita, H. Ohsawa, and M. Ohtsu, A photon scanning tunneling microscope using an AlGaAs laser, *Jpn. J. Appl. Phys.* **30**: 2107–2111 (1991).

23. C. J. Chen, Attractive interatomic force as a tunneling phenomenon, *J. Phys. Condens. Matter* **3**: 1227–1245 (1991).

24. E. A. Synge, A suggested method for extending microscopic resolution into the ultra-microscopic region, *Phil. Mag.* **6**: 356–362 (1928).

25. J. A. O'Keefe, Resolving power of visible light, *J. Opt. Soc. Am.* **46**: 359 (1956).

26. E. Ash and G. Nicholls, Super-resolution aperture scanning microscope, *Nature* **237**: 510–512 (1972).

27. M. Fee, S. Chu, and T. W. Hänsch, Scanning electromagnetic transmission line microscope with sub-wavelength resolution, *Opt. Commun.* **69**: 219–224 (1989).

28. F. Keilmann, Laser-driven corrugation instability of liquid metal surfaces, *Phys. Rev. Lett.* **51**: 2097–2100 (1983).

29. F. Keilmann, K. W. Kussmaul, and Z. Szentirmay, Imaging of optical wavetrains, *Appl. Phys. B* **47**: 169–176 (1988).

30. T. Pangaribuan, K. Yamada, S. Jiang, H. Ohsawa, and M. Ohtsu, Reproducible fabrication technique of nanometric tip diameter fiber probe for photon scanning tunneling microscope, *Jpn. J. Appl. Phys.* **31**: L1302–L1304 (1992).

31. S. Jiang, H. Ohsawa, K. Yamada, T. Pangaribuan, M. Ohtsu, K. Imai, and A. Ikai, Nanometric scale biosample observation using a photon scanning tunneling microscope, *Jpn. J. Appl. Phys.* **31**: 2282–2287 (1992).

32. M. Naya, R. Micheletto, S. Mononobe, R. Uma Mahesuwari, and M. Ohtsu, High resolution near-field optical imaging of biological samples in water with optical feedback control, *Appl. Opt.* **36**: 1681–1683 (1997).

33. T. Pangaribuan, S. Jiang, and M. Ohtsu, Two-step etching method for fabrication of fiber probe for photon scanning tunneling microscope, *Electron. Lett.* **29**: 1978–1979 (1993).

34. S. Mononobe and M. Ohtsu, Fabrication of pencil-shaped fiber probe for near-field optics by selective chemical etching, *J. Lightwave Technol.* **14**: 2231–2235 (1996).

35. S. Mononobe, T. Saiki, M. Naya, and M. Ohtsu, Reproducible fabrication of a fiber probe with a nanometric protrusion for near-field optics, *Appl. Opt.* **36**: 1496–1500 (1997).

36. H. A. Bethe, Theory of diffraction by small holes, *Phys. Rev.* **66**: 163–182 (1944).

37. C. J. Bowkamp, Diffraction theory, *Rep. Progr. Phys.* **17**: 35–100 (1954).

38. Y. Leviatan, Study of near-zone fields of a small aperture, *J. Appl. Phys.* **60**: 1577–1583 (1986).

39. A. Roberts, Small-hole coupling of radiation into a near-field probe, *J. Appl. Phys.* **70**: 4045–4049 (1991).

40. J. M. Vigoureux, F. Depasse, and C. Girard, Superresolution of near-field optical microscopy defined from properties of confined electromagnetic waves, *Appl. Opt.* **31**: 3036–3045 (1992).

41. C. Girard and M. Spajer, Model for reflection near field optical microscopy, *Appl. Opt.* **29**: 3726–3733 (1990).

42. J. M. Vigoureux, C. Girard, and D. Courjon, General principle of scanning tunneling optical microscopy, *Opt. Lett.* **14**: 1039–1041 (1989).

43. B. Labani, C. Girard, D. Courjon, and D. Van Lebeke, Optical interaction between a dielectric tip and a nanometric lattice: Implications for near-field microscopy, *J. Opt. Soc. Am. B* **7**: 936–943 (1990).

44. O. J. F. Martin, C. Girard, and A. Dereux, Generalized field propagator for electromagnetic scattering and light confinement, *Phys. Rev. Lett.* **74**: 526–529 (1995).

45. B. Chen and D. F. Nelson, Wave-vector-space method for wave propagation in bounded media, *Phys. Rev. B* **48**: 15365–15371 (1993).

46. Z. Zhang and S. Satpathy, Electromagnetic wave propagation in periodic structures: Bloch wave solution of Maxwell's equations, *Phys. Rev. Lett.* **65**: 2650–2653 (1990).

47. D. Maystre, A new kind of surface wave: The localiton, in *Near Field Optics*, D. W. Pohl and D. Courjon, eds., Kluwer, Dordrecht, 1993, pp. 367–376.

48. H. Hori, Quantum optical picture of photon STM and proposal of single atom manipulation, in *Near Field Optics*, D. W. Pohl and D. Courjon, eds., Kluwer, Dordrecht, 1993, pp. 105–114.

49. K. Cho, Nonlocal theory of radiation–matter interaction: Boundary-condition-less treatment of Maxwell equations, *Progr. Theor. Phys. Suppl.* **106**: 225–233 (1991).

50. K. Cho, Y. Ohfuti, and K. Arima, Study of scanning near-field optical microscopy (SNOM) by nonlocal response theory, *Jpn. J. Appl. Phys.* **34**: 267–270 (1994).

51. U. Ch. Fischer and D. W. Pohl, Observation of single-particle plasmon by near-field optical microscopy, *Phys. Rev. Lett.* **62**: 458–461 (1989).

52. M. Specht, J. D. Pedaring, W. M. Heckl, and T. W. Hänsch, Scanning plasmon near-field microscope, *Phys. Rev. Lett.* **68**: 476–479 (1992).

53. P. Dawson, F. de Fornel, and J.-P. Goudonnet, Imaging of surface plasmon propagation and edge interaction using a photon scanning tunneling microscope, *Phys. Rev. Lett.* **72**: 2927–2930 (1994).

54. T. Saiki, S. Mononobe, and M. Ohtsu, Nanometric integrating tip: Enhanced sensitivity of fluorescence detection in photon STM, in *Technical Digest, Quantum Electronics and Laser Science Conference*, Baltimore, 1995, pp. 84–85.

55. R. Kopelman, W. Tan, Z.-Y. Shi, and D. Birnbaum, Near field optical and exciton imaging, spectroscopy and chemical sensors, in *Near Field Optics*, D. W. Pohl and D. Courjon, eds., Kluwer, Dordrecht, 1993, pp. 17–24.

56. R. Uma Maheswari, H. Tatsumi, Y. Katayama, and M. Ohtsu, Observation of subcellular nanostructure of single neurons with an illumination mode photon scanning tunneling microscope, *Opt. Commun.* **120:** 325–334 (1995).

57. T. Saiki, S. Mononobe, M. Ohtsu, N. Saito, and J. Kusano, Statially-resolved photoluminescence spectroscopy of lateral $p–n$ junctions prepared by Si-doped GaAs using a photon scanning tunneling microscope, *Appl. Phys. Lett.* **67:** 2191–2193 (1995).

58. R. D. Grober, T. D. Harris, J. K. Trautman, E. Betzig, W. Wegscheider, L. Pfeiffer, and K. West, Optical spectroscopy of a GaAs/AlGaAs quantum wire structure using near-field scanning optical microscopy, *Appl. Phys. Lett.* **64:** 1421–1423 (1994).

59. U. Mohideen, M. J. Yoo, H. Hess, W. S. Hobson, F. Ren, R. Kopf, and R. E. Slusher, GaAs/AlGaAs quantum-dot near-field scanning optical microscopy, *Tech. Digest Quantum Electron. Laser Sci.* **16:** 85 (1995).

60. E. Betzig, P. L. Finn, and J. S. Weiner, Combined shear force and near-field scanning optical microscopy, *Appl. Phys. Lett.* **60:** 2484–2486 (1992).

61. R. Toledo-Crow, P. C. Yang, Y. Chen, and M. Vaez-Iravani, Near-field differential scanning optical microscope with atomic force regulation, *Appl. Phys. Lett.* **60:** 2957–2959 (1992).

62. M. Abe, T. Uchihashi, M. Ohta, H. Ueyama, Y. Sugawara, and S. Morita, Measurement of evanescent field using noncontact mode atomic force microscopy, *Opt. Rev.* **4:** 232–235 (1997).

63. J. K. Trautman, J. J. Macklin, L. E. Brus, and E. Betzig, Near-field spectroscopy of single molecules at room temperature, *Nature* **369:** 40–42 (1994).

64. W. P. Ambrone, P. M. Goodwin, J. C. Martin, and R. A. Keller, Single molecule detection and photochemistry on a surface using near-field optical excitation, *Phys. Rev. Lett.* **72:** 160–163 (1994).

65. X. S. Xie and R. C. Dunn, Probing single molecule dynamics, *Science* **265:** 361–364 (1994).

66. E. Betzig, J. K. Trautmann, R. Wolfe, E. M. Gyorgy, P. L. Finn, N. H. Kryder, and C.-H. Chang, Near-field magneto-optics and high density data strage, *Appl. Phys. Lett.* **61:** 142–144 (1992).

67. S. Jiang, J. Ichibashi, H. Monobe, M. Fijihira, and M. Ohtsu, Highly localized photochemical process in LB films of photochromic material by using a photon scanning tunneling microscope, *Opt. Commun.* **106:** 173–177 (1994).

68. B. D. Terris, H. J. Mamin, and D. Rugar, Near-field optical data storage, *Appl. Phys. Lett.* **68:** 141–143 (1996).

69. S. Hosaka, T. Shintani, Y. Miyamoto, A. Hirotsune, M. Terao, M. Yoshida, K. Fujita, and S. Kramer, Nanometer-sized phase-change recording using a scanning near-field optical microscope with a laser diode, *Jpn. J. Appl. Phys.* **35:** 443–447 (1996).

70. S. Kawata and T. Sugiura, Movement of micrometer-sized particles in the evanescent field of a laser beam, *Opt. Lett.* **16:** 772–774 (1992).

71. F. Depasse and D. Courjon, Inductive force generated by evanescent light fields: Application to local probe microscopy, *Opt. Commun.* **87:** 79–83 (1992).

72. T. Funatsu, Y. Harada, M. Tokunaga, K. Saito, and T. Yanagida, Imaging of single fluorescent molecules and individual ATP turnovers by single myosin molecules in aqueous solution, *Nature* **374:** 555–559 (1995).

73. V. I. Balykin, V. S. Letokhov, Yu. B. Ovchinnikov, and A. I. Sidorov, Quantum-state-selective mirror reflection of atoms by laser light, *Phys. Rev. Lett.* **60:** 2137–2140 (1988).

74. H. Hori, S. Jiang, M. Ohtsu, and M. Ohsawa, A nanometric-resolution photon scanning tunneling microscope and proposal of single atom manipulation, in *Technical Digest of the 18th International Quantum Electronics Conference, Vienna*, June 1992, Vol. 9, pp. 48–49.

75. C. G. Aminoff, A. M. Steane, P. Bouyer, P. Desbiolles, J. Dalibard, and C. Cohen-Tannoudji, Cesium atoms bouncing in a stable gravitational cavity, *Phys. Rev. Lett.* 71: 3083–3086 (1993).

76. W. Seifert, C. S. Adams, V. I. Balykin, C. Heine, Yu. Ovchinnikov, and J. Mlynek, Reflection of metastable argon atoms from an evanescent wave, *Phys. Rev. A* 49: 3814–3823 (1994).

77. S. Feron, J. Reinhardt, M. Ducroy, O. Gorceix, S. Nic Chormaic, Ch. Miniatura, J. Robert, J. Baudon, V. Lorent, and H. Haberland, Doppler-tuned multiphoton resonances in an atom reflection by a standing evanescent wave, *Phys. Rev. A* 49: 4733–4741 (1994).

78. M. A. Ol'Shanii, Yu. B. Ovchinnikov, and V. S. Lethokov, Laser guiding of atoms in a hollow optical fiber, *Opt. Commun.* 98: 77–79 (1993).

79. W. Jhe, M. Ohtsu, H. Hori, and S. R. Freiberg, Atomic waveguide using evanescent wave near optical fibers, *Jpn. J. Appl. Phys.* 33: L1680–L1682 (1994).

80. M. J. Renn, D. Montgomery, O. Vdovin, D. Z. Anderson, C. E. Wieman, and E. A. Cornell, Laser-guided atoms in hollow-core optical fibers, *Phys. Rev. Lett.* 75: 3253–3256 (1995).

81. H. Ito, T. Nakata, K. Sakaki, M. Ohtsu, K. I. Lee, and W. Jhe, Laser spectroscopy of atoms guided by evanescent waves in micron-sized hollow optical fibers, *Phys. Rev. Lett.* 76: 4500–4503 (1996).

82. C. Kittel, *Introduction to Solid State Physics*, 6th ed., Wiley, New York, 1986.

83. E. Yablonovitch and T. J. Gmitter, Photonic band structure: The face-centered-cubic case, *Phys. Rev. Lett.* 63: 1950–1953 (1989).

84. K. M. Leung and Y. F. Liu, Full vector wave calculation of photonic band structures in a face-centered-cubic dielectric media, *Phys. Rev. Lett.* 65: 2646–2649 (1990).

85. O. Costa De Beauregard, Ch. Imbert, and J. Ricard, Energy-momentum quanta in Fresnel's evanescent wave, *Int. J. Theor. Phys.* 4: 125–140 (1971).

86. J. P. Gordon, Radiation force and momenta in dielectric media, *Phys. Rev. A* 8: 14–21 (1973).

87. D. F. Nelson, Momentum, pseudomomentum, and wave momentum: Toward resolving the Minkowski–Abraham controversy, *Phys. Rev. A* 44: 3985–3996 (1991).

88. R. Peierls, *More Surprise in Theoretical Physics*, Princeton University Press, Princeton, New Jersey, 1991, Sections 2.4–2.6, pp. 30–42.

89. M. Kristensen and J. P. Woerdman, Is photon angular momentum conserved in dielectric medium? *Phys. Rev. Lett.* 72: 2171 (1994).

90. H. He, M. E. Friese, N. R. Heckenberg, and H. Rubinsztein-Dunlop, Direct observation of transfer of angular momentum to absorptive particles from a laser beam with a phase singularity, *Phys. Rev. Lett.* 75: 826–829 (1995).

91. C. K. Carniglia, L. Mandel, and K. H. Drexhage, Absorption and emission of evanescent photons, *J. Opt. Soc. Am.* 62: 476–486 (1972).

92. D. Suter, J. Äbersold, and J. Mlynek, Evanescent wave spectroscopy of sublevel resonances near a glass/vapor interface, *Opt. Commun.* 84: 269–274 (1991).

93. M. Chevrollier, M. Fichet, M. Orisa, G. Rahmat, D. Bloch, and M. Ducloy, High resolution selective reflection spectroscopy as a probe of long-range surface interaction: Measurement of the surface van der Waals attraction exerted on excited Cs atoms, *J. Phys. II France* 2: 631–657 (1992).

94. A. Sommerfeld, *Partial Differential Equations in Physics*, Academic Press, New York, 1949.

95. E. Wolf and M. Niet-Vesperinas, Analyticity of the angular spectrum amplitude of scattered fields and some of its consequence, *J. Opt. Soc. Am. A* 2: 886–890 (1985).

96. Th. Martin and R. Landauer, Time delay of evanescent electromagnetic waves and the analogy to particle tunneling, *Phys. Rev. A* 45: 2611–2617 (1992).

97. A. M. Steinberg and R. Chiao, Tunneling delay times in one and two dimensions, *Phys. Rev. A* **49:** 3283–3295 (1994).

98. C. Cohen-Tannoudji, B. Diu, and F. Laloë, *Quantum Mechanics*, Vol. 1, Wiley, New York, 1977, Chapter 1, p. 71.

99. P. R. Holland, *The Quantum Theory of Motion*, Cambridge University Press, Cambridge, 1993.

100. J. P. Fillard, M. Castagne, and C. Prioleau, Atomic force microscopy silicon tips as photon tunneling sensors: A resonant evanescent coupling experiment, *Appl. Opt.* **34:** 3737–3742 (1995).

101. H. Levine and J. Schwinger, On the theory of diffraction by an aperture in an infinite plane screen. I, *Phys. Rev.* **74:** 958–974 (1948).

102. G. Mie, *Ann. Phys.* (Leipzig) **25:** 377 (1908).

103. B. B. Dasgupta, and R. Fuchs, Polarizability of a small sphere including nonlocal effects, *Phys. Rev. B* **24:** 554–561 (1981).

104. R. Fuchs and F. Claro, Multipolar response of small metallic spheres: Nonlocal theory, *Phys. Rev. B* **35:** 3722–3727 (1987).

105. F. Hache, D. Richard, and C. Girard, Optical nonlinear response of small metal particles: A self-consistent calculation, *Phys. Rev. B* **38:** 7990–7996 (1988).

106. R. K. Bullough, Many-body optics I. Dielectric constants and optical dispersion relations, *J. Phys. A (Proc. Phys. Soc.) Ser. 2* **1:** 409–430 (1968).

107. R. K. Bullough, Many-body optics II. Dielectric constant formulation of the binding energy of a molecular fluid, *J. Phys. A (Gen. Phys.) Ser. 2* **2:** 477–486 (1969).

108. R. K. Bullough, Many-body optics III. The optical extinction theorem and ε_l (k,ω), *J. Phys. A (Gen. Phys.)* **3:** 708–725 (1970).

109. R. K. Bullough, Many-body optics IV. The total transverse response and ε_t(k,ω), *J. Phys. A (Gen. Phys.)* **3:** 726–750 (1970).

110. R. K. Bullough, Many-body optics V. Virtual-mode theory, and phenomenological binding energies in the complex-dielectric-constant approximation, *J. Phys. A (Gen. Phys.)* **3:** 751–773 (1970).

111. C. Girard and C. Girardet, Self-consistent interaction potential for a molecule absorbed on a dielectric surface: A symmetric top molecule on an ionic crystal, *J. Chem. Phys.* **86:** 6531–6539 (1987).

112. C. Girard, Theoretical atomic-force-microscopy study of a stepped surface: Nonlocal effects in the probe, *Phys. Rev. B* **43:** 8822–8828 (1991).

113. S. John, H. Sompolinsky, and M. J. Stephen, Localization in a disordered elastic medium near two dimensions, *Phys. Rev. B* **27:** 5592–5603 (1983).

114. S. John and M. J. Stephen, Wave propagation and localization in a long-range correlated random potential, *Phys. Rev. B* **28:** 6358–6368 (1983).

115. S. John, Electromagnetic absorption in a disordered medium near a photon mobility edge, *Phys. Rev. Lett.* **53:** 2169–2172 (1984).

116. S. John, Localization and absorption of waves in a weekly dissipative disordered medium, *Phys. Rev. B* **31:** 304–309 (1985).

117. S. John, Strong localization of photons in certain disordered dielectric sphere-lattices, *Phys. Rev. Lett.* **58:** 2486–2489 (1987).

118. S. John and J. Wang, Quantum electrodynamics near a photonic band gap: Photon bound states and dressed atoms, *Phys. Rev. Lett.* **64:** 2418–2421 (1990).

119. C. K. Carniglia and L. Mandel, Quantization of evanescent electromagnetic waves, *Phys. Rev. D* **3:** 280–296 (1971).

120. J. A. Wheeler and R. P. Feynman, Interaction with the absorber as the mechanism of radiation, *Rev. Mod. Phys.* **17:** 157–181 (1945).

121. R. P. Feynman, *Quantum Electrodynamics*, Benjamin/Cummings, Reading, Massachusetts, 1961.

122. R. P. Feynman, *The Theory of Fundamental Processes*, Benjamin/Cummings, Reading, Massachusetts, 1962, Section 20, pp. 95–100.

123. G. Torardo di Francia, On the theory of some Cerenkovian effects, *Nuovo Cimento* **16**: 1085–1101 (1960).

124. D. A. Tidman, A quantum theory of radiative index, Cerenkov radiation and the energy loss of a fast charged particle, *Nucl. Phys.* **2**: 289–346 (1956/1957).

125. G. S. Agarwal, Quantum electrodynamics in the presence of dielectrics and conductors. I. Electromagnetic-field response functions and blackbody fluctuations in finite geometries, *Phys. Rev. A* **11**: 230–242 (1975).

126. G. S. Agarwal, Quantum electrodynamics in the presence of dielectrics and conductors. II. Theory of dispersion forces, *Phys. Rev. A* **11**: 243–252 (1975).

127. G. S. Agarwal, Quantum electrodynamics in the presence of dielectrics and conductors. III. Relations among one-photon transition probabilities in stationary and non stationary fields, density of states, the field-correlation functions, and surface-dependent response functions, *Phys. Rev. A* **11**: 253–264 (1975).

128. H. Chew, P. J. McNulty, and M. Kerker, Model for Raman and fluorescent scattering by molecules embedded in small particles, *Phys. Rev. A* **13**: 396–404 (1976).

129. H. Chew, Transition rates of atoms near spherical surfaces, *J. Chem. Phys.* **87**: 1355–1360 (1987).

130. D. Mechede, W. Jhe, and E. A. Hinds, Radiative properties of atoms near a conducting plane: An old problem in a new light, *Phys. Rev. A* **41**: 1587–1596 (1990).

131. E. A. Hinds and V. Sandoghdar, Cavity QED level shifts of simple atoms, *Phys. Rev. A* **43**: 398–403 (1991).

132. W. Jhe and J. W. Kim, Atomic energy-level shifts near a dielectric microsphere, *Phys. Rev. A* **51**: 1150–1153 (1995).

133. M. Janowicz and W. Żakowicz, Quantum radiation of a harmonic oscillator near the planar dielectric–vacuum interface, *Phys. Rev. A* **50**: 4350–4364 (1994).

134. J. J. Hopfield, Theory of the contribution of excitons to the complex dielectric constant of crystals, *Phys. Rev.* **112**: 1555–1567 (1958).

135. S. M. Barnett, B. Huttner, and R. Roudon, Spontaneous emission in absorbing dielectric media, *Phys. Rev. Lett.* **68**: 3698–3701 (1992).

136. P. A. M. Dirac, Classical theory of radiating electrons, *Proc. Roy. Soc. Lond. A* **167**: 148–169 (1938).

137. G. N. Plass, Classical electrodynamic equations of motion with radiative reaction, *Rev. Mod. Phys.* **33**: 37–62 (1961).

138. H. Schwarz and H. Hora, Modulation of an electron wave by a light wave, *Appl. Phys. Lett.* **15**: 349–351 (1969).

139. H. Hora, Coherence of matter waves in the effect of electron waves modulation by laser beams in solids, *Phys. Stat. Sol.* **42**: 131–136 (1970).

140. J. Bae, H. Shirai, T. Nishida, T. Nozokido, K. Furuya, and K. Mizuno, Experiantal verification of the theory on the inverse Smith–Purcell effect at a submillileter wavelength, *Appl. Phys. Lett.* **61**: 870–872 (1992).

141. J. Bae, S. Okuyama, T. Akizuki, and K. Mizuno, Electron energy modulation with laser light using a small gap circuit: A theoretical consideration, *Nucl. Instrum. Meth. Phys. Res. A* **331**: 509–512 (1993).

142. P. K. Tien and J. P. Gordon, Multiphoton process observed in the interaction of microwave fields with the tunneling between superconductor films, *Phys. Rev.* **129**: 647–651 (1962).

143. L. P. Kouwenhoven, S. Jauhar, K. McCormic, D. Dixon, P. L. McEuen, Yu. V. Nazarov, N. C. van der Vaart, and C. T. Foxon, Photon-assisted tunneling through a quantum dot, *Phys. Rev. B* **50**: 2019–2022 (1994).

144. L. P. Kowenhoven, S. Jauhar, J. Orenstein, P. L. McEuen, Y. Nagamune, J. Motohisa, and H. Sakaki, Observation of photon-assisted tunneling through a quantum dot, *Phys. Rev. Lett.* **73:** 3443–3446 (1994).

145. P. Johansson, R. Monreal, and P. Appel, Theory of light emission from a scanning tunneling microscope, *Phys. Rev. B* **42:** 9210–9213 (1990).

146. R. Berndt, J. K. Gimzewski, and P. Johansson, Inelastic tunneling excitation of tip-induced plasmon modes on noble-metal surface, *Phys. Rev. Lett.* **37:** 3796–3799 (1991).

147. R. Berndt and J. K. Gimzewski, Injection luminescence from CdS(1120) studied with scanning tunneling microscopy, *Phys. Rev. B* **45:** 14095–14099 (1992).

148. M. A. Paesler and P. J. Moyer, *Near-Field Optics: Theory, Instrumentation, and Applications,* Wiley, New York, 1996.

149. J. P. Fillard, *Near Field Optics and Nanoscopy,* World Scientific, Singapore, 1996.

TWO

PRINCIPLES OF NEAR-FIELD OPTICAL MICROSCOPY

2.1. AN EXAMPLE OF NEAR-FIELD OPTICAL MICROSCOPY

Let us investigate an example of a practical nanometer-resolution scanning near-field optical microscope (NOM). An NOM setup is shown in Fig. 2.1, and a recipe for preparing and operating the system is listed below. Several examples of such a system are given in Chapter 3, optical probes are described in Chapter 4, and experimental results showing evidence of nanometer resolution are given in Chapter

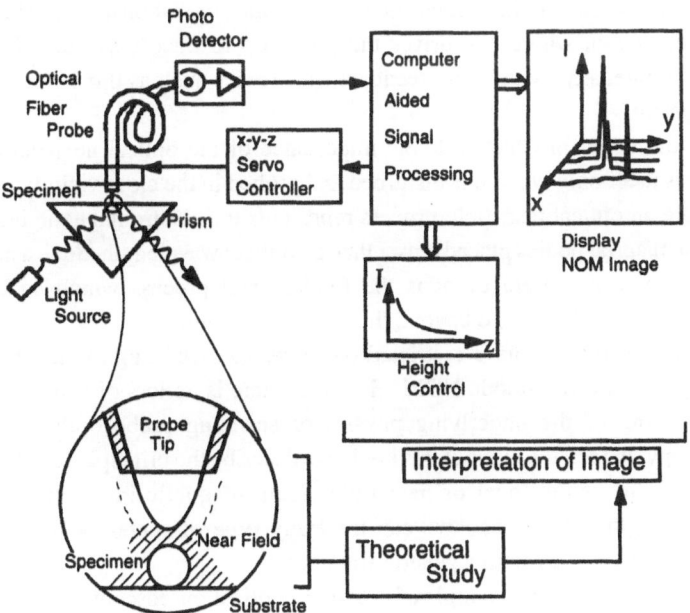

Figure 2.1.   Typical setup of a near-field optical microscope (NOM). An optical fiber probe and collection-mode operation are employed in this case.

5. In this chapter, we consider *collection-mode NOM*, but we will see in later chapters that the ideas and basic setup are almost the same for *illumination-mode NOM*.

A nanometer-size specimen is put on a flat, clean dielectric substrate and illuminated by an evanescent wave produced by light incident at the angle of total internal reflection. An optical fiber probe with a nanometer-size sharpened probe tip is scanned by a three-dimensional piezoelectric driver on a plane several tens of nanometers above the substrate. The optical power picked up and propagated by the optical fiber probe is measured by a very sensitive photodetector. Mapping the signal intensity detected as a function of the planar position of the probe tip with computer-aided signal processing yields the NOM image of the specimen.

The NOM system described above is a *collection-mode, constant-height* system, since, as the light source illuminates the specimen, the probe tip collects the light while being scanned at a constant height above the substrate. There exist several versions of near-field optical microscopes in which light illumination induces a local specimen–probe tip interaction at close distances and the resulting scattered light wave is measured by a photodetector. One could, for instance, reverse the role of light source and photodetector to prepare an illumination-mode NOM experiment. Or one might be interested in a *feedback* operation instead of constant-height operation, where one drives the probe tip to trace a surface of constant picked-up intensity by utilizing feedback electronics such as those in an electron STM system.

As discussed in Chapter 1, the importance of the optical near-field regime manifests itself only when it is disturbed and probed in the close vicinity of matter. Therefore an optical near-field process represents the electromagnetic interaction between material bodies placed closer than an optical wavelength. Such a near-field detection of a near-field event is the fundamental process which enables this microscopy technique to go beyond the diffraction limit.

In this section we have restricted ourselves to describing the technique and principle of collection-mode NOM. However, there is no loss of generality in our understanding of the underlying physics of scanning probe optical near-field microscopy. In fact, throughout this book, we describe the principles of NOM from the viewpoint of the local or near-field electromagnetic interaction between subwavelength objects, and we can treat any type of scanning optical probe near-field microscopy in this general framework.

The following is an example of near-field optical microscopy.

- Prepare a sharpened optical fiber probe having a specially fabricated nanometer-size probe tip on one end of a single-mode optical fiber.

- Mount the optical fiber probe on a three-dimensional piezoelectric transducer, which makes it possible to scan the probe tip in the x, y, and z directions with nanometric precision using an electronic servocontroller.
- Connect the other end of the optical fiber to a very sensitive optical detector such as a photoelectron multiplier.
- Prepare a very clean prism as a substrate and put a nanometric-size dielectric specimen on it.
- Illuminate the specimen by an evanescent wave on the prism surface produced via light incident at the angle of total internal reflection.
- Drive the probe tip very close to the prism surface, monitoring the exponentially increasing optical power detected as the probe tip enters the evanescent field further.
- Keeping the height of the probe tip several nanometers above the prism surface, scan the tip in a submicrometer-size rectangular plane above the specimen.
- Map the picked-up optical power on the scanning plane drawn on an adequate display such as a cathode ray tube monitor with the aid of a computer and image-forming software.
- Discard the relatively flat background signal and observe the NOM image of the nanometer-size specimen as a map of signal height even though visible light of submicrometer wavelength was used for illumination.
- Take care in the interpretation of the image since the NOM image provides information about the optical properties of the specimen as well as its geometrical shape; consult the appropriate theories.

There are many types of NOM recipes; one chooses an approach based on what one wants to see.

2.2. CONSTRUCTION OF THE NOM SYSTEM

2.2.1. Building Blocks of the NOM System

The NOM system is constructed from several fundamental building blocks, each of which has a different characteristic scale with respect to the electromagnetic interaction. The function of each building block as part of the overall NOM system is shown in Fig. 2.2 together with its relation to the contents of this book. An interpretation of any NOM image requires an appropriate evaluation for all of the following characteristics of building blocks of the specific NOM system:

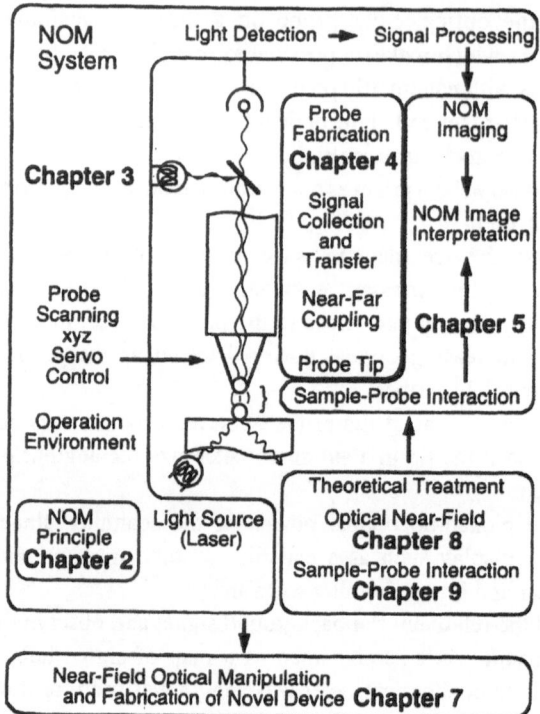

Figure 2.2. Building blocks of the NOM system, with the chapters in this book which describe the details of each block.

- **Local electromagnetic interaction**
  Electromagnetic interaction occurs on a subwavelength scale: *near-field process* $\ll \lambda_{optical}$.
- **Coupling from near- to far-field**
  Electromagnetic interaction mediates signal transfer from subwavelength-scale electromagnetic interaction to propagating light waves: *intermediate process* $\sim \lambda_{optical}$.
- **Signal collection and transfer**
  Light waves are guided both from source to specimen and from probe to photodetector; the quantity relevant to the near-field process is the optical power or the photon number detected: *far-field process* $\gg \lambda_{optical}$.

The NOM system is not complete when one has simply specified the optical process and its detection. The following parts are indispensable components in the function of NOM in order to produce the image of a specimen:

- **Production of NOM image**
  The probe tip is scanned and the signal is mapped to form an image of the sample object; averaging the signal and eliminating the background signal are part of the image processing. Signal processing and image-forming techniques are similar to those developed for electron STM.
- **Interpretation of image**
  The signal intensity cannot be simply related to the topographic image. Careful consideration must be given to the optical properties of the specimen, the polarization dependence of the signal, the way waves are guided from the probe tip to the photodetector, and so on. Appropriate theoretical consideration is required.

These are the characteristics that distinguish NOM or general interaction-type microscopes from interference types. In the latter, the optical process can in principle produce a magnified image of a specimen directly on the eye or photographic plate without elaborate signal processing and careful interpretation.

2.2.2. Environmental Conditions

In an NOM experiment, the following auxiliary conditions are also required, involving the fabrication of the probe, sample preparation, and so on (detailed discussions are presented in Chapters 3–5):

- **Preparation of probe tip**
  A wide variety of probe-tip and operation modes is available according to the specimen's optical properties, size, and environment. The size of the probe tip provides the measure of attainable resolution provided that the other conditions are adequate, such as those on the screening of the background signal, precision in movement, and residual noise in the servocontrol of the probe tip.
- **Light source and photodetector**
  Stability and intensity of the light source and sensitivity of the photodetector determine the quality and resolution of the NOM image. For a nanometer-sized specimen the pickup power is inevitably weak and a very high sensitivity up to the photon counting level is required depending on the nature of the probe tip, the means of signal collection and signal processing, and so on.
- **Fabrication of sample**
  Careful sample preparation is required to obtain nanometer resolution; a good specimen should be prepared so that it is only a protrusion on a

relatively flat background and is attached tightly in a thin layer. The specimen should also be isolated enough not to interfere with its neighbors.
- **Operating environment**
  The operating environment can be vacuum, air, or liquid.
- **Appropriate laboratory conditions**
  An acoustic shield of a certain level is required depending on the resolution expected, stiffness of the probe, stability of the substrate, and so on.

2.2.3. Functions of the Building Blocks

Let us study the functions of each of the building blocks of the NOM system.

2.2.3.1. Local Electromagnetic Interaction

The primary consideration is the local electromagnetic interaction between the nanometer-size specimen and the probe tip, which exhibits a short-range nature, that is, the interaction is effective only when the specimen and probe are very close to each other. Such a short-range electromagnetic interaction corresponds to a very local perturbation of the electromagnetic background due to the presence of subwavelength-size material objects, i.e., the specimen and probe tip. The optical properties and intrinsic field distribution of the probe–tip system should be simple in order to measure the geometrical character of the specimen.

The near-field criterion for the local electromagnetic interaction is discussed in Chapter 1.

2.2.3.2. Coupling from Near- to Far-Field

Local deformations in electromagnetic fields are manifested when one takes a microscopic electromagnetic process into account in the very near vicinity of matter. However, we usually observe a macroscopic average at least over a wavelength-size region when we observe a light–matter interaction such as a light source plus detector system. In order to observe the local interaction between the specimen and probe tip one has to couple the local event to some propagating field which extends to the light source and photodetector. That is, only by using propagating light can one extract information about the local interaction. The coupling schemes in the system described in the recipe use an evanescent wave to irradiate the specimen and the connection of the probe tip to a single-mode optical fiber.

The probe tip plus the coupling scheme should have the character of a spatial-frequency filter which picks up the spatial Fourier component lying within

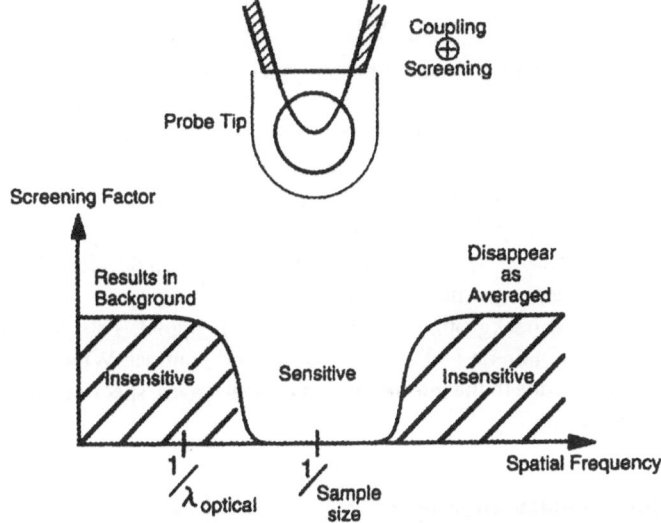

Figure 2.3. Basic characteristics required for an NOM probe tip as a spatial frequency filter. Both the coupling of the near- to the far-field region and the screening of large-scale interactions are of fundamental importance for NOM probes.

a window fit to the geometrical shape of the specimen, as shown in Fig. 2.3. Only when this condition is satisfied can we interpret the NOM image readily.

2.2.3.3. Signal Collection and Transfer

In order to send a message to a local specimen and to observe its reaction, we use a light source and a photodetector placed at either end of the optical system, as usual for any type of optical measurement. Two important aspects of the NOM system are a scheme for isolating the source field from the detector field and an extremely sensitive photodetection technique. The former reduces the signal background, which contains information not about the local event, but about the averaged optical properties of the system. To achieve such isolation the evanescent wave is incident at the angle of total internal reflection and a small aperture with a metal-cladded probe tip or a steeply tapered waveguide is used. Together these give rise to spatial frequency filtering in the lower frequency regime.

A high signal-to-noise ratio for detection is required in order to sense a local event taking place in a very small volume of several cubic nanometers. One technique for signal processing uses a modulated source using laser light and phase-sensitive detection, as well as a highly sensitive detector at the photon-counting level.

2.2.3.4. Production of the NOM Image

As is usual for general scanning probe-type microscopes, such as the electron STM and atomic force microscope (AFM), the NOM image is provided by means of a computer-aided mapping of the signal for the positions of the probe tip scanned in two dimensions. One can use either constant-height or constant-intensity operation, depending on the purpose of the measurement. In the latter scheme an electronic feedback loop is installed with the height-controlling driver of the three-dimensional piezotransducer. The technique has been developed extensively for general scanning probe microscopes.

A relatively flat background still remains in the signal. This should be subtracted in order to obtain an NOM image with good contrast. When a corrugated background appears due to the interference of optical waves, specific remedies need to be programmed in.

2.2.3.5. Interpretation of Images

In order to interpret the NOM signal, one needs to employ theoretical analysis regarding, for example, the nature of the interaction between the specimen and the probe tip, the polarization dependence on the illumination and signal collection schemes, and the dependence of the signal transfer of the optical fiber probe on the spatial modulation of the local field described in the spatial Fourier frequency basis. We will discuss these points in detail in the following sections.

2.3. THEORETICAL DESCRIPTION OF NEAR-FIELD OPTICAL MICROSCOPY

Proper theoretical understanding is indispensable for image interpretation with the NOM system. One has to deconvolute the specimen–probe tip interaction so as to determine the unperturbed state of the specimen. This requirement is general for any kind of interaction-type scanning probe microscopy. Thus the theoretical tools take up a large portion of this book.

2.3.1. Basic Character of the NOM Process

From a global point of view, the NOM system is a very complicated version of a light scattering problem in which the electromagnetic interaction of matter occurs on several different characteristic scales at once (Fig. 2.4a). However, in principle, if NOM is to work, it needs to provide a simple understanding and theoretical description of the local electromagnetic interaction, otherwise we cannot discuss any NOM image in relation to the topographic nature of an observed

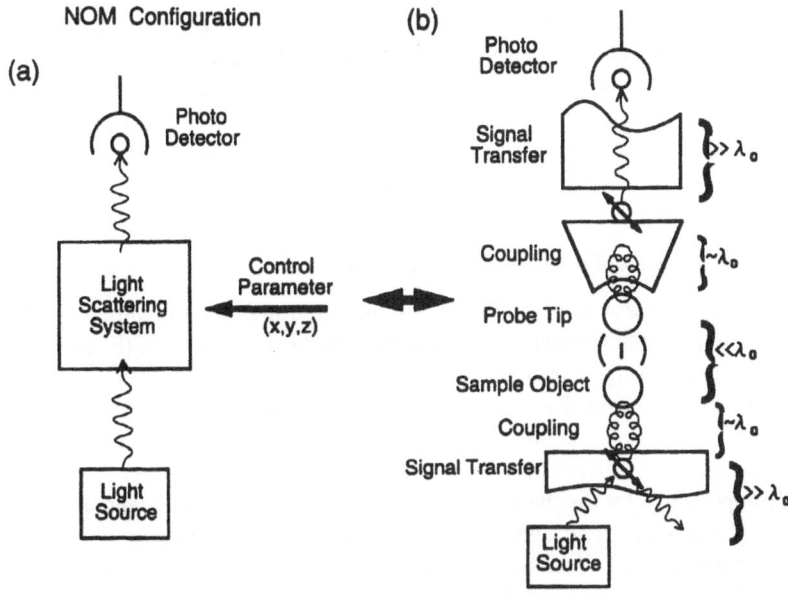

Figure 2.4. (a) The NOM configuration as an optical process and (b) characteristic interaction processes with several distinguishing scales. There are three fundamental parts: the local electromagnetic interaction in the near-field regime ($\ll \lambda_0$), near- to far-field coupling on the mesoscopic scale ($\sim\lambda_0$), and illumination and light transfer at the macroscopic scale ($\gg \lambda_0$).

specimen. Experimentally we have techniques for fabricating well-defined probe tips as well as sensitive signal collection and processing schemes. From the theoretical point of view, the problem is to identify the local interaction of a specimen and a probe tip as a well-defined subsystem involved in the global light scattering process.

The important point is that the NOM system involves several subsystems with different characteristic scales with respect to electromagnetic interaction (Fig. 2.4b), and we can identify the function of each subsystem with a certain clarity by virtue of the differences in scale. In this case we can describe the most important subsystem of the specimen plus probe tip system in terms of a local theoretical treatment, or nonglobal theory, which in turn allows us to interpret the NOM image. The remainder of the NOM system is considered as the means of communication between the macroscopic light source, photodetector, and subwavelength-size subsystem. There are two important schemes with different scales by which we can connect the scales from the nanometer to the micrometer ranges, namely near-to-far coupling and light wave transmission (Fig. 2.4b). This is how we give meaning to the process of near-field detection of near-field events within the global light scattering process of the NOM system. On this basis we can separate the NOM system into several components as discussed in the previous section.

It should be noted that such a nonglobal theoretical treatment is a general approach for mesoscopic systems and an appropriate model is an indispensable component in understanding the nature of the phenomena as well as their applications.

2.3.2. Extraction of a Meaningful Subsystem

Let us break down the NOM system into the following fundamental subsystems, as in Fig. 2.5: (a) local interaction of sample and probe tip, (b) coupling of local events to the signal transfer system, and (c) elimination of signal background. It is important to find the conditions for the appropriateness of this subsystem description from both the theoretical and experimental points of view.

Each of these subsystems is concerned with a different type of electromagnetic mode appropriate for the theoretical description of the physical event taking place in the specific subsystem, such as a planar, axial, or spherical mode.

It is therefore essential to have transformations between these mode descriptions, although each one of them spans a complete set of mode functions. For instance, a scattering problem concerned with spherical dielectric bodies involves a local specimen–probe tip interaction which is irradiated by an evanescent wave at a planar dielectric–air boundary and produces a signal coupled to an optical fiber probe with transfer modes of axial or cylindrical symmetry. The polarization dependence is also related to the axial symmetry of the local specimen–probe tip

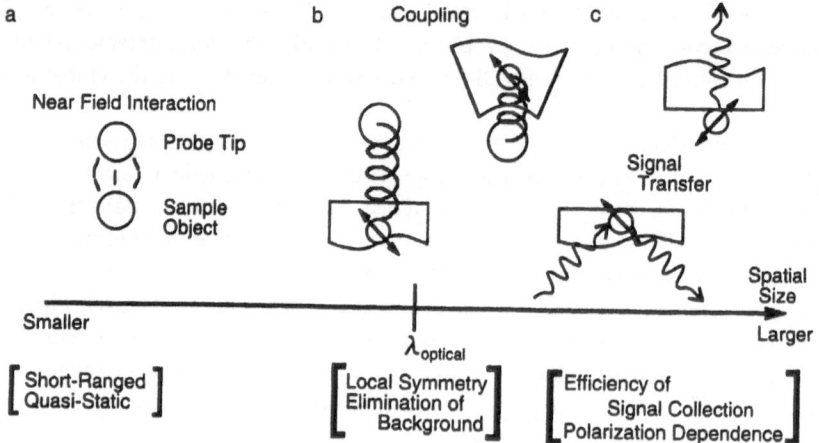

Figure 2.5. Building blocks of the NOM and their characteristics: (a) near-field interaction, (b) near- to far-field coupling, (c) illumination and signal transfer.

system. The vector plane waves and their coupling to spherical modes are concerned with the illumination of the specimen and the evaluation of signal background due to the direct coupling of the illuminating field to the probe tip. Several convenient formulations are provided in Chapters 8 and 9, including a description of the vectorial nature of the electromagnetic field in terms of the rotation group. In this book, we use extensively theoretical techniques from quantum mechanics which were in fact originally developed from work in optics. For example, to account for the polarization dependence of the near-field interaction we introduce a generalized cavity-QED concept to consider a representative dipole which produces a propagating light wave into the signal transmission line extended to the photodetector.

2.3.3. Demonstration of Localization in the Near-Field Interaction

To some extent, we can describe the NOM process in terms of conventional electromagnetic theories, such as boundary value problems, scattering problems, and so on. Let us introduce an instructive result derived from conventional electromagnetic theory by Jang and Jhe [1], which demonstrates the short-range nature and spatial localization of the optical interaction of two closely spaced objects such as the sample-plus-probe system of the NOM. The important point is that the result is scaled not by the optical wavelength, but by the sizes of the object and the probe.

It should be emphasized that a field calculation alone is not sufficient for the theoretical description of the NOM process. Instead we should consider an interacting sample–probe system as an important subsystem separated from a global NOM process. The model employed here consists of a small dielectric sphere as a sample placed on a flat dielectric surface and a small aperture as a probe swept above the sample, as shown in Fig. 2.6a. This models the basic structure of the sample–probe system in the NOM.

The sample–probe system is irradiated by an evanescent wave with wavelength larger than the sample–probe system. The transferred optical intensity is calculated on the basis of conventional Mie scattering and Kirchhoff's integral, and therefore the result is related to the far-field observation of the sample–probe scattering. Sample results are given in Figs. 2.6b and 2.7. The transferred intensity is shown versus the vertical sample–probe distance in Fig. 2.6 for the case when the aperture radius is equal to the radius of the sphere.

One clearly sees a rapid decay of the transferred intensity as the normalized sample–probe distance is increased. This size-dependent localization of the near-field optical interaction is one of the most important features of a general NOM system. In Fig. 2.7, the transferred optical intensity is shown as a function of the ratio of the aperture radius to the sphere radius. Transmission reaches a maximum when the size of the aperture and that of the sphere coincide. This sharp size–resonance

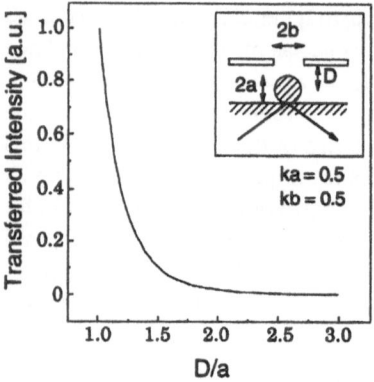

Figure 2.6. Size-dependent decay characteristics of optical near-field interaction between a small dielectric sphere (specimen) and a small aperture (probe) calculated by means of a non-global theoretical treatment developed by Jang and Jhe [1]. Total intensity scattered into the far-field region behind the aperture is indicated as a function of sphere–aperture distance. This shows evidence of the size-dependent localization of the optical near-field interaction between subwavelength-size objects.

behavior demonstrates another important characteristic of the general NOM process, that the optical near-field interaction between a sphere and an aperture exhibits a kind of resonance character with respect to spatial frequency.

These results provide theoretical background for NOM experiments. In order to obtain a clear image of a subwavelength-size specimen we need to use a probe tip with a size similar to that of the specimen, and in sweeping it, also keep the vertical distance close to the size of the specimen.

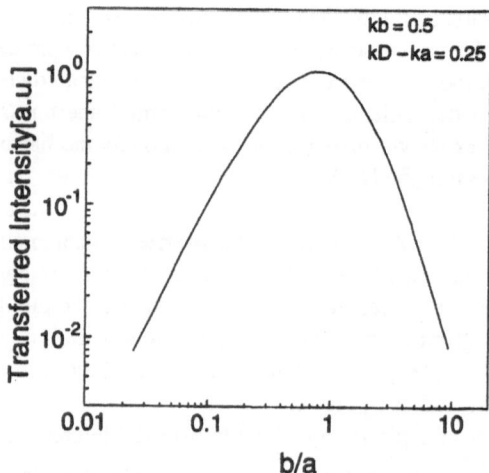

Figure 2.7. Size–resonance characteristics in the optical near-field interaction calculated by the same scheme as in Fig. 2.6 by Jang and Jhe [1]. The total scattered intensity is calculated as a function of the size of the probe aperture. This shows evidence of a size–resonance effect in the optical near-field interaction between subwavelength-size sphere and aperture.

2.3.4. Representation of the Spatial Localization of an Electromagnetic Event

To support experimental work with the NOM, the theoretical evaluation of the scattered intensity from the sample–probe system is not sufficient. First we need to understand the origin of the ultrahigh spatial frequency which enables the NOM to go beyond the diffraction limit. Second we must evaluate the spatial resolution of the NOM based on the idea of the signal transfer function which describes the spatial frequency filtering characteristics of the NOM measurement.

One of the most useful means of theoretical description is the angular spectrum representation of the scattered field. As discussed in Chapter 1, the angular spectrum provides a way to evaluate the electromagnetic interaction of the near-field regime in terms of waves with very large spatial frequency and corresponding short decay length. The scattered light field from an object of arbitrary shape is represented in terms of plane waves with a spectrum of angular distribution of wavevectors extending into the entire complex space. In other words, the scattered wave is represented as the sum of propagating (homogeneous) plane waves in all spatial directions and evanescent (inhomogeneous) waves with an entire set of values of the penetration depth. The importance of the plane wave expansion is related to the spatial translation, or displacement, which provides the measure of the interaction propagator. The propagating plane wave simply describes the parallel displacement in space, and the evanescent wave represents the short-range interaction with a rapid spatial frequency in the lateral direction. For instance, one can find the dominant spatial frequency from the peak of the angular spectrum and the lateral localization of the interaction from the spectral width around the peak. This can be compared with the Fourier analysis of the modulated signal. Knowing the peak in the Fourier spectrum, one can find the dominant temporal frequency contained in the signal, and from the spectral width one can find the duration for which the interaction is sustained. Such a theoretical basis enables us to analyze the NOM characteristics in terms of a transfer function with respect to spatial frequencies.

2.3.5. Model Description of a Local Electromagnetic Interaction

For further development of NOM STM theory one needs to establish an empirical or intuitive model describing the quasistatic and short-range local optical processes related to the nanometer-size specimen–probe tip interaction.

One instructive way to do this is to discuss the NOM process in the general framework of scanning probe microscopes including the electron STM and AFM. Such a model would be especially useful if it could provide interesting physical information on the local interacting subsystem, or more generally optical near-field problems, such as the excitation transfer or tunneling of quasiparticles. As an example, we will discuss in Chapter 8 a quasiparticle model of the local specimen–probe tip interaction described in terms of a Yukawa-type screened potential. Such a model provides us with a convenient way of evaluating the signal transfer function in terms of the spatial frequency, and thus to understand the NOM process as a spatial frequency filter. The model description has the potential to let us extract some physical information from the localized interaction or optical near-field problems. A quantum mechanical view and related problems also are discussed in Chapter 8. We also investigate the potential of the NOM as a near-field optical manipulator of atom- and molecule-size particles including nanometer-size biological samples.

Requirements for NOM theory are summarized:

- An *electromagnetic field calculation* demonstrating the existence of a localized electromagnetic field in the near-field regime and providing rigorous support for a more convenient description of the NOM.
- Expression of the specimen–probe tip interaction in terms of a *mode description* which enables us to understand the origin of high-spatial-frequency filtering and resulting resolution beyond the diffraction limit. This provides the basis to construct a signal transfer function of the NOM system with the character of a spatial frequency filter.
- Preparation of the mathematical background for *mode transforms* between planar, axial, and spherical electromagnetic descriptions consisting of the important subsystems of the overall NOM system.
- Empirical and intuitive *model* of a quasistatic and short-range local optical process related to specimen–probe tip interaction in a nanometer-size space.

Note: The following should be evaluated for the interpretation of an NOM image:

Near-Field Behavior of Sample Field

$\otimes$

Sample–Probe Interaction

$\otimes$

Illumination, Signal Collection, and Transfer

2.4. NEAR-FIELD PROBLEMS AND THE TUNNELING PROCESS

We conclude this introduction to near-field optical microscopy with a brief survey of the relation between tunneling phenomena and optical near-field problems. First we introduce the important idea of the tunneling current as developed by Bardeen as a theoretical description of the STM process, which is relevant to the electron near-field problem. The STM process is based on the electronic interaction of an atomic-size specimen and a probe tip, which needs to be treated theoretically as a coupled system. Bardeen's idea of the tunneling current will show us how to describe a sample–probe problem in terms of a coupled system. Then we will compare the tunneling problem of electrons with optical near-field problems.

2.4.1. Bardeen's Description of Tunneling Current in STM

According to Green's theorem, any field distribution can be reproduced from boundary values in terms of an integral of all the contributions from each segment of the source at the boundary to each point of interest inside the closed boundary (see Fig. 2.8). Now we would like to describe separate sample and probe fields in the theory of the tunneling process in the STM, as shown in Fig. 2.8.

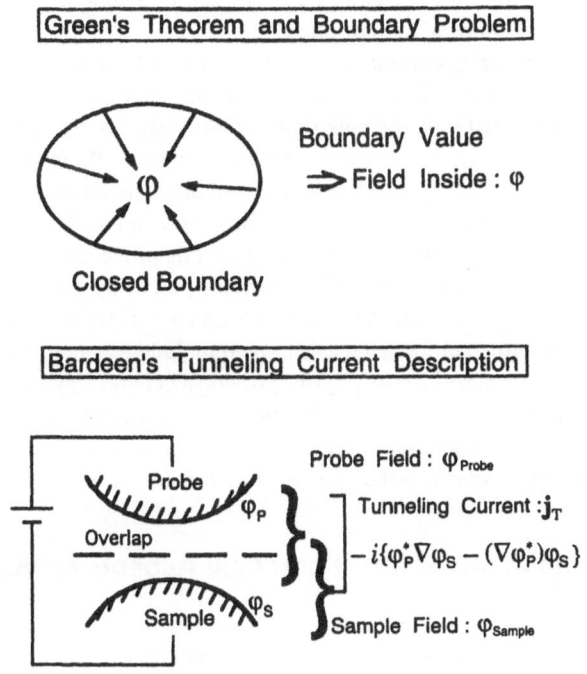

Figure 2.8. Boundary-value problem for disconnected surfaces, and Bardeen's tunneling current. One can see the distinguishing feature of the tunneling current description when one compares it with the ordinary boundary-value problem for a closed boundary based on Green's theorem.

The basic idea in Bardeen's picture to describe the tunneling process in STM is the separation of the sample wave function ψ_s, which extends to the source, and the probe wave function ψ_p, which extends to the detector. The sample wave function is defined to be an appropriate function only in the half-space including the source and the tunneling gap, and the probe wave function is the appropriate function in the half-space including the detector and the tunneling gap. From the wave function of the whole system, Bardeen extracted the most fundamental cross-term of the tunneling current,

$$j_{Bardeen} = -ie\{\psi_p^* \nabla \psi_s - (\nabla \psi_p^*)\psi_s\} \qquad (2.4.1)$$

and provided a comprehensive description of the STM process [2].

Our aim is to extend Bardeen's idea by studying the conditions under which we can relate the separate sample and probe evanescent fields to the idea of a tunneling current calculated from the overlap of these separate evanescent fields. Once we establish the analogy between NOM and STM, we can discuss many characteristics in the general framework of tunneling microscopy. We also can give such tunneling photons or excitations an interpretation which would allow further development of optical near-field physics.

Bardeen's tunneling current also offers a theoretical approach to general surface physics problems. In order to describe an electronic state at a material surface one usually start with the bulk states of a crystal. Subtracting the contribution of the crystalline layer just above the surface from the bulk electron correlations in the Hamiltonian of the system and quantizing the residual bulk terms, one can obtain the theoretical description of the surface state and analyze its nature. The upper half of the bulk material is then cut off by this procedure. Next, one puts a local probe on the material surface which recovers part of the electron correlations in the close vicinity of the surface to the probe-tip contact region. Then the part of the electron correlation subtracted needs to be reincluded as a tunneling term.

Such insights from solid-state physics provide a general way to understand the NOM as well as general near-field optical phenomena. In turn, the techniques and physics of near-field optics can provide useful methods to diagnose and produce novel electronic phenomena at the surface of materials.

2.4.2. Comparison of the Theoretical Aspects of NOM and STM

It will be shown in Chapter 8 that a light scattering problem, by a stepwise variation of the refractive index, is equivalent to that of the scattering of Schrödinger waves (Fig. 2.9a) when the problem is reduced to a one-dimensional one. In particular, the case of the total internal reflection of a light wave corresponds to the case of a wave incident with initial energy below the potential barrier and the

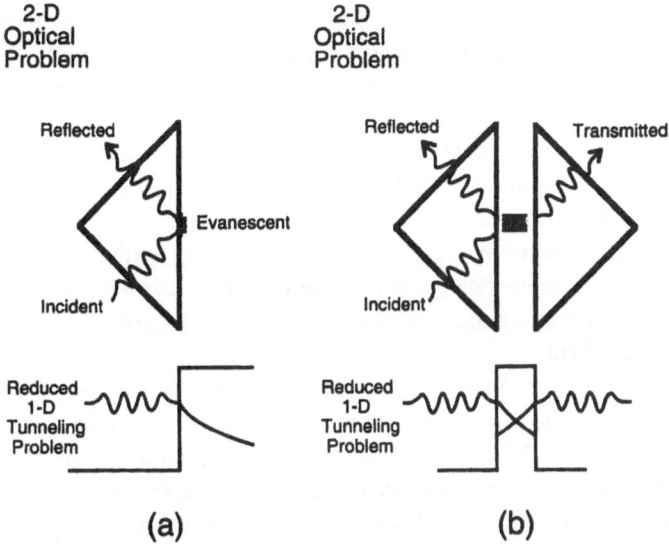

Figure 2.9. Evanescent waves and their relation to tunneling phenomena. (a) Evanescent wave at total internal reflection in the reduced one-dimensional potential barrier problem. (b) Evanescent waves and tunneling effect in the reduced one-dimensional potential barrier problem.

solution involves a wave exponentially decaying into the barrier region. This problem is extended to the case of the reduced one-dimensional refractive index barrier problem shown in Fig. 2.9b, where two planar dielectric interfaces are placed with a narrow gap of air, and light waves are incident at the angle of total internal reflection. The air gap between the dielectrics provides a reduced one-dimensional potential barrier for the light waves. The waves propagating in the medium and decaying from the surface correspond to good solutions for both source side and detector side. The overlap integral of evanescent waves describes the coupling of the decaying wave from the source side surface to the exponentially increasing wave on the detector side surface. Such surface-to-surface transmission of evanescent light waves is similar to the tunneling of Schrödinger waves [3, 4]. It has been shown that the angular spectrum representation of scattered waves and the introduction of evanescent waves are not just mathematical tricks, but exert physical effects on atomic particles [5].

In general, any scattered light wave from a material of arbitrary shape can be expressed in terms of plane waves when both homogeneous and inhomogeneous waves are included [6]. Such an expansion of an arbitrary form of scattered waves into plane waves with a complex wavenumber is called an angular spectrum representation. One must assume planar boundaries at the front and rear surfaces of the scatterer and expand the scattered wave in the form of a complex integral of plane waves with respect to the directional angle of the complex wavevector (Fig. 2.10a).

When the scattered field is described in terms of the plane waves, it is straightforward to translate spatially to another point in real space. Depending on the shape of the scatterer, the scattered wave involves the angular spectrum corresponding to the evanescent waves with short penetration depth and large wavevector along the boundary surface. The amplitude distribution in the angular spectrum then provides a measure of the range of penetration and localization of the optical near-field.

The interaction of two closely spaced scatterers in the near-field zone is described in terms of an overlap integral of the corresponding angular spectrum components (see Fig. 2.10b). Therefore the angular spectrum representation provides a measure of the range and localization of the near-field interaction. We will see in Chapter 9 that it is possible to develop an angular spectrum representation of a two-boundary interaction propagator. By comparing these theoretical descriptions with the case of electron waves we can achieve a tunneling view of the NOM

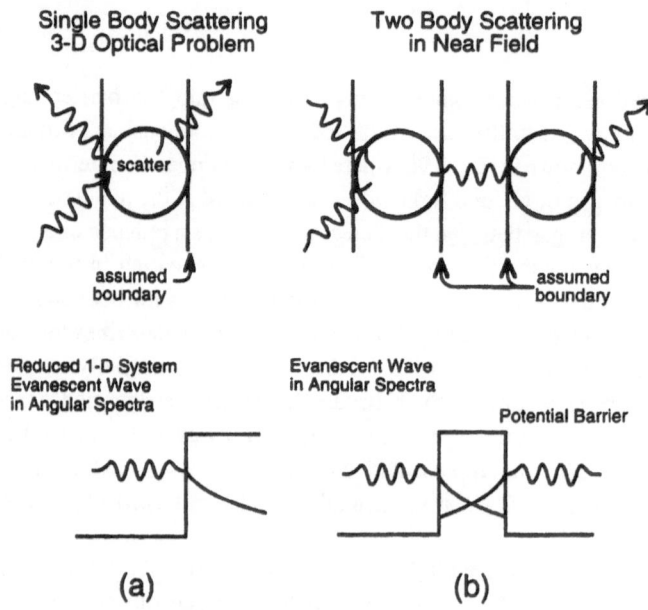

Figure 2.10. Angular spectrum representation of scattered fields and its relation to the reduced one-dimensional tunneling problem.

process and understand it in the general framework of scanning probe-type microscopy.

2.5. REFERENCES

1. K. Jang and W. Jhe, Nonglobal model for a near-field scanning optical microscope using diffraction of the optical near field, *Opt. Lett.* **21:** 236–238 (1996).
2. J. Bardeen, Tunneling from a many-particle point of view, *Phys. Rev. Lett.* **6:** 57–59 (1961).
3. Th. Martin and R. Landauer, Time delay of evanescent electromagnetic waves and the analogy to particle tunneling, *Phys. Rev. A* **45:** 2611–2617 (1992).
4. A. M. Steinberg and R. Chiao, Tunneling delay times in one and two dimensions, *Phys. Rev. A* **49:** 3283–3295 (1994).
5. T. Matsudo, H. Hori, T. Inoue, H. Iwata, Y. Inoue, and T. Sakurai, Direct detection of evanescent electromagnetic waves at a planar dielectric surface by laser atomic spectroscopy, *Phys. Rev. A* **55:** 2406–2412 (1997).
6. E. Wolf and M. Niet-Vesperinas, Analyticity of the angular spectrum amplitude of scattered fields and some of its consequence, *J. Opt. Soc. Am. A* **2:** 886–890 (1985).

THREE

INSTRUMENTATION

3.1. BASIC SYSTEMS OF A NEAR-FIELD OPTICAL MICROSCOPE

Figure 3.1 explains the basic concept of a near-field optical microscope (NOM). Figure 3.1a shows how an evanescent field is generated on the surface of a subwavelength-size sample sphere. The volume of the evanescent field depends on the size of the sample, as described in Section 2.4.3 (see Fig. 2.6). Since an evanescent field is nonradiative, it cannot be detected in the far-field region. By placing a subwavelength-size sphere in the near-field region to scatter the evanescent field, the scattered light intensity can be detected in the far-field. The relation between the probe position and the scattered light intensity, obtained by scanning the sphere, displays the spatial profile of the evanescent field. The optical response of the sample surface is thus obtained. This relation corresponds to the image taken by the NOM. Characteristics of the probe (size, shape, structure) and the sample–

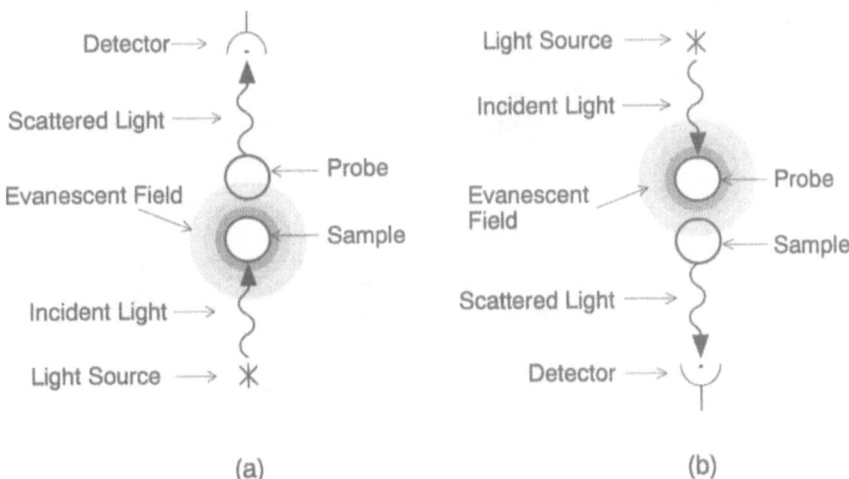

Figure 3.1. Basic NOM configurations: (a) Collection mode (C-mode). (b) Illumination mode (I-mode).

probe separation determine the image resolution, which can be higher than that of the conventional diffraction-limited optical microscope. Since the evanescent field on the sample surface is scattered and collected by the probe, this mode of operation (shown in Fig. 3.1a) is called the *collection-mode* (C-mode).

Not only is the evanescent field scattered by the probe, but the resulting scattered field is in turn scattered by the sample itself and also by the probe. Therefore, the NOM is operated by utilizing the short-range electromagnetic interaction between the sample and probe via an evanescent field. Since this interaction system is the same even if the sample and probe positions are inter-changed, the sample can alternatively be illuminated by the evanescent field on the probe surface. This is called the *illumination mode* (I-mode), as illustrated in Fig. 3.1b.

More practical structures for the C- and I-modes are shown in Figs. 3.2a and 3.2b, respectively. Here, the top of a metal-coated sharpened fiber is used as a probe, whose representative shape is shown in Fig. 4.23. Instead of the spherical probe,

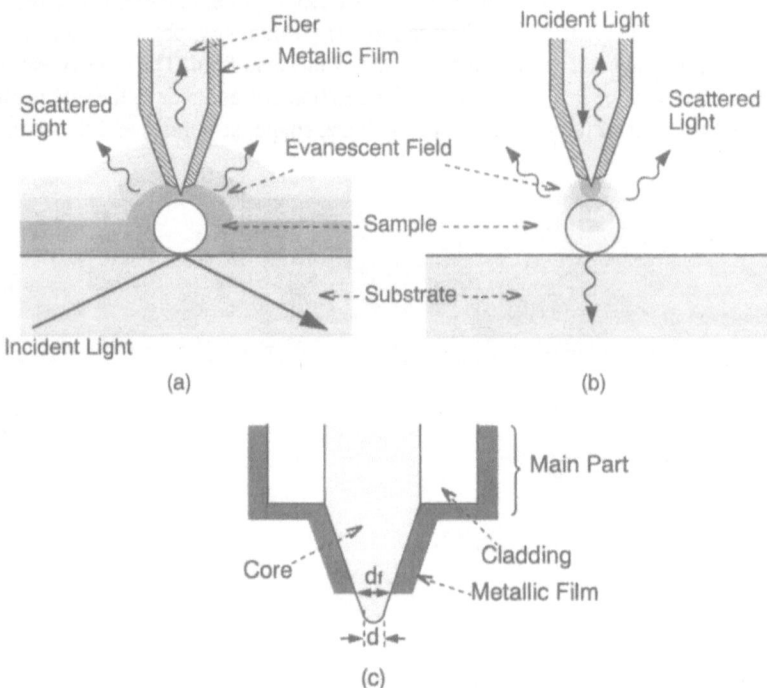

Figure 3.2. Practical basic structure of the NOM using an inverted conical probe, which is a sharpened fiber core protruding from the metal film. (a) Collection mode (C-mode). (b) Illumination mode (I-mode). (c) Definition of the foot diameter d_f and apex diameter d.

an inverted conical probe is used, which is a sharpened fiber core protruding from the metal film. This type of inverted conical probe is called a *protruded probe*. A schematic view of the metal-coated sharpened fiber probe is shown in Fig. 3.2c. The foot diameter d_f and apex diameter d are defined as shown in the figure. (Definition of the tip diameter d for more practical fiber probes is given in Fig. 4.2) The main part of the fiber of which the probe tip is formed is used to transmit the light scattered by the probe tip.

Due to the resonance phenomenon related to the sizes of the sample and the probe (see Fig. 2.7), it is essential to use a probe whose size is equivalent to that of the sample in order to have the most effective scattering of the evanescent field on the sample surface, and hence the most sensitive imaging of the sample. Furthermore, due to this resonance phenomenon and the size-dependent decay length of the evanescent field intensity away from the sample surface, the detection efficiency of the evanescent field by the inverted conical probe exhibits bandpass filtering characteristics in the spatial Fourier frequency domain (Fig. 3.3a). Although the bandpass filtering characteristic depends also on the sample–probe separation, its high and low cutoff frequencies are given by the inverse of the apex diameter and

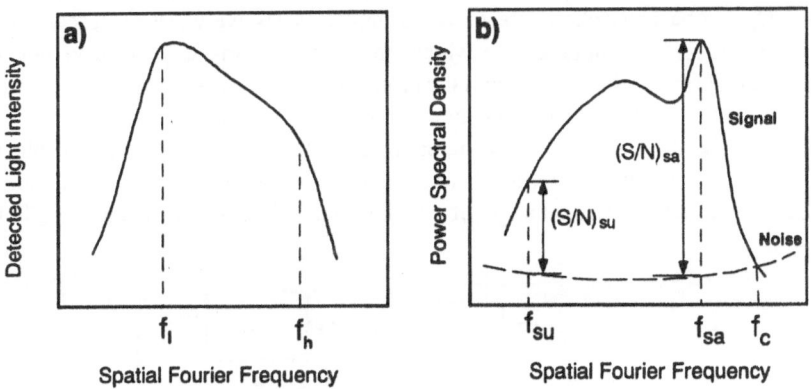

Figure 3.3. (a) Bandpass filtering characteristics of the detection efficiency of the scattered light intensity by the inverted conical fiber probe of the C-mode NOM, displayed as a function of the spatial Fourier frequency. The high and low cutoff frequencies f_h and f_l are given by the inverses of the apex diameter d and the foot diameter d_f of the protruded probe, respectively. (b) An alternative measure of resolution. The solid curve is the spatial power spectral density of the image intensities. The broken curve represents the noise magnitude of the NOM system. The resolution can be defined by the inverse of the spatial Fourier frequency f_c at 0-dB signal-to-noise ratio (S/N), i.e., the frequency at which the two curves cross. However, this measure is determined not only by the deterministic characteristics of the NOM, but also by the characteristics of the sample and noise. The contrast of the image can be defined as the ratio of the values of $(S/N)_{sa}$ and $(S/N)_{su}$ at the spatial Fourier frequencies f_{sa} and f_{su}, which correspond to the sizes of the sample and substrate, respectively.

the foot diameter of the protruded probe, respectively. The high cutoff frequency can be used as a measure of the resolution of the NOM.

Figure 3.3b shows an alternative definition of resolution. Here, resolution is given by the spatial Fourier frequency f_c at 0-dB signal-to-noise ratio (S/N), i.e., the frequency at which the spatial power spectral density of the image intensity intersects that of the noise associated with the detection system. It should be noted that this measure of resolution is not determined solely by the deterministic characteristics of the NOM, but depends also on the characteristics of the sample and noise.

The contrast of the image can be defined as the ratio between the values of S/N at the spatial Fourier frequencies which correspond, respectively, to the size of the observed smallest feature in the sample and the variation of the surface.

3.1.1. Modes of Operation

3.1.1.1. Collection Mode

Figure 3.4 shows a typical C-mode NOM system, in which a light source, a sample, a probe, and a photodetector are arranged successively. An evanescent field is generated on the sample surface, which mediates the short-range electromagnetic interaction between the sample and the probe. This interaction generates scattered light from the probe, whose intensity is measured by a photodetector.

To facilitate scanning, either the prism-mounted sample or the fiber probe is placed on an actuator. The light is incident into the substrate prism with an incident angle larger than the critical angle of total internal reflection. The evanescent field

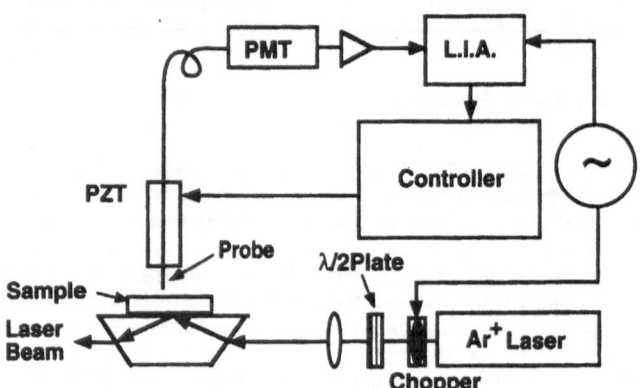

Figure 3.4. The experimental setup of the C-mode NOM. L.I.A., Lock-in amplifier; PMT, photomultiplier tube; PZT, pieozoelectric transducer.

generated on the sample surface is scattered by the probe. The main part of the fiber probe collects some of the scattered light, converting it into the guided mode of the fiber. At the photodetector this light is converted into an electrical signal. The signal intensity is then plotted as a function of the position of the probe which is scanned over the sample surface. This is the image of the sample. The required scanning control and the image processing units are the same as used for a scanning probe microscope (SPM) [1].

The detected light power depends on the output power of the light source and the shape, structure, and material of the probe. As the detected power is very low (picowatt or nanowatt range), a highly sensitive detection technique is required (see Section 3.2). For example, Fig. 3.5 shows the variation of detected power plotted on a semilog scale as a function of the sample–probe separation for different cone angles of the probe without metal coating. This figure shows that for an incident light power of 30 mW the typical detected light power is only several picowatts when the sample–probe separation is fixed at about 10 nm. This figure also shows that the sharper the probe, the lower is the detected intensity. This suggests that spatial resolution and detection sensitivity must be traded off against each other.

Characteristics of the C-mode NOM which have been empirically found by the authors are as follows:

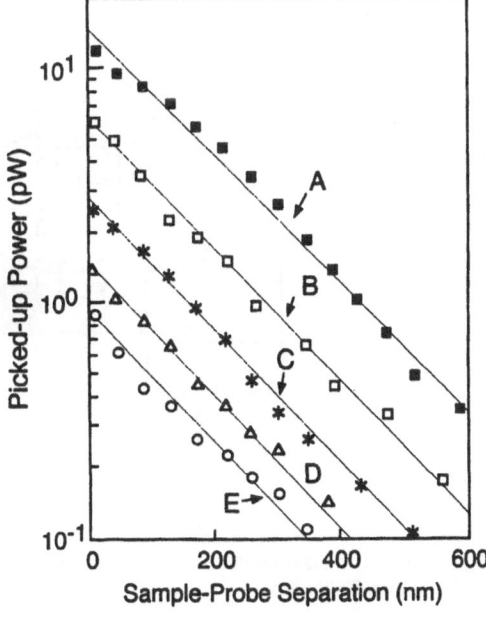

Figure 3.5. The detected power as a function of the sample–probe separation for different cone angles of the probe. A semiconductor laser of 780 nm wavelength and 30 mW output power was used as a light source. Five sets of results were obtained by using fibers sharpened by one-step etching (Section 4.1.1) without coating the metallic film. Their cone angles are (A) 100, (B) 45, (C) 34, (D) 25, and (E) 20 deg.

1. The sample–probe separation can be controlled by monitoring the rapid change of the detected signal intensity as a function of the sample–probe separation, as shown by Fig. 3.2.
2. The polarization state of the incident light can be easily controlled by using a waveplate.
3. Operation in a liquid environment is easy because of the capability of optical feedback given by characteristic 1.
4. A transparent or semitransparent material has to be used for the substrate of the sample. It is difficult to observe thick, strongly absorbing samples.

3.1.1.2. Illumination Mode

Figure 3.6 shows a typical I-mode NOM system, in which the roles of the probe and sample are exchanged, i.e., the evanescent field is generated on the probe surface and is scattered by the sample. Part of the scattered light is detected and is displayed as a function of the position of the scanned probe.

The scattered light can be detected by conventional collection optics. In some cases, however, e.g., photoluminescence (PL) spectroscopy of a semiconductor (see Section 5.3.2), the PL signal is collected by the probe (as in the C-mode) in order to avoid the limit placed on spatial resolution by the diffusion lengths of electrons and holes.

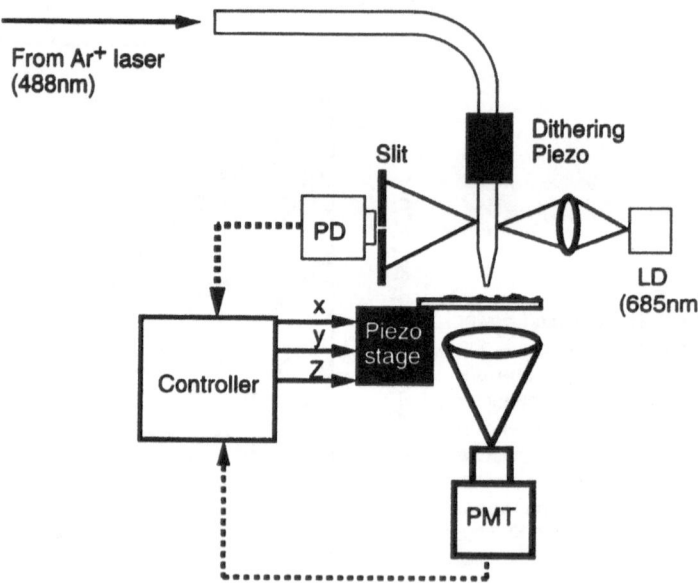

Figure 3.6.   The experimental setup of the I-mode NOM. LD, Diode laser; PD, photodiode; PMT, photomultiplier tube.

Characteristics of the I-mode NOM which have been empirically found by the authors are as follows:

1. High image contrast of the sample is obtained as compared with the image of the substrate because the sample is selectively illuminated by the evanescent field. However, the contrast depends on the nature of the sample.
2. Contrast enhancement is possible when the sample exhibits strong absorption.
3. The signal-to-noise ratio can be relatively high.
4. Such a system can be used for fabrication and manipulation due to the use of local illumination.
5. An auxiliary method of monitoring the sample–probe separation is required for the separation control. One popular method is to monitor the shear force between the sample and the probe by dithering the probe laterally (Section 3.1.2).
6. Due to the lateral dithering, there are chances for the probe to scratch the surface of fragile and soft samples such as biological samples.
7. Operation flexibility depends on the feedback control used. Under shear-force feedback, operation in water is difficult due to the decrease in the Q-factor of the resonance dithering.
8. Interpretation of the characteristics of the obtained image is difficult. This is because the probe is scanned by monitoring the equipotential contour of the shear force and is not scanned by monitoring the equienergy contour of the evanescent field. This difficulty can cause artifacts in the image [2].
9. Under very small sample–probe separation, the dithering motion of the probe introduces cross-talk between the I-mode and shear-force topographic images.
10. It is difficult to determine *a priori* the exact polarization state of the light emitted from the probe.
11. Too much incident light power into the probe will induce residual heating at the apex of the probe, which will damage the probe. The temperature increase due to this residual heating has been measured by using a micrometer-size thermocouple [3].

3.1.2. Position Control of the Probe

3.1.2.1. Collection Mode

The sample–probe separation can be regulated by monitoring the rapid change of the detected scattered light intensity as a function of the sample–probe separation (see Fig. 3.5). That is, the monitored signal is negatively fed back to the actuator to control the probe position. By this servocontrol, the probe can be scanned on the equipower contour of the evanescent field.

3.1.2.2. Illumination Mode

An auxiliary method of monitoring the sample–probe separation is required for position control of the probe. One popular method is to monitor the shear force between the sample and the probe [4, 5]. This is accomplished by dithering the probe laterally (horizontally in Fig. 3.6) at its resonant frequency. The dithering amplitude is measured by shining laser light on the side of the probe and measuring the scattered light power using phase-sensitive detection. The additional light source and the detector used for this measurement are shown in Fig. 3.6. Figure 3.7 shows the resonance curve of the lateral dithering for a fiber probe with a reduced-diameter cladding (fabricated by two-step etching Section 4.1.2). The resonance frequency and Q-factor are 24.9 kHz and 360, respectively, sufficiently large for the detection of shear force. Values of this order can be realized when the length of the supported probe is 2 mm. For probes with the original diameter of the cladding, these values were much smaller. The two-step etching process is indispensable for realizing probes capable of shear-force detection.

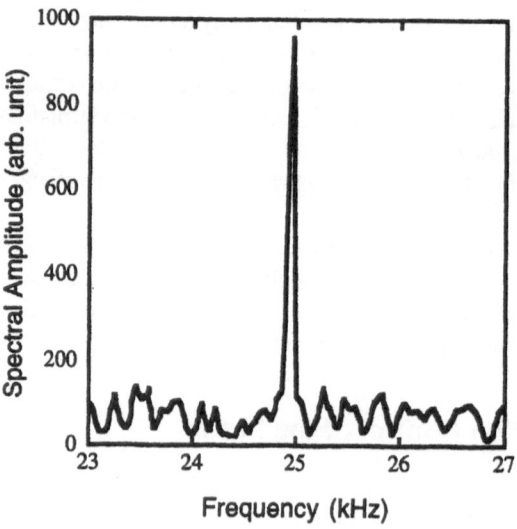

Figure 3.7. The resonance curve of the lateral dithering for a fiber probe with a reduced-diameter cladding. Resonance frequency and the Q-value are 24.9 kHz and 360, respectively.

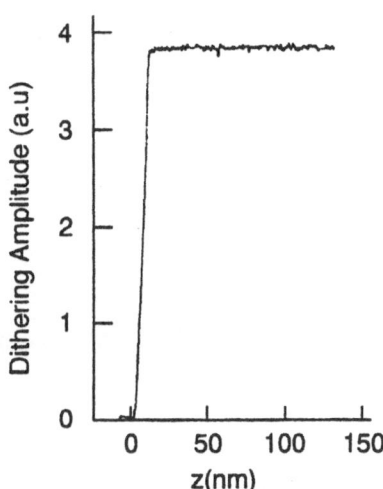

Figure 3.8. The relation between the sample–probe separation z and the measured dithering amplitude.

When the sample–probe separation is smaller than 10 nm, the dithering amplitude decreases rapidly due to shear force, as shown in Fig. 3.8. This signal is negatively fed back to the actuator (cf. Fig. 3.11) to control the probe position. Thus, the sample–probe separation can be regulated by keeping the shear force constant. It should be noted that the equipotential contour on which the probe is scanned is usually different from the equienergy contour of the evanescent field. This difference can induce artifacts in the optical near-field image [2]. As a new method of monitoring shear force, a nonoptical method has been recently developed by using a tuning fork [6].

An alternative method of detecting the atomic force between the sample and probe is required when imaging fragile samples such as biological specimens, as they can be easily damaged by the laterally dithered probe. One such method is shown in Fig. 3.9a. A probe fabricated by two-step etching is scanned by tilting its axis about 5 deg relative to the vertical. The magnitude of deflection of the tilted probe due to the atomic force can be easily measured by detecting the scattered laser power. The reason this tilted probe is sensitive to the atomic force is that its reduced cladding diameter allows it to bend easily. Figure 3.9b shows the dependence of the measured atomic force on the sample–probe separation. Since this force exists in the vicinity of the sample surface, very high gain can be obtained for servocontrol. One problem is the coupling between the x and y components of the deflection characteristics due to the axially symmetric shape and spring constant of the fiber, resulting in fluctuations of the probe along the y axis when it is scanned along the x axis of the sample surface.

Characteristics 6–9 given in Section 3.1.1.2 could impose some technical problems when using shear force or atomic force for position control of the probe.

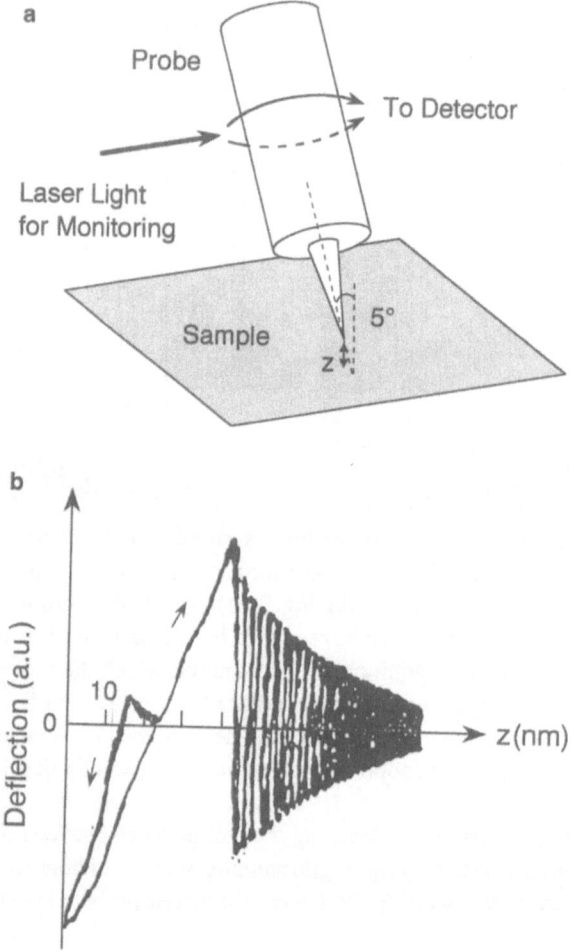

Figure 3.9. (a) Detection of an atomic force between the sample surface and fiber probe tip with decreased cladding diameter, whose axis was tilted about 5 deg. (b) Dependence of measured atomic force on the sample–probe separation z.

In order to solve these problems, optical feedback techniques can be employed by using a secondary light whose wavelength is different from that used for the imaging. As shown in Fig. 3.10, this light is incident into the rear surface of the substrate of the sample under the condition of total internal reflection in order to generate the evanescent field on the sample surface. This evanescent field is used for position control of the probe, as in the case of the C-mode. Proper choice of

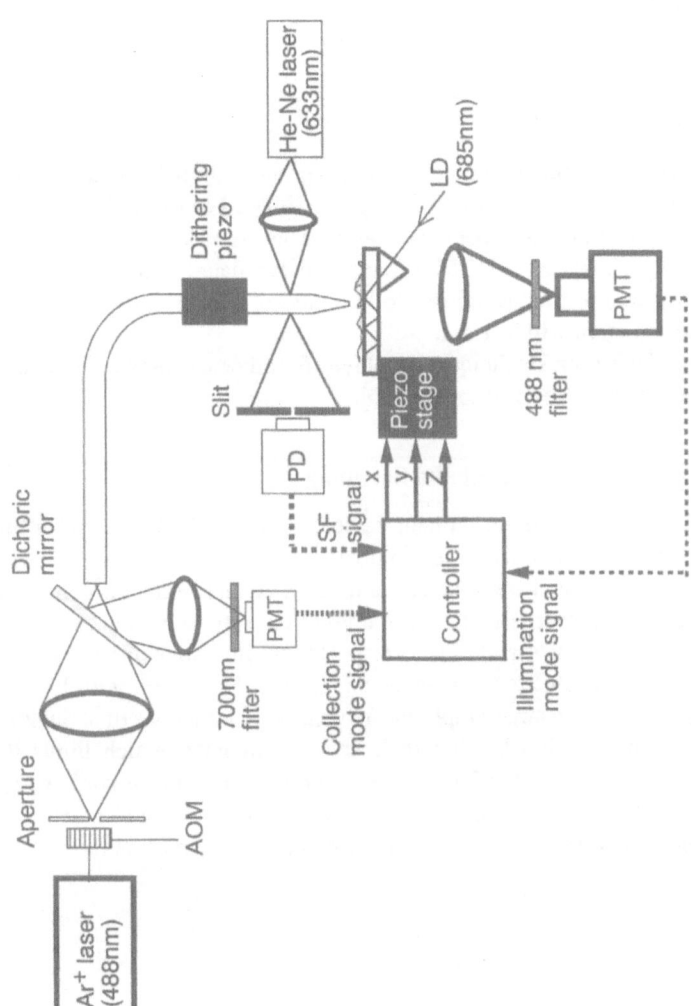

Figure 3.10. Experimental setup of the I-mode NOM utilizing the secondary evanescent field generated from a light source with different wavelength for controlling the sample–probe separation. LD, Diode laser; PD, photodiode; PMT, photomultiplier tube; AOM, acoustooptic modulator.

wavelength of the secondary light will sufficiently reduce the cross-talk problem. Although this control method cannot be applied for an opaque substrate, the probe can be scanned on the equienergy contour of the evanescent field for the imaging. This is because the intensity distribution of the evanescent field on the sample surface is independent of wavelength if the sample material is not resonant to the light.

3.1.3. Mechanical Components

Special devices and techniques for scanning a probe in the near-field region on the sample surface are essential for realizing the NOM. For example, since the thermal expansion coefficient of conventional materials is as large as $1 \times 10^{-5}/°C$ m, care must be taken when the NOM is made of these materials. For such cases, the ambient temperature fluctuation has to be reduced to 0.01°C to regulate the sample–probe separation within 1 nm.

Mechanical devices and techniques employed for other scanning probe microscopes [1] have also been utilized for the NOM.

3.1.3.1. Devices for Fine Control and Scanning

Figure 3.11a shows the basic actuator device, a tripod, which is used for fine control and scanning of the probe. It is composed of three columns several square millimeters in cross-sectional area and tens of millimeters in length. It is composed of a piezoelectric transducer (PZT), whose conversion efficiency and maximum excursion are about 1 nm/V and 1 μm, respectively. Its length is changed by applying a voltage to the electrodes attached on its sides. Nonlinearity is observed when it is modulated with a large-amplitude ac voltage. A hysteresis of about 20% is also seen. The modulation bandwidth is several kilohertz, which limits the maximum scanning speed. Figure 3.11b shows a more advanced actuator in which the temperature drift of the horizontal scanning is reduced by fixing four actuators to form a symmetric cross. The fifth actuator for vertical scanning is fixed at the

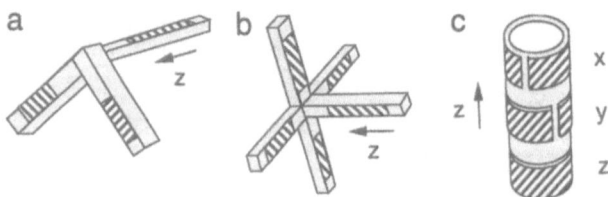

Figure 3.11. (a) The tripod-type actuator. (b) An advanced actuator to compensate for the temperature drift. (c) A tube-type actuator.

center of the symmetric cross. The length of each component is about 10 mm and the resonance frequency is as high as 20 kHz. Figure 3.11c shows a popular tube-type actuator. It utilizes the expansion and bending of a cylindrical PZT device. Two of its advantageous properties are its high symmetry and high resonance frequency (20 kHz). However, the motions along the two horizontal axes (x and y axes) are not completely independent of each other.

3.1.3.2. Reduction of Mechanical Vibration

Techniques for reducing the mechanical vibration from external sources are required in order to control the sample–probe separation accurately. As a reference, the amplitude of mechanical vibrations of a conventional laboratory floor is as large as 1 μm in the frequency range lower than 100 Hz. In order to isolate the NOM system from these vibrations, the resonant frequency of the vibration-isolating table should be lower than 1 Hz and that of the NOM unit should be greater than 10–20 kHz.

Since a single vibration-isolation table may not be sufficient for some NOM systems, additional small isolation tables may be installed between the main isolation table and the NOM unit for further vibrational reduction.

3.1.3.3. Reduction of Acoustic Vibration

Reduction of acoustic vibration is necessary to reduce low-frequency noise because the fiber probe is sensitive to acoustics. Reduction of acoustic power as high as 40 dB can be obtained at 125 Hz by using a shield box of glass wool. The magnitude of the reduction increases monotonically with increasing acoustic frequency. A reduction of 60 dB is possible at 4 kHz. Further effective reduction can be expected by fixing a lead sheet on the inner wall of the box.

3.1.4. Noise Sources Internal to the NOM

To avoid unwanted scattered light sources, a thin sample should be carefully fixed on the substrate. Processes of fixing samples on the substrate should be similar to those used in atomic force microscopy (AFM). Even when the sample is carefully prepared and steps are taken to reduce mechanical and acoustic vibrations, there can be several noises mixed into the output signal of the NOM system. Such noises have to be carefully rejected to allow the very weak signal light to be detected. Possible noise sources are summarized in the following.

3.1.4.1. Light Scattering

As described in Section 8.3.2.2, physical information which depends solely on the size of the sample can only be obtained when the sample size is smaller than about 100 nm. If either the sample or probe is larger than 100 nm, interpretation of the obtained image becomes complicated since scattered light intensity can be mixed into the near-field components. For example, Fig. 3.12 shows the relation between the sample–probe separation and the light power detected by the C-mode NOM, for which a diffraction grating (500 nm pitch and 100 nm corrugation depth) is used as a sample. Superimposed on the expected rapid decrease of the detected light power is a successive periodic variation due to the interference of the scattered light generated around the edges of the diffraction grating because of the large pitch and depth of the corrugations. This shows that the contribution of the scattered light cannot be neglected in the near-field region, making the interpretation of the measured image very complex. Thus, in order to study the intrinsic features of short-range electromagnetic interaction between the sample and probe via an evanescent field, it is essential to carry out high-spatial-resolution imaging experiments using a sample and probe much smaller than 100 nm.

3.1.4.2. Mechanical/Acoustic Vibrations and Thermal Drift

Typical magnitudes of the noises due to mechanical/acoustic vibration and thermal drift are shown in Fig. 3.13. Curve A shows the power spectral density of the relative noise intensity associated with a low-optical-power detection of an evanescent field on a flat glass plate by a C-mode NOM, where the semiconductor laser power incident into the glass plate is 10 mW. Curve B is the reference level representing the relative noise intensity of –40 dB, which corresponds to an error variance of 1 nm^2/Hz in measuring the sample–probe separation; in other words, the normal resolution (i.e., discrimination sensitivity of intensity variation of the

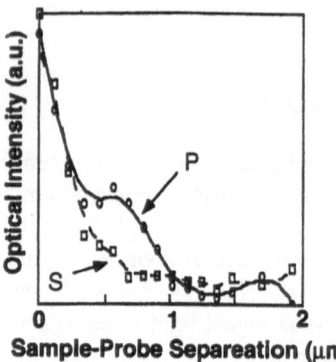

Figure 3.12. The relation between the sample–probe separation and the light power detected by the C-mode NOM, for which the surface of a diffraction grating was used as a specimen. P and S, p-polarized and s-polarized incident light, respectively.

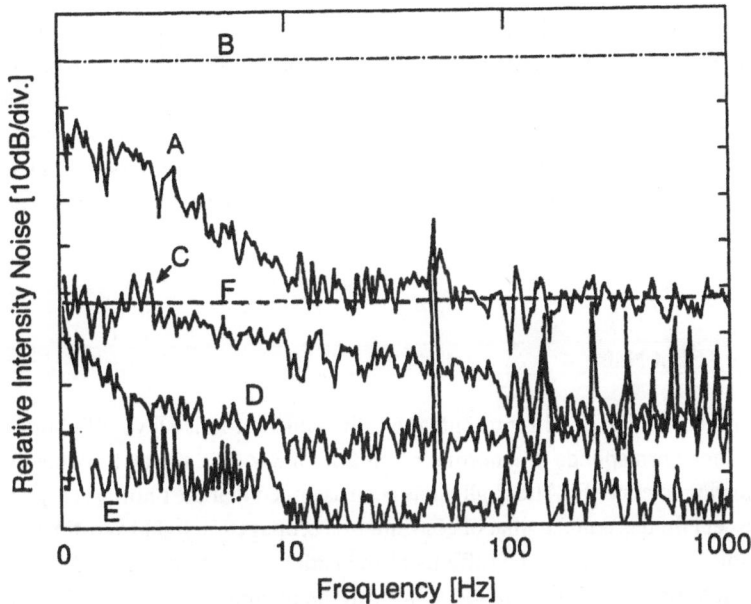

Figure 3.13. The power spectral density of the relative intensity noise associated with the C-mode NOM. The laser power incident into the prism was 10 mW. (A) Measured value, (B) reference level representing the relative intensity noise of −40 dB, which corresponds to an error variance of 1 nm^2/Hz in measuring the prism–probe separation, (C) intensity fluctuation of the incident laser light, (D) noise from the power supply for the PZT actuator, (E) excess noise from the photodetector, (F) estimated shot-noise level.

evanescent field normal to the sample surface) is 1 nm at 1 Hz bandwidth of the measurement. The magnitudes of laser power fluctuation (curve C) and electrical noises (curves D and E) have been reduced to be lower than the value of curve A.

By comparing the curves, it is found that the principal noise source at Fourier frequencies lower than 10 Hz is due to mechanical/acoustic vibrations and thermal drift of the fiber, while that for $f > 10$ Hz is the shot noise of detected photons. Curve A shows that the mechanical/acoustic vibration- and thermal drift-limited normal resolution at $f = 1$ Hz is 0.1 nm at 1 Hz bandwidth. Careful isolation of mechanical vibration, use of an acoustic shield, and temperature control are required for further improvements in resolution.

3.1.4.3. Electrical Noise

Care should be taken to reject the electrical noise which is mixed into the very weak signal of the NOM system. The actuator made of a PZT and its power supply can pick up electrical noises because of the high electrical impedance of the PZT. The optical power fluctuations of the laser caused by the electrical noise of the power supply should be carefully reduced. Furthermore, a low-noise postdetector amplifier should be used in order to reduce the excess noise from the photodetector. Typical power spectral densities of these noises are shown by curves C, D, and E of Fig. 3.13. It should be noted that these noises have been carefully reduced to a level lower than that of curve A.

3.1.4.4. Shot Noise

Curve F of Fig. 3.13 represents the estimated shot-noise level of photodetection, which corresponds to a normal resolution of 3×10^{-3} nm at 1 Hz bandwidth by referring to curve B. Although it has been assumed that the photon energy of the evanescent field is equal to that of conventional propagating light for estimating the shot-noise level, it may not be sufficient to consider only the zero-point fluctuations of the light field for this estimation. Instead, the magnitude of zero-point fluctuations of a small number of elementary excitations need to be estimated because the shot noise is generated at the point where the evanescent field and the optically induced polarizations couple to each other at the sample surface. In this sense, it should be noted that the specific noise is included in the evanescent field detection, which is different from the noise properties in detecting conventional propagating light.

3.1.5. Operation under Special Circumstances

One of the advantageous properties of the NOM is its capability of operation in water as well as in air. In contrast to the STM, which is used in high vacuum, during operation in air, water vapor molecules can contaminate the probe and sample surfaces. In particular, when the probe is made of porous silica glass, water vapor is easily adsorbed on the probe surface. The effect of surface contamination will be observed if the spatial resolution of the NOM is improved further in the future. Furthermore, one should reduce the water capillary effect due to water vapor molecules in the gap between the sample and probe, not only to reduce the effect of water, but also to avoid the attractive force due to the electric charges on dielectric surfaces.

Additional technical considerations required when the NOM is operated in water or at low temperature are reviewed in the following.

3.1.5.1. NOM Operation in Water

It is essential to be able to operate the NOM in a liquid environment in order to obtain the image of a living biological sample. The most important technique for this operation is to control the sample–probe separation. The scanning speed of the probe should also be decreased due to the viscous drag of water. Furthermore, the probe position may fluctuate randomly due to the Brownian motion of the water molecules. This may be observed in the future with improved spatial resolution.

Especially when shear force is used to control the sample–probe separation in the I-mode NOM, the Q-value of the resonant dithering of the probe is decreased drastically in water due to viscous drag. The screening of the atomic force by water molecules further decreases the slope of the force curve, reducing the gain of the servocontrol loop regulating the sample–probe separation. High-resolution imaging is, however, possible with the C-mode NOM because shear force detection is not used (Section 5.2.1.2).

3.1.5.2. NOM Operation at Low Temperature

Low-temperature (i.e., liquid He) operation is required for spatially resolved photoluminescence spectroscopy of semiconductors, as their emission efficiencies increase due to the reduction of nonradiative relaxation processes at low temperature. In addition, a narrowing of the emission spectral linewidth can also be expected. For low-temperature operation, several technical problems need to be solved: (1) The conversion efficiency of the PZT actuator is temperature dependent. For example, the conversion efficiency of a tube-type PZT actuator (200 nm/V at room temperature) is decreased to 24 nm/V at 10 K. This decrease requires careful control of the gain of the servocontrol circuit in order to maintain both constant sample–probe separation and maximum excursion of the scanning. (2) The thermal expansion coefficients differ among the components. For example, the resin used to fix a fiber probe can crack. Mechanical supports for the sample, probe, and PZT actuators need to be designed using materials whose thermal expansion coefficients are close to that of the PZT [an alloy of Fe + Ni (42%) is one such material]. (3) Mechanical vibration due to the flow of liquid He is an extra noise source in the shear-force signal used for control in the I-mode NOM. Careful isolation from the source of mechanical vibration is required.

Figure 3.14 shows a low-temperature I-mode NOM system which has solved the three problems mentioned above [7]. (A similar system has also been constructed by Grober et al. [8].) The system is installed in a helium-flow-type cryostat of 3.5 inch inner diameter. Its height and base area are 200 mm and 50×50 mm^2, respectively. Liquid He is introduced through a transfer tube. Temperature sensors and heaters are fixed at the input port and a mounting table for the sample. Controlling electronics can vary the temperature of the mounting table down to 4.2

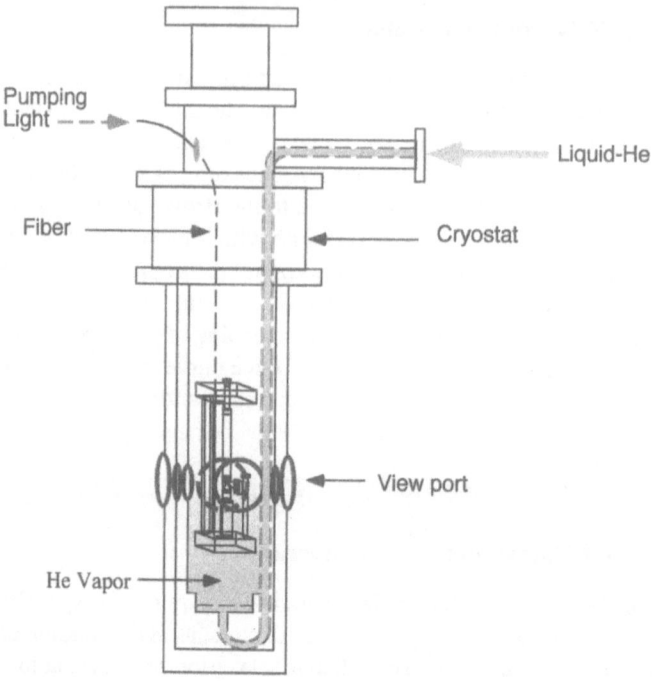

Figure 3.14. Structure of a low-temperature I-mode NOM system installed in a helium-flow-type cryostat.

K with a drift as low as 0.1 K/hr. Four view ports of the cryostat are used for monitoring shear force and detecting the emitted photoluminescence. Pumping light for photoluminescence is introduced through an optical fiber which is connected to the probe.

The main body of the system in the cryostat is shown schematically in Fig. 3.15. Three PZT actuators are used for this system. The first is a tube-type PZT actuator which exhibits a three-dimensional scan length of $10 \ \mu m \times 10 \ \mu m \times 0.6 \ \mu m$ at 10 K, corresponding to 12% that at room temperature. It is used for scanning the sample. The second tube-type PZT actuator is used for dithering the probe for shear-force monitoring. The third is a stacked-type PZT actuator which is used for auxiliary scanning along the vertical axis (z axis). These PZT actuators are mounted on plates and rods made of Fe + Ni (42%) alloy. A stepping motor is used for coarse scanning along the z axis.

Residual temperature drift is mainly due to that of the PZT actuator because the sample, probe, and their holders are fixed on a common frame of the system. The estimated drift is lower than 5 nm/hr at liquid He temperature because the

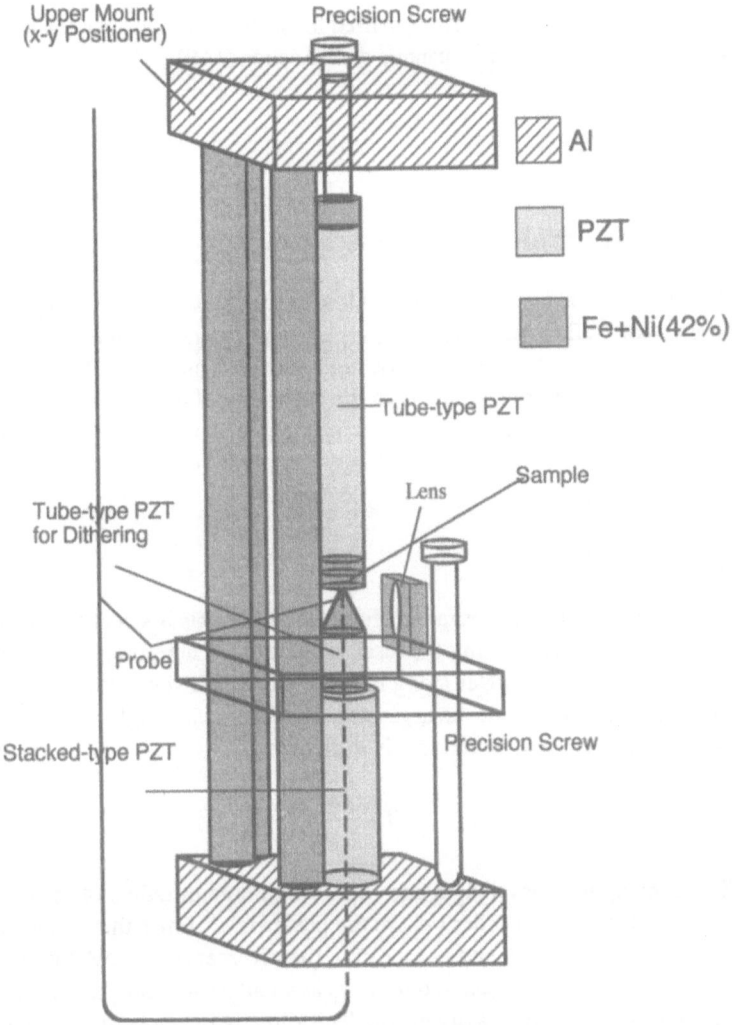

Figure 3.15. Structure of the main body of a low-temperature I-mode NOM system.

thermal expansion coefficient of the PZT is about 50 nm/K and the temperature drift is lower than 0.1 K/hr. Although the actual magnitude of the thermal drift is found to be larger than the estimated value, it is lower than 100 nm/hr, to which unexpected thermal drifts at other parts of the system contribute. In spite of this residual thermal drift, the system is confirmed to be sufficiently stable even though the shear force and photoluminescence are detected through view ports by using external optical components. It has been used for spatially resolved photoluminesence spectroscopy of semiconductor quantum dots (Section 5.3.3).

3.2. LIGHT SOURCES

3.2.1. Basic Properties of Lasers

Both continuous wave (CW) and pulsed lasers have been used as a coherent light source for NOM. Advantageous properties of lasers are their high temporal and spatial coherence, beam directivity, high-energy density, high brightness, and well-defined polarization, coupled with the capability of external modulation and control. The laser oscillation, however, can become unstable due to external perturbation. This section describes the basic properties of laser light.

3.2.1.1. Directivity

The intracavity light propagating along the cavity axis is selectively amplified by stimulated emission, starting the laser oscillations. Therefore, the output light beam from the laser propagates along the cavity axis. However, laser beam divergence due to diffraction cannot be neglected. The divergence angle (half-width) $\Delta\theta$, representing the degree of directivity, is expressed as

$$\Delta\theta = \frac{\lambda}{\pi w_0} \tag{3.2.1}$$

where λ is the optical wavelength and w_0 is the minimum radius of the Gaussian cavity mode of the lowest order (where the power is e^{-1} times that of the center). For example, for $\lambda = 1$ μm and $w_0 = 1$ mm, the divergence angle $\Delta\theta$ is about 0.3 mrad. This means that the beam radius stays as small as 3.2 mm even at a position 10 m away from the output cavity mirror. This directivity can be further improved by expanding the beam radius.

3.2.1.2. Monochromaticity

Quantum fluctuations (i.e., fluctuations due to spontaneous emission) are the basic source of phase and frequency fluctuations of laser light. The field spectral linewidth (full-width at half-maximum) of CW laser oscillation has been used as a measure to represent the magnitude of the quantum fluctuation and can be expressed as

$$\Delta\nu = \frac{h\nu}{8\pi P_o}\left(\frac{c}{nL}\right)^2\left(\alpha_l L + \ln\frac{1}{R}\right)\left(\ln\frac{1}{R}\right)\frac{N_a}{(N_a - N_b)_{th}} \tag{3.2.2}$$

where h is Planck's constant, ν is the optical frequency, P_o is the CW laser output power, c is the speed of light in vacuum, n is the refractive index of the matter in the cavity, L is the cavity length, α_l is the intracavity optical loss coefficient, R is the power reflectivity of the cavity mirror, N_a is the population density of the upper level of the laser transition, and $(N_a - N_b)_{th}$ is the population difference between the upper and lower levels of the laser transition at the threshold.

This is the so-called Schawlow–Townes formula. The linewidth estimated by this formula is several millihertz for a He–Ne laser with an output power of several milliwatts. A smaller value of $\Delta\nu$ corresponds to a higher temporal coherence.

3.2.1.3. Energy Density

Although the average output power of the laser is not always higher than that of other, incoherent light sources because of the low efficiency of laser oscillation, the laser light energy is concentrated spatially, temporally, and spectrally so that it has a very high energy density.

By focusing the laser light using a convex lens of focal length f, the cross-sectional beam radius at the focal plane is $f\Delta\theta$, and thus the power density of the laser beam is $P/\pi(f\Delta\theta)^2$, where P is the CW laser output power. Thus, the power density of a focused Gaussian beam of radius w_0 is given by

$$I_0 = \frac{\pi w_0^2}{f^2\lambda^2} P \tag{3.2.3}$$

This value can be as high as 3.1 MW/cm^2 for $w_0 = 1$ mm, $f = 1$ cm, $\lambda = 1$ μm, and $P = 1$ W.

In the case of a pulsed laser, the peak output power P_p is approximated as $W_p/\Delta t$, where W_p and Δt represent the pulse energy and pulse width, respectively. A value of P_p as high as 1 GW is possible for $W_p = 1$ mJ and $\Delta t = 1$ psec. It corresponds to a peak power density as high as 3.1 PW/cm^2 at the focal plane.

Even though the actual spectral linewidth of a laser is in practice much larger than the theoretical limit given by Eq. (3.2.2), the brightness of the laser light is still very high. The effective brightness temperature (defined as the blackbody temperature radiating the same light intensity at the relevant wavelength) is given by

$$T_B = P/(k\Delta\nu) \tag{3.2.4}$$

where P is the CW laser power, $\Delta\nu$ is the spectral linewidth, and k is Boltzmann's constant (= 1.38×10^{-23} J/K). A value of T_B as high as 10^{10}–10^{20} K is possible for most practical CW lasers.

3.2.1.4. Capability of External Modulation and Control

An outstanding property of the laser is its capability of modulation and control by an external signal. The power, wavelength, and optical phase of the laser light can be controlled and varied by applying an external signal. Even if the laser power or wavelength fluctuates, external modulation can reduce these fluctuations. For example, wideband and highly efficient phase modulation of a semiconductor laser has realized an accurate negative feedback loop in order to control the injection current so as to reduce the field spectral linewidth from 1 MHz to 7 Hz [9]. By using the capability of external modulation and control, a very short optical pulse can also be generated.

3.2.2. Characteristics of CW Lasers

3.2.2.1. Gas Lasers

Although a wideband external modulation is difficult, gas lasers have several advantages such as high coherence and small divergence angle (several milliradians). The characteristics of several CW gas lasers in the visible region are described below.

The He–Ne laser utilizes the population inversion among the metastable levels of neutral Ne atoms for laser oscillation at several wavelengths such as 1.15 μm, 633 nm, and 612 nm. Output power ranges from 0.1 mW to several milliwatts. In order to obtain an output power higher than 1 mW, the cavity length needs to be sufficiently long to allow oscillation on many longitudinal modes.

The Ar^+ laser oscillates by using the population inversion of Ar ions generated by electron bombardment. The most popular oscillation wavelengths are 477, 488, and 515 nm. Although the basic structure of the laser cavity is almost the same as other gas lasers, a large discharge current density is required for generating gaseous ions, which results in a high output power. The total output power for multiline oscillation can be as high as 1–20 W.

The He–Cd laser oscillates by using the population inversion of Cd ions generated by the discharge of metallic Cd vapor. This laser is useful as a light source for providing near-ultraviolet light. Oscillation wavelengths of 325 and 442 nm can be realized by using a glow discharge. Output powers can be as high as 10 and 100 mW, respectively. Red, green, and blue light with wavelength 636, 538, and 442 nm, respectively, can be simultaneously generated by using a hollow cathode. This has been called a "white light laser."

3.2.2.2. Dye Lasers

A dye solution with an organic solvent can be used as a gain medium for laser oscillation. The principal advantage of the dye laser is its wide range of wavelength tuning in the visible region. For example, using a 4-W Ar^+ laser as a pump source, wavelength tuning between 560 and 700 nm is possible with an output power of several hundred milliwatts.

A ring cavity configuration is employed to eliminate spatial hole burning, allowing high-power and single-longitudinal-mode operation. Etalons and birefringence filters are used for wavelength tuning. A thin film of dye flow is formed by spraying it out from a nozzle at the beam waist of the ring cavity. The pumping light is incident on this dye jet. Fluctuations in the optical pumping power, in the thickness and flow rate of the dye jet, and in the microbubbles contained in the dye jet can cause the power and phase of the output light to fluctuate. The spectral linewidth due to these fluctuations ranges from several to several tens of megahertz.

3.2.2.3. Solid-State Lasers

Laser oscillation is realized by doping impurities such as neodymium ions (Nd^{3+}) into a transparent host material (crystal or amorphous glass). The impurities function as the gain medium. A high output power is expected because the active ion density in the crystal is much higher than that in the gas lasers.

The Nd:YAG (yttrium aluminum garnet) laser supports laser oscillation at a number of wavelengths in the 1-μm region. The crystal is optically pumped by a tungsten lamp, a xenon arc lamp, or a krypton arc lamp. Recently AlGaAs lasers of 810 nm wavelength have been employed as a pumping source to reduce the power consumption and the volume of the device. Among laser devices are the nonplanar ring oscillator (NPRO) and the monolithic isolated single-mode end-pumping ring (MISER) laser, in which a specially polished crystal surface forms a ring cavity, and end pumping by a semiconductor laser is employed [10]. Laser oscillation is very stable, as the monolithic cavity reduces mechanical vibration. Output power of several hundred milliwatts is obtained. The spectral linewidth of the NPRO is narrower than 10 kHz, while the theoretical limit given by Eq. (3.2.2) is 1 Hz for an output power of 1 mW. The modulation bandwidth is about 25 MHz, which is limited by the relaxation oscillation.

In the $Ti:Al_2O_3$ laser, titanium ions (Ti^{3+}) in a sapphire crystal are used as the gain medium to support oscillation in the wavelength region of 0.7–1.1 μm. The cavity configuration is equivalent to that of the ring dye laser. Output power as high as 2 W is realized using a 13-W Ar^+ laser pump. The actual spectral linewidth is narrower than 100 kHz.

In addition to the lasers described above, a variety of oscillation wavelengths from the near-infrared to the infrared region have been realized utilizing a combi-

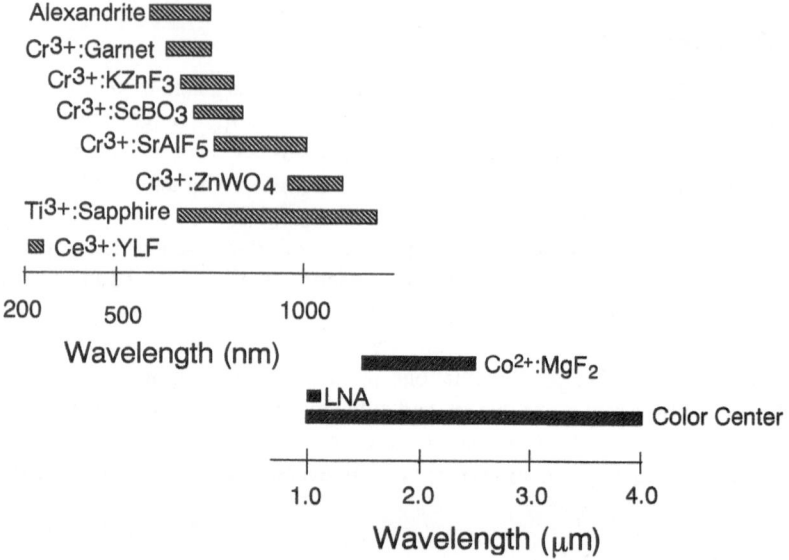

Figure 3.16. Oscillation wavelengths of solid-state lasers.

nation of active ions and host crystals (Fig. 3.16). Development of stable solid-state lasers pumped by semiconductor lasers has been progressing steadily.

3.2.2.4. Semiconductor Lasers

Direct transition-type semiconductors are used as the gain media. In such types of semiconductors, an electron in the conduction band is deexcited to the valence band and then recombines with a positive hole while conserving its momentum, emitting a photon. Direct transition-type semiconductors are found among composite semiconductors, which are composed of more than two elements. Since the value of the energy gap between the conduction and valence bands can be varied by changing the mole fraction of each element in ternary or quarternary composite semiconductors, oscillation wavelengths covering the infrared to visible regions can be realized, as shown in Fig. 3.17. Among semiconductor laser devices, the AlGaAs laser (0.8 μm) for optical disk memory, the InGaAsP laser (1.3, 1.5 μm) for optical fiber transmission, and the AlGaInP laser (0.6 μm) for optical sensing are used as light sources for the NOM. Recent developments in device fabrication for the ZnSe laser have been remarkable. A compact coherent blue light source is expected in the near future.

In addition to employing a double-hetero structure to confine the electrons and photons in the active layer, an index-guided structure has been used to emit a

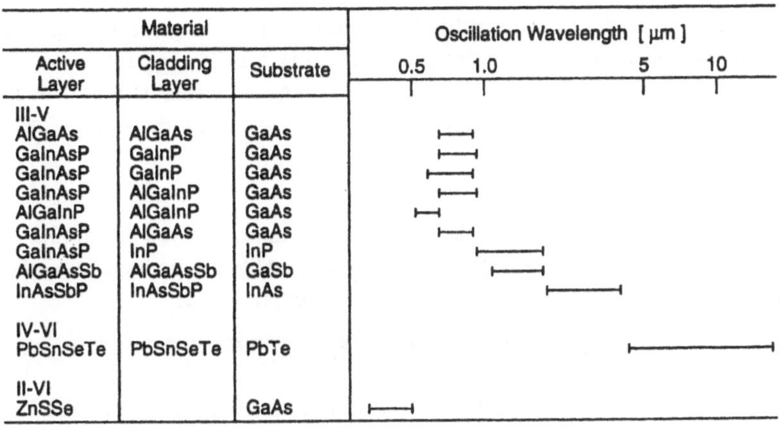

Material			Oscillation Wavelength [μm]			
Active Layer	Cladding Layer	Substrate	0.5 1.0		5	10
III-V						
AlGaAs	AlGaAs	GaAs				
GaInAsP	GaInP	GaAs				
GaInAsP	GaInP	GaAs				
GaInAsP	AlGaInP	GaAs				
AlGaInP	AlGaInP	GaAs				
GaInAsP	AlGaAs	GaAs				
GaInAsP	InP	InP				
AlGaAsSb	AlGaAsSb	GaSb				
InAsSbP	InAsSbP	InAs				
IV-VI						
PbSnSeTe	PbSnSeTe	PbTe				
II-VI						
ZnSSe		GaAs				

Figure 3.17. Oscillation wavelengths of III–V, VI–VI, and II–VI compound semiconductor lasers.

Gaussian beam (the lowest order lateral mode). Distributed feedback (DFB) cavities have been fabricated allowing single-longitudinal-mode oscillation. These have been used for optical fiber transmission systems. It should be noted that such advanced structures are not always employed for high-power semiconductor lasers. Although a typical output power is several tens of milliwatts, power as high as 10 W has been realized by an array-type laser in which parallel active layers are grown on a substrate.

Since the active layer works as an optical waveguide, Eq. (3.2.1) does not apply in the crystal, and the divergence angle of the output beam depends on the waveguide structure. Its value is much larger than the one given by Eq. (3.2.1). The values parallel and perpendicular to the active layer surface are about 15 and 30 deg, respectively.

Power and phase/frequency of the output light can be directly modulated by injection current, via modulation of the gain and refractive index. Modulation bandwidth is limited by the relaxation oscillation frequency of electrons, which is expressed as

$$f_r = \frac{1}{2\pi} \sqrt{\frac{I/I_{th} - 1}{\tau_s \tau_p}} \qquad (3.2.5)$$

where I, I_{th}, τ_p, and τ_s are the dc injection current, the threshold value, the photon lifetime, and the intraband relaxation time of electrons, respectively. The value of f_r is as large as several gigahertz because the typical values of τ_p and τ_s are about 1 psec and 3 nsec, respectively. A novel device with f_r as high as several tens of gigahertz is also possible by employing a quantum well structure.

In addition to the theoretical limit of spectral linewidth given by Eq. (3.2.2), the contribution of the carrier density fluctuation results in a refractive index fluctuation which is induced by the spontaneous emission fluctuations. The magnitude of this contribution is represented by α^2, and the linewidth is expressed as

$$\Delta\nu_{LD} = \Delta\nu(1 + \alpha^2) \tag{3.2.6}$$

where $\Delta\nu$ is the linewidth due to the spontaneous emission given in Eq. (3.2.2), in which the quantity $N_a/(N_a - N_b)_{th}$ should be replaced by the spontaneous emission factor n_{sp} in a semiconductor laser. This factor is expressed as

$$n_{sp} = 1/[1 - \exp[(h\nu + E_{Fv} - E_{Fc})/kT]] \tag{3.2.7}$$

where E_{Fv} and E_{Fc} are the quasi-Fermi levels of the valence and conduction bands, respectively, k is Boltzmann's constant, and T is the temperature. The value of n_{sp} falls between 1 and 2. Equation (3.2.6) is the modified Schawlow–Townes formula for a semiconductor laser, and α is the linewidth enhancement factor [11]. Since a typical value of α is about 2–9, the value of $\Delta\nu_{LD}$ can be as wide as several hundred kilohertz to several tens of megahertz. A novel device with a linewidth as narrow as several kilohertz has been fabricated recently by introducing a quantum well structure to reduce the value of α [12, 13]. The actual linewidth is wider than the quantum noise-limited value because the contributions from the competition effect between longitudinal and lateral modes, the injection current fluctuation, and the ambient temperature fluctuation must be considered.

3.2.3. Additional Noise Properties of CW Lasers

3.2.3.1. Gas, Dye, and Solid-State Lasers

Fluctuations in the refractive index and the cavity length can induce additional noise since the frequency of the laser oscillation is close to the cavity resonance frequency [14]. For example, additional noise in a He–Ne laser is caused by air flows in the laser cavity, mechanical vibrations, plasma instability in the He–Ne discharge gas, noise in the current source, and ambient temperature fluctuation. The magnitude of such noise is usually much larger than that of the quantum noise. The noise in an Ar$^+$ laser is larger than that in a He–Ne laser because of the large fluctuations in refractive index and temperature caused by large discharge current fluctuations, temperature fluctuations due to water cooling, and gradual decreases in gas pressure by sputtering.

In dye lasers, the principal noise sources are fluctuations in the optical pumping power, thickness and flow rate of the dye jet, and microbubbles contained in the dye jet. An optimal nozzle structure for the dye jet, a filter to purify the dye solution, and the addition of solvents (e.g., ethylene glycol, to increase viscosity) can reduce the size and number of these microbubbles.

In the case of a Nd:YAG laser, the principal noise sources are the power fluctuation of the lamp and temperature fluctuations in the crystal. Recently developed semiconductor laser-pumped Nd:YAG lasers have contributed greatly to the stability of the Nd:YAG laser system. Very stable laser oscillation has been realized using the NPRO architecture, as the Nd:YAG crystal itself forms a laser cavity, thus reducing the contributions from additional external noise sources.

3.2.3.2. Semiconductor Lasers

Semiconductor lasers exhibit specific noise characteristics because their cavity losses are large, the adiabatic approximation of carrier density variation usually used in the calculations is not valid due to the large and fast carrier density variation, and the width of the gain spectral profile is more than 100 times larger than the longitudinal mode separation [15]. Because of the very wide-gain spectral profile, unwanted vestigial side modes exist even in a longitudinal-mode-controlled laser (e.g., a DFB laser). Consequently, the concept of single-longitudinal-mode semiconductor lasers is still an approximation.

In order to simplify the following discussion, we will consider only two longitudinal modes of an index-guided laser, without longitudinal mode control. The characteristics of the mode hopping and mode partition phenomena described below depend mainly on the intraband relaxation of carriers. That is, they are quantum phenomena driven by spontaneous emission fluctuations coupled to each longitudinal mode and based on gain saturation due to the intraband relaxation of carriers.

The linear gains of the two longitudinal modes can be set almost equal by adjusting the injection current and temperature. Figure 3.18a shows the time dependence of each mode intensity observed under this condition, which shows a clear switching phenomenon [16] called *mode hopping*. This occurs because of the suppression of one mode due to a decrease in carrier density as a result of oscillating the other mode. It has been shown that the switching frequency decreases exponentially with increasing bias [17].

Nearly single longitudinal mode oscillation can be realized by fixing the linear gain of one mode far larger than that of the other by proper choice of the injection current and temperature. Figure 3.18b shows the time dependences of two mode powers under such conditions [16]. The transient decrease in the main mode power is called a *power dropout*. The power exchange between the main and submodes shown in this figure is called *mode partitioning*. This intensity noise has the same origin as that in mode hopping; the gain of the main mode decreases with a decrease in carrier density due to a transient increase in submode power driven by spontaneous emission. It has been confirmed experimentally as well as theoretically that

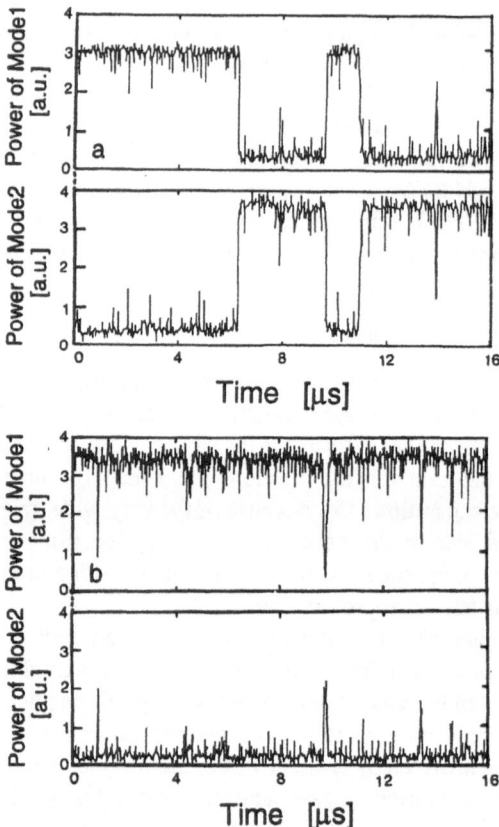

Figure 3.18. Measured temporal variation of the mode power of a semiconductor laser when two longitudinal modes oscillate. (a) Mode hopping. (b) Mode partitioning.

the pulse width of the power dropout, the probability of occurrence of power dropout, and the variance of power fluctuations decrease exponentially with increasing injection current. The phenomena described above are attributed to the strong coupling between the two modes, which is due to the very short intraband relaxation time. In contrast to semiconductor lasers, most gas lasers show weak coupling, which means that each mode oscillates independently.

Figure 3.19 shows a schematic configuration in which further significant intensity noise occurs. Laser oscillation becomes unstable when reflected light feeds back into the laser from an external reflecting surface. The probability of occurrence of this instability induced by the injection of phase-delayed light is large in semiconductor lasers because their facet reflectivities are low. Thus carriers are easily optically pumped by the injected light. Figure 3.20 shows results of computer

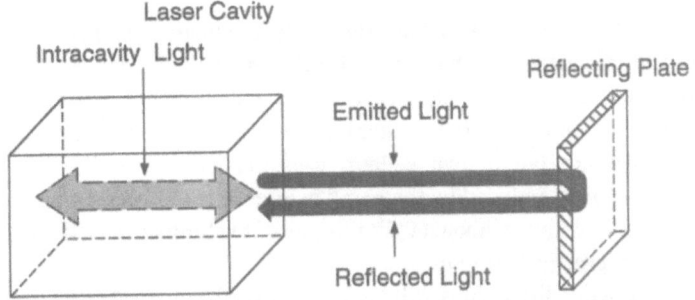

Figure 3.19. Noise generation induced by injection of reflected light.

simulations to estimate the increase in intensity noise by increases in reflected light intensity. It is easily seen from this figure that the effect of reflected light appears even when 0.003% of the emitted intensity is reflected back into the laser; finally, at 5%, a pulsed oscillation takes place.

This phenomenon is a deterministic instability which occurs when a phase-delayed light wave is injected into a self-sustained oscillator (i.e., a CW laser) and finally leads

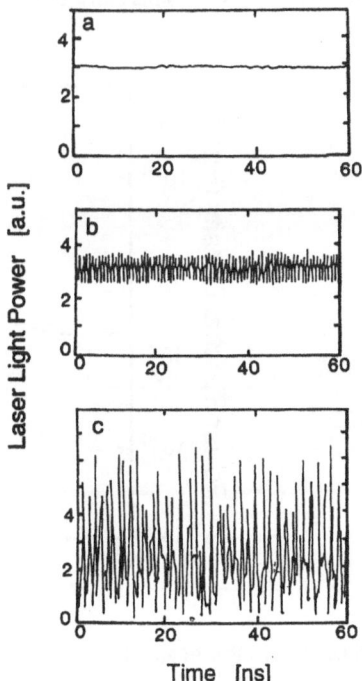

Figure 3.20. Calculated result of the temporal power variation of a semiconductor laser caused by the injection of reflected light. The distance between output and reflection surfaces is 5 cm. The ratio between the emitted power and the injected power after reflection is (a) 0%, (b) 0.003%, (c) 5%.

to chaos [18]. In an actual system, randomly fluctuating temperature, injection current, and the position of the external reflector make the characteristics of this instability very complicated and cause further increases in the magnitude of the instability. It is thus necessary to either increase the laser facet reflectivity or to use an optical isolator (an optical isolator uses the nonreciprocal rotation of the polarization of light induced by the Faraday effect). It has been empirically found that an optical isolation of about 60 dB is required to eliminate the effect of reflected light waves in practical systems.

The following discussions will concentrate on the characteristics of noise in a single-longitudinal-mode laser (more strictly, a nearly single-longitudinal-mode laser) by assuming that the additional noise sources presented above have been removed. In addition to the quantum noise induced by the spontaneous emission, intensity noise can also be induced by carrier density fluctuation which arises out of optical pumping from spontaneous emission. This carrier-induced noise can also be regarded as a type of quantum noise. Figure 3.21 shows the power spectral density of this noise source. The carrier density fluctuation gives a resonant peak at the relaxation oscillation frequency [19]. This relaxation oscillation frequency determines the noise bandwidth. Furthermore, this carrier density fluctuation also causes current fluctuations leading to temperature fluctuation due to self-heating of the laser device. This can also be regarded as a type of quantum noise. In addition,

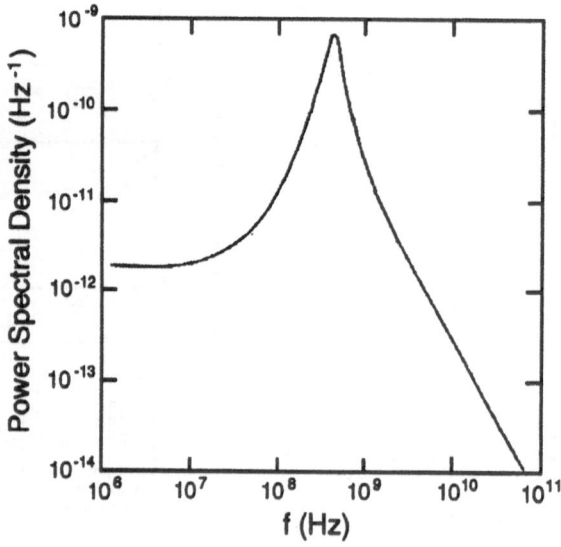

Figure 3.21. Power spectral density of the intensity fluctuation of a single-longitudinal-mode semiconductor laser.

there is also intensity noise due to current fluctuation in the current source and fluctuation in the threshold current or quantum efficiency due to ambient temperature fluctuation.

A similar discussion can be given for the sources of frequency noise. Quantum noise is composed of spontaneous emission, carrier density fluctuation, and temperature fluctuation. That is, in addition to the quantum noise due to spontaneous emission, the longitudinal mode frequency can fluctuate due to fluctuations in the refractive index caused by fluctuation in the carrier density due to optical pumping by spontaneous emission. The ratio between the power spectral density due to carrier-induced fluctuations and that by spontaneous emission is represented by α^2, as in Eq. (3.2.6). The intensity and frequency noises are correlated with each other via the carrier density fluctuation. That is, the linewidth enhancement factor is given by the ratio between changes in the real and imaginary parts of the complex susceptibility.

Figure 3.22 shows the power spectral density of the frequency noise. The contribution of the carrier density fluctuation gives a resonant peak at the relaxation oscillation frequency. Below this frequency, the ratio between the values due to carrier density fluctuations and spontaneous emission corresponds to α^2. In addition to these contributions, slowly fluctuating components are observed due to the drift in longitudinal mode frequency caused by self-heating due to current fluctuations induced by carrier density fluctuation. These contributions should also be regarded as quantum noise. The Fourier transform of these power spectral densities gives the

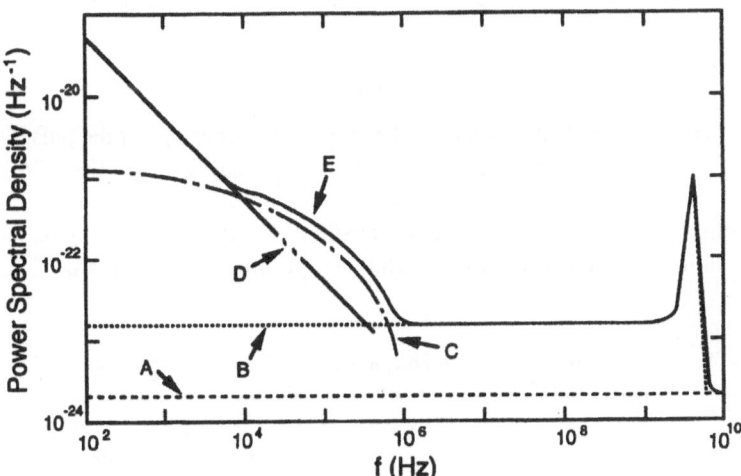

Figure 3.22. Power spectral density of the quantum frequency fluctuation of a single-longitudinal-mode semiconductor laser. (A) Fluctuation of spontaneous emission, (B) fluctuation of carrier density, (C) fluctuation of temperature, (D) 1/f fluctuation, (E) total quantum fluctuation.

field spectral profile, whose half linewidth is given by Eq. (3.2.6). It also shows FM sidebands originating from the relaxation oscillation of the carrier.

As additional noise source, flicker noise, can sometimes be observed and is independent of the laser power. This has been attributed to several origins, such as fluctuation in the carrier density [20]. However, its precise origin has not yet been fully identified. Due to this flicker noise, a power-independent linewidth of about 2–8 MHz can be observed even within the limit of infinite laser power ($P \to \infty$) [20].

Furthermore, as an external source of noise, current fluctuation from the current source and ambient temperature fluctuations cause frequency noise. Table 3.1 summarizes the rates of frequency drift due to slow changes in current and ambient temperature.

3.2.4. Short-Pulse Generation

Measurements of fluorescence and photoluminescence lifetimes require ultrashort optical pulses, which can be generated by controlling the temporal behavior of lasers. The technology of ultrashort optical pulse generation has developed rapidly. Pulses as short as 6 fsec have been generated [21]. This value is already very close to one period of a sinusoidal light wave in the visible region.

By assuming a constant-phase monochromatic optical pulse with Gaussian envelope, its Fourier transform gives a Gaussian spectral shape. By defining their full-widths at half-maximum as the pulse width Δt and spectral width $\Delta \nu$, respectively, we obtain their product as

$$\Delta t \, \Delta \nu = (2/\pi) \ln 2 \qquad (3.2.8)$$

The value of this product depends on the shape of the envelope of the pulse. In the case of a pulse with a time-dependent phase, this product is larger than $(2/\pi) \ln 2$. When (3.2.8) holds, the pulse is said to be *Fourier transform limited*.

When a pulse propagates through a medium, it suffers a frequency-dependent phase shift. The second-order phase shift, proportional to the square of the fre-

Table 3.1. Variations of Lasing Frequency of a Semiconductor Laser with a Change in DC Injection Current and Ambient Temperature

	AlGaAs laser (0.8 μm wavelength)	InGaAsP laser (1.5 μm wavelength)
Variation of frequency		
With dc injection current (GHz/mA)	−2.8	−1.1
With ambient temperature (GHz/K)	−28	−11

quency, is referred to as the *group velocity dispersion* and deforms the pulse shape and causes the phase to vary. As a result, the optical frequency varies. This is called *chirping*. This phenomenon can also be used to reduce the pulse width down to the Fourier transform-limited value.

The problem to be solved for generating Fourier transform-limited pulses is to form a constant-phase pulse over a wide optical frequency bandwidth. The technique of mode locking has been popularly used for this purpose. Representative mode locking techniques are summarized in the following. Table 3.2 summarizes the results obtained by these techniques.

3.2.4.1. Active Mode-Locking

A loss or phase modulator is installed in the laser cavity, and the modulation frequency is tuned to the longitudinal mode separation in order to phase-lock successively the adjacent longitudinal modes to the modulation sideband. Since its capture time is long, this technique has been applied only to narrow-gain-bandwidth lasers, such as Nd:YAG and Ar$^+$ lasers. A pulse width of several tens of picoseconds has been achieved. Because of its high operation stability, it has been applied to the pulsed laser for the synchronous excitation of a dye laser.

3.2.4.2. Passive Mode-Locking

The head and tail of the pulse are suppressed by an intracavity saturable absorber and gain saturation of the laser medium, respectively. Although the lifetime of the upper state of the absorber is as long as several nanoseconds, a pulse width of about 100 fsec can be obtained because of gain saturation. Although this is a simple technique, it suffers from low operation stability.

Table 3.2. Ultra-Short-Pulse Lasers

	Wavelength (nm)	Bandwidth (THz)	Pulse width (fsec)	
			Theoretical limit	Measured value
Ti:Al$_2$O$_3$	790	120	3	8.2
Cr:LiSAF	850	91	4	33
Cr:Mg$_2$SiO$_4$	1280	35	12	25
Cr:YAG	1450	30	12	60
Nd:glass	1053	7.5	42	40
Dye (Rh6G)	600	40	10	27

3.2.4.3. Synchronous Excitation of a Dye Laser

Mode locking is realized by tuning the cavity length of the dye laser to the pulse separation of the mode-locked laser used for pumping. A pulse width of about 100 fsec has been realized for a wide wavelength range by using the second harmonics of a Nd:YAG laser of 3.5 psec pulse width for pumping. A shorter pulse width has been obtained by installing a saturable absorber or a dispersion-compensating prism in the laser cavity. This laser has been popular because of its high stability and high output power.

3.2.4.4. Colliding Pulse Mode-Locking

Two counterpropagating pulses are generated at the moment when they collide with each other in a saturable absorber installed in a ring cavity (Fig. 3.23). This is because the cavity loss due to the absorber decreases when the two pulses collide. A pair of prisms is used to compensate for the frequency chirping introduced by the laser gain medium and the saturable absorber to minimize pulse width. A pulse width of 27 fsec has been realized with this technique. A pulse width of 6 fsec has been obtained by using an optical fiber as a pulse width compressor. This technique has not become popular due to its strict optical alignment requirements.

3.2.4.5. Additive Pulse Mode-Locking

A laser cavity is coupled to an external cavity in which an optical fiber is installed. The output pulse is transmitted through the fiber and is injected back into the laser. Self-phase modulation is induced in the fiber. By adjusting the external cavity length so as to optimize the magnitude of the induced phase shift of the pulse, pulse compression is obtained. Ultrashort pulses can be generated by this technique,

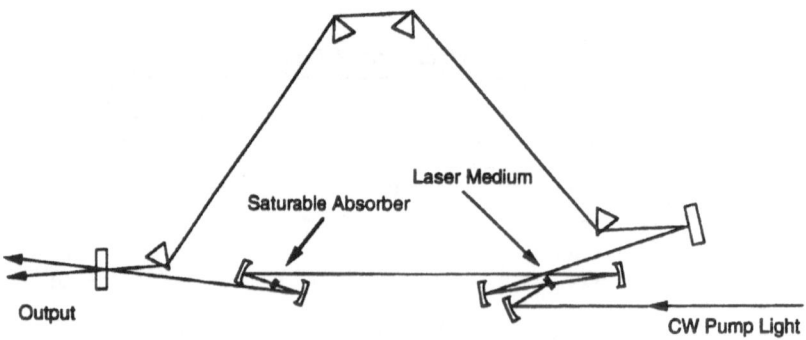

Figure 3.23. The experimental setup of colliding-pulse mode-locking.

as the response time for self-modulation in the fiber is very short. This technique is applicable for high-power lasers (i.e., solid-state lasers) where techniques relying on saturable absorbers fail. A pulse width of 40 fsec has been realized for a Nd:glass laser. A problem of this technique is mechanical instability: The cavity length has to be adjusted accurately to utilize the interference between the two counterpropagating pulses.

3.2.4.6. Kerr Lens Mode-Locking

Cavity loss in a $Ti:Al_2O_3$ laser is modulated by self-focusing due to the Kerr effect in the $Ti:Al_2O_3$ crystal. Self-focusing allows the peak of the pulse to pass through an aperture installed in the cavity, while the tails of the pulse cannot. The pulse width is decreased by repeating this selective transmission. This technique is very stable because self-stabilization takes place, as in the case of the optical soliton when group velocity dispersion is properly compensated. It has been applied to several solid-state lasers as summarized in Table 3.2.

3.2.4.7. Pulse Compression

Since an optical medium generally exhibits positive group velocity dispersion, pulse compression is possible by compensating for this dispersion by providing negative group velocity dispersion. Use of a pair of diffraction gratings or a pair of prisms is popular for providing the required negative group velocity dispersion. A pulse width as short as 10 fsec can be realized by this dispersion-compensation technique if the pulse is close to the Fourier transform limit and the gain bandwidth of the laser is sufficiently wide. An optical fiber is used for further compression of the pulse if its spectral width is limited by the gain bandwidth. When a short pulse is transmitted through a fiber, the Kerr effect changes its refractive index, resulting in a change in instantaneous optical frequency. Since the upchirping induced by this phenomenon is in the visible wavelength region, successive dispersion compensations by a pair of diffraction gratings or prisms can compress the pulse. The shortest recorded pulse width of 6 fsec was realized by this method [21].

3.2.5. Nonlinear Optical Wavelength Conversion

In order to generate coherent light at frequencies not directly realizable by laser oscillation, techniques of nonlinear optical wavelength conversion (second harmonic, sum-frequency, and difference-frequency generation) have been used. An inorganic oxide or organic material is often used for this purpose. Conversion techniques using a semiconductor laser or a semiconductor laser-pumped solid-state laser as a primary light source have been recently developed. Wavelengths as

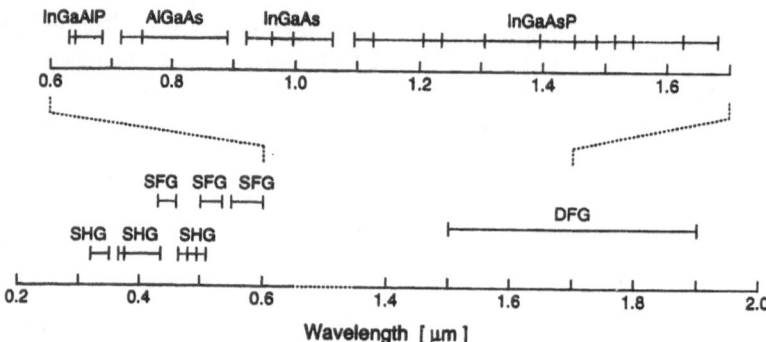

Figure 3.24. Oscillation wavelength ranges of commercially available semiconductor lasers (top) and of the nonlinearly converted light from these lasers (bottom). SHG, Second harmonic generation; SFG, sum-frequency generation; DFG, difference-frequency generation.

short as about 0.2 μm have been realized. Conversion efficiencies up to 50% have been obtained by installing the nonlinear optical crystal in a buildup cavity. A wideband tunable light source covering the whole visible range is possible by mixing the output of several semiconductor lasers in a nonlinear crystal to generate their sum or difference frequencies. This is because the oscillation wavelengths of semiconductor lasers range from the visible to the near-infrared. The frequency tuning range has been estimated to be as large as 1 PHz (Fig. 3.24) [22].

3.3. LIGHT DETECTION AND SIGNAL AMPLIFICATION

3.3.1. Detector

When light is shone on the photocathode of a vacuum tube, electrons are emitted and collected by the anode. A photocurrent is thus produced. This is called *photoelectric emission* or the *external photoelectric effect*. The lowest optical frequency for this emission, i.e., the cutoff frequency ν_c, is given by

$$h\nu_c = \phi \tag{3.3.1}$$

where h and ϕ represent Planck's constant and the work function of the photocathode material, respectively.

By applying a voltage to the anode, the value of the photocurrent I_p from the photocathode is proportional to the incident power P of the monochromatic light. In order to represent this proportionality quantitatively, the quantum efficiency η has been defined:

$$\eta = \frac{I_p/e}{P/h\nu} \qquad (3.3.2)$$

where e and ν are the electron charge and the optical frequency, respectively.

An internal photoelectric effect is observed in a semiconductor when an electron–hole pair is created by photon absorption in an interband transition. The cutoff optical frequency ν_c for this transition is given by $\nu_c = E_g/h$, where E_g represents the bandgap energy. An external circuit is connected to utilize such photoconductivity for photon detection.

Doping of the semiconductor with impurities can be used to lower the bandgap energy and thus the cutoff optical frequency. By applying an inverse-bias voltage to the p–n junction, a photocurrent is produced when light is incident on the junction. By increasing the reverse-bias voltage, an electron avalanche is induced, thus increasing the detection sensitivity. Such a device is called an *avalanche photodiode*.

Spectral sensitivity, minimum detectable power, signal-to-noise ratio, and response time are the most essential measures to evaluate the performances of the detector. Figure 3.25 shows the relation between the working wavelength regions and response times for several detectors. Minimum detectable power is represented by the noise equivalent power (NEP) in units of $W/Hz^{1/2}$. The NEP is the signal level that produces a unity signal-to-noise ratio, and is expressed as

$$\text{NEP} = P(V_n/V_s)/B^{1/2} \qquad (3.3.3)$$

where V_n, V_s, and B are the output noise voltage, output signal voltage, and the noise bandwidth, respectively. Typical NEP values for photodiodes are on the order of several hundred nanowatts. Alternatively, the normalized detectivity D^* is also used

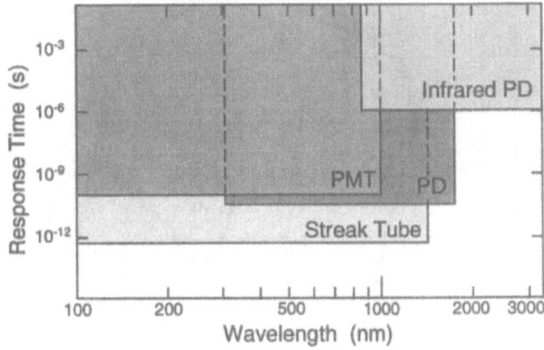

Figure 3.25. Relation between the working wavelength regions and response times for several detectors.

as a figure of merit for a detector, which is the reciprocal of the NEP normalized by the detector area A_d and bandwidth B:

$$D^* = (A_d B)^{1/2}/\text{NEP} \qquad (3.3.4)$$

Basic performances of each detector type are reviewed in the following.

3.3.1.1. Photomultiplier Tube

The photomultiplier tube (PMT) consists of a photoemissive detector and a low-noise amplifier packaged in a vacuum jacket. When an incident photon is absorbed by the photocathode, it produces a photoelectron, which is accelerated toward the first dynode by the voltage drop across the resistor, gaining high kinetic energy. Upon striking the dynode, this electron excites a number of secondary electrons, which are accelerated toward the second dynode. In this way a large number of electrons are collected by the anode.

A variety of PMTs have been produced commercially; their spectral sensitivity as a function of optical wavelength is shown in Fig. 3.26. The advantages of the PMT are high gain (10^6) high dynamic range (10^6), fast response time (0.1–10 nsec) low noise (NEP = 10^{-16} W/Hz$^{1/2}$), and high sensitivity (detection of a single photoelectron is possible).

3.3.1.2. Photodiode

Silicon, GaAsP, InGaAs, and Ge are used to form p–n junction diodes, Schottky-junction diodes, etc. By applying a reverse-bias voltage, these can be used

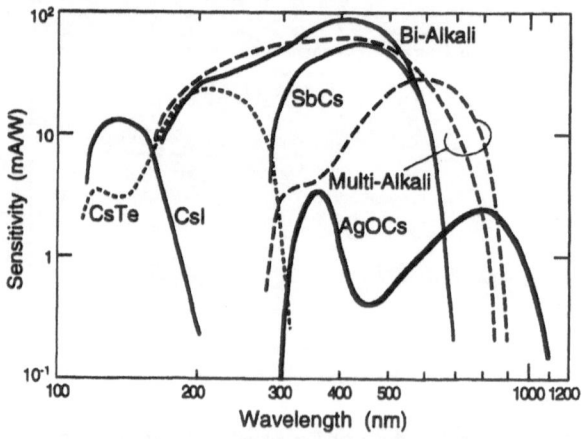

Figure 3.26. Spectral sensitivity of photomultiplier tubes as a function of optical wavelength.

for light detection in the visible to near-infrared region. Figure 3.27 shows the spectral sensitivity for several photodiodes (PDs) as a function of the optical wavelength. In the case of a p–n junction, a "pin-PD" has been popularly used in order to realize fast response (a high-resistance intrinsic layer is grown between the p and n layers). This is due to the presence of a strong electric field in the reverse-biased intrinsic layer.

Response time of the PD is limited by the traversal time of the carrier through the depletion layer, the diffusion time of a carrier generated outside the depletion layer, and the junction capacity. In the case of the pin-PD, a typical response bandwidth, which is inversely proportional to the response time, is several hundred megahertz, although it can be as high as several gigahertz.

The signal-to-noise ratio of the PD can be expressed as

$$S/N = 2I_p^2/[2e(I_p + I_d)B + 4kTFB/R] \tag{3.3.5}$$

where I_p is the photocurrent given by Eq. (3.3.2), I_d is the dark current due to thermally excited carriers, and k, T, F, and R are the Boltzmann constant, temperature, noise figure, and input resistor of the postdetector amplifier, respectively. The numerator of this equation corresponds to the signal power. The first term of the denominator represents the shot-noise power due to the temporal number fluctuation of the generated photoelectrons, and the second term is the thermal noise generated in the resistance R. The incident power corresponding to $S/N = 1$ is about 0.5 μW.

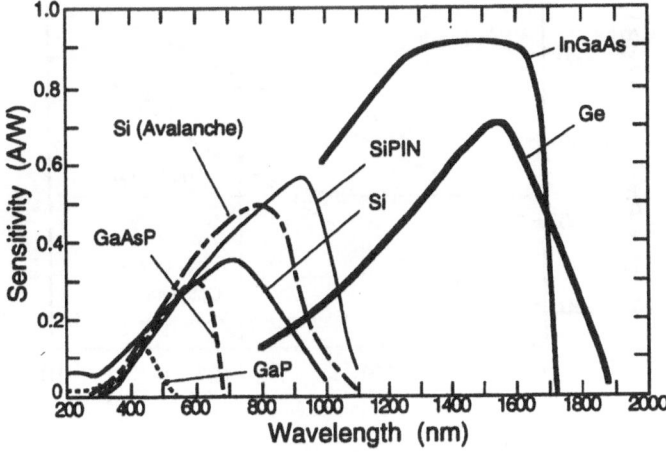

Figure 3.27. Spectral sensitivity of several photodiodes as a function of optical wavelength.

3.3.1.3. Avalanche Photodiode

Avalanche buildup of the number of carriers can be realized by applying sufficiently high reverse-bias voltage to the PD. Generated electrons or holes are accelerated by the high electric field of the depletion layer and successively generate a number of electrons and holes. The photocurrent is amplified by this carrier multiplication, as in the case of the PMT. A PD based on this avalanche buildup is called an avalanche photodiode (APD).

The multiplication factor M, depending on the bias voltage, can be as large as several hundred for the Si APD. This diode is widely used because of its high sensitivity. The response time is determined not only by the limiting factors for the PD, but also by the avalanche buildup time. Since the latter is proportional to the multiplication factor, the gain–bandwidth product is used as a measure to evaluate the performance of the APD. A gain–bandwidth product as large as several hundred gahertz has been realized for the Si APD.

The signal-to-noise ratio is expressed as

$$\frac{S}{N} = \frac{2I_p^2 M^2}{2e(I_p + I_d)M^2 M^x B + 2eBI_l + 4kTFB/R} \tag{3.3.6}$$

where M^x is the *excess noise figure* and has a value depending on the structure. The value of x for the Si APD is about 0.35. The first term of the denominator corresponds to the shot noise induced by the photocurrent, dark current, and temporal fluctuation of the avalanche buildup. The second term represents the shot

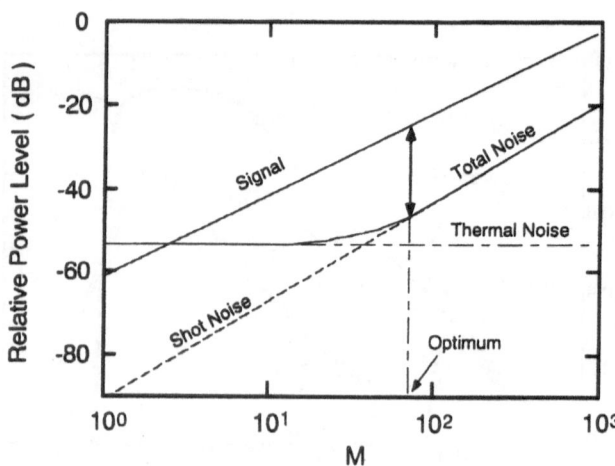

Figure 3.28. Dependence of the relative power levels of signal and noise of the avalanche photodiode on the value of the multiplication rate M.

noise due to the leak current I_l in the vicinity of the p–n junction. The third term is for the thermal noise. Figure 3.28 shows that the value of S/N becomes maximum for a certain value of M. The bias voltage should be adjusted to maximize the value of S/N.

3.3.2. Signal Detection and Amplification

3.3.2.1. Video Detection

Video detection generates the output signal by coupling the light wave directly to the photodetector. The output current is given by Eq. (3.3.2), and the values of S/N for the PD and APD are given by Eqs. (3.3.5) and (3.3.6), respectively.

3.3.2.2. Heterodyne Detection

Equations (3.3.5) and (3.3.6) give the minimum detectable power

$$P_{min} = \frac{h\nu B}{\eta} \tag{3.3.7}$$

for $M^x = 1$ and in the absence of dark current, leak current, and thermal noise. This is the *quantum detection limit*. It is very difficult to realize this limit by video detection. However, it is possible by heterodyne detection, as explained in Fig. 3.29. In this detection scheme, high-power local light is incident on the photodiode simultaneously with the signal light. The total amplitude of the electric field incident on the detector is

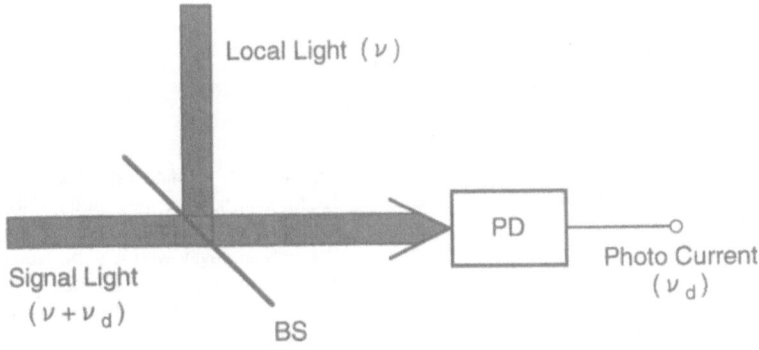

Figure 3.29. Heterodyne detection. BS, Beam splitter.

$$E(t) = E_s \cos 2\pi(\nu + \nu_d)t + E_l \cos 2\pi\nu t \qquad (3.3.8)$$

where E_s, E_l, $\nu + \nu_d$, and ν are the electric field amplitudes and frequencies of the signal and local light, respectively. The photocurrent $i_p(t)$ including dc and ac components is thus proportional to

$$i_p(t) \propto E_s^2 + E_l^2 + 2E_lE_s \cos 2\pi\nu_d t \qquad (3.3.9)$$

By noting that $E_l \gg E_s$ and referring to Eq. (3.3.2), one can derive an expression by using the power P_s and P_l of the signal and local light waves, respectively:

$$i_p(t) = (e\eta P_l/h\nu) (1 + 2\sqrt{P_s/P_l} \cos 2\pi\nu_d t) \qquad (3.3.10)$$

According to this equation, the time average of the output signal power is $2(P_s/P_l) (e\eta P_l/h\nu)^2 M^2$. On the other hand, by comparing this equation and the denominator of Eq. (3.3.6), it is also found that the shot-noise power corresponds to $2e((e\eta P_l/h\nu) + I_d)M^2M^xB$. Thus, the value of S/N is given by

$$\frac{S}{N} = \frac{2(P_sP_l)(e\eta/h\nu)^2 M^2}{2e(e\eta P_l/h\nu + I_d) M^2M^xB + 2eBI_l + 4kTFB/R}$$

$$(3.3.11)$$

Here, since the first term of the denominator is much larger than the other terms, this equation is approximated as

$$S/N = P_s/(h\nu BM^x/\eta) \qquad (3.3.12)$$

From this equation it can be seen that the quantum detection limit can be realized if $M^x = 1$ since the shot noise of the high-power local light is the dominant noise source of the system. A PMT or pin-PD can be used for this sensitive detection because their values of M^x are close to unity.

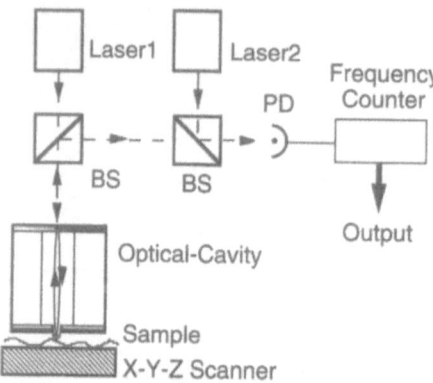

Figure 3.30. Reflection-resonance-type NOM, in which the magnitude of the frequency shift due to the phase change of the optical cavity with a subwavelength-size aperture used as a probe is measured by the heterodyne detection method. Laser1: the laser which is phase-locked to the optical cavity. Laser2: reference laser. BS, Beam splitter.

An example of applying the heterodyne detection to the NOM is shown as follows. In order to improve the detection sensitivity of the NOM, the phase change of the light wave can be measured instead of the power. Figure 3.30 explains schematically the principle of operation of the reflection-resonance-type NOM (RR-NOM) [23]. The evanescent field generated by a subwavelength-size aperture on a small optical cavity facet is scattered by the sample. Due to this scattering, the complex optical impedance of the cavity facet varies slightly, shifting the cavity resonance frequency. This small frequency shift can be measured by the heterodyne detection method using a reference laser and one which is phase-locked to the cavity. The magnitude of the frequency shift obtained as a function of the position of the scanned optical cavity represents the three-dimensional image of the sample surface.

The magnitude of the induced resonant frequency shift can be estimated by using a simplified theoretical model [23]. The electric field leaking out from the aperture including the evanescent field is expressed in terms of its spatial plane-wave spectrum. The tunneling of the leaked electric field to the sample is decomposed into separate plane-wave components. One can then obtain an analytical expression for the intracavity reflected field from the apertured cavity facet. From this expression, the resonant frequency shift can be obtained, which depends on the size and topography of the aperture as well as the sample–probe separation.

Figure 3.31 shows numerically calculated results of the relationship between the aperture diameter and the frequency shift normalized to the free spectral range (FSR) of the cavity. Curves A–D are the values of the shot-noise limit depending on the detected power P and the finesse F of the cavity. In the case of $P = 10$ mW and $F = 100$ (curve A), the shot-noise-limited lateral resolution, defined as the diameter of the aperture for simplicity, is evaluated as $\lambda/100$ (λ is the wavelength). For $\lambda = 800$ nm, this corresponds to a frequency shift of 14 Hz for an FSR of 100 GHz and a resolution of 8 nm. A heterodyne optical phase-locked loop linking the two lasers with a tracking accuracy of 4×10^{-4} Hz can guarantee the detection of such a small magnitude of the frequency shift [24]. On the other hand, since a supercavity with finesse higher than 10^5 and a single-mode diode laser with an output power on the order of 1 W are available, shot-noise-limited lateral resolution higher than $2.3 \times 10^{-3} \lambda$ ($= 1.7$ nm) (curve D) can be expected.

As a preliminary experiment, the resonant frequency shift of a high-reflection (HR) coated Fresnel-rhomb prism was detected by changing the prism–probe separation. Figure 3.32a shows the experimental setup. An AlGaAs laser (830 nm wavelength) with linewidth narrowed by the optical feedback technique is used as a light source. A sharpened fiber fabricated by the technique described in Section 4.1.2 is used to shift the resonant frequency of the HR-coated Fresnel-rhomb prism cavity (FSR = 1.75 GHz, $F = 10$). The shift is induced by disturbing the evanescent field generated on a side facet of the cavity. Phase-sensitive detection is used for a high sensitivity.

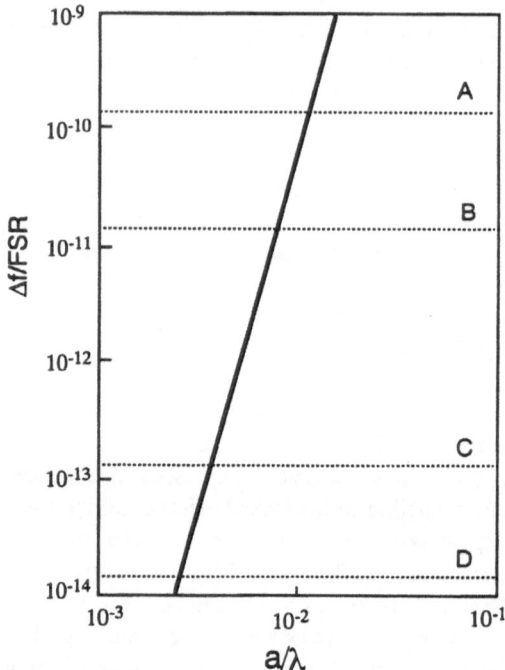

Figure 3.31. Numerically estimated results of the relationship between the aperture diameter and the frequency shift. FSR, Free spectral range of the cavity; λ, optical wavelength. Sample–probe separation is assumed to be $1 \times 10^{-3}\lambda$. Curves A–D are the shot-noise-limited values for (A) $P = 10$ mW, $F = 10^2$, (B) $P = 1$ W, $F = 10^2$, (C) $P = 10$ mW, $F = 10^5$, (D) $P = 1$ W, $F = 10^5$.

Figure 3.32b shows the relationship between the prism–probe separation and normalized frequency shift Δf/FSR. From the result shown in this figure and considering that the experimental resolution of the phase-sensitive detection is higher than 10^{-3}FSR/F to 10^{-5}FSR/F, a normal resolution as high as 0.1–0.001 nm can be expected with a prism–probe separation less than 10 nm. Furthermore, since a frequency shift as small as 4×10^{-4} Hz can be detected by the optical phase-locked loop, normal resolution higher than 10^{-3} nm can be expected by using the setup shown in Fig. 3.30.

As it is not straightforward to fabricate a subwavelength-size aperture on the facet of a high-finesse cavity, an additional demonstration was carried out in the microwave region ($\lambda = 4.4$ cm, frequency $f_0 = 6.8$ GHz) to confirm the operation of the present RR-NOM. A microwave cavity for the TE_{111} mode was used as a probe, and the resonant frequency shift was measured by phase-sensitive detection

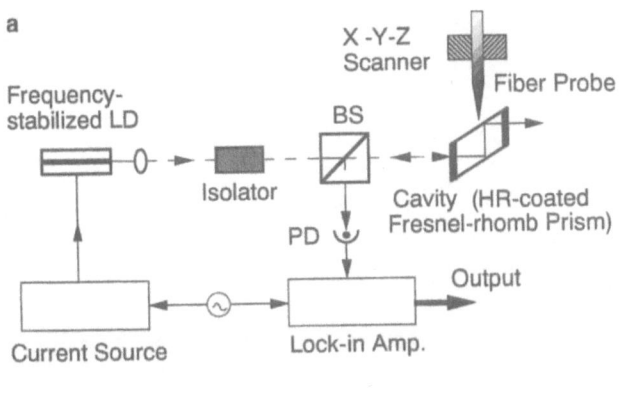

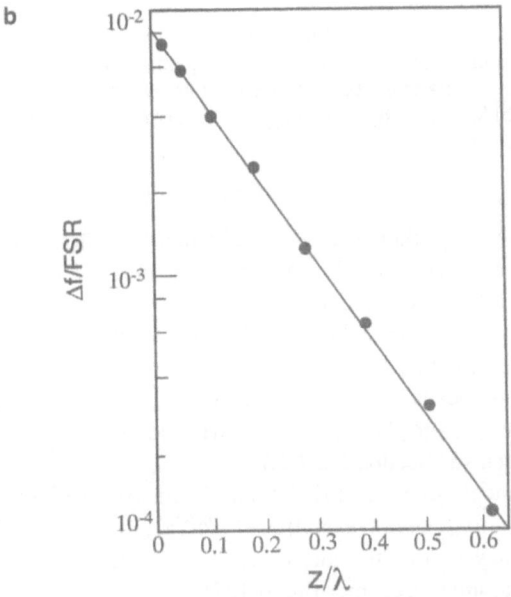

Figure 3.32. (a) Experimental setup and (b) result for the resonant frequency shift Δf of the HR-coated Fresnel-rhomb prism. LD, Diode laser; BS, beam splitter; PD, photodiode. The wavelength, FSR, and finesse of the prism are 830 nm, 1.75 GHz, and 10, respectively. The horizontal axis in panel (b) is the prism–probe separation z normalized to the wavelength λ.

[25, Fig. 1]. Figure 3.33a shows the cross-sectional profile of an Al plate having subwavelength structure that was used as a sample. Figure 3.33b shows the measured frequency shift when the probe was scanned laterally. It can be seen that the separation $b = 0.13\lambda$ of two corrugation elements, corresponding to the

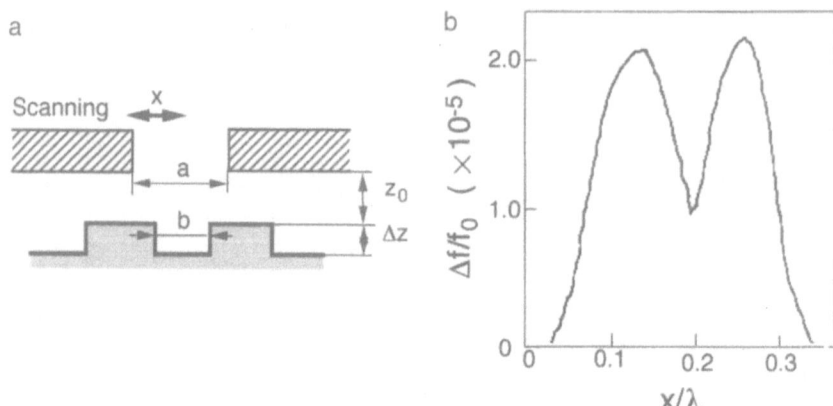

Figure 3.33. (s) Cross-sectional profile of the aperture on the microwave cavity wall and corrugated sample. The microwave wavelength λ and frequency f_0 are 4.4 cm and 6.8 GHz, respectively. The aperture diameter a is 0.2λ. Values of other parameters are $b = 0.13\lambda$, $\Delta z = 0.05\lambda$, and $z_0 = 0.02\lambda$. (b) Measured resonant frequency shift Δf of the microwave cavity as a function of its lateral position x.

separation between the two peaks of the resonant frequency shift, is resolved clearly. The value of the resonant frequency shift $\Delta f/f_0$ is on the order of 10^{-5}. Considering that the FSR of this microwave cavity is equal to f_0, it can be seen that such a resonant frequency shift agrees with the calculated result. This result also confirms that the RR-NOM can resolve structure (0.13λ) which is even smaller than the aperture diameter (0.2λ). This is because of the inhomogeneous intensity distribution of the evanescent field on the aperture plane, partly due to the boundary effect. (For further discussion of this effect, see Section 5.2.2.2.)

A similar experiment has been carried out in the microwave frequency region using an interferometer to achieve a resolution of 1/4000 times the wavelength [26]. In the optical frequency region, the interferometric technique has been used in conjunction with the atomic force microscopy [27].

3.3.2.3. Phase-Sensitive Detection

Flicker noise is often associated with the signal. Since the magnitude of its power spectral density decreases with the increase of the Fourier frequency f, this noise is also called $1/f$ noise. Even when the signal is in the low-frequency region so as to be superposed with the $1/f$ noise (see Fig. 3.34), the value of S/N can be drastically improved by high modulation of the signal and successive demodulation after it has been shifted to a region in which the magnitude of the $1/f$ noise is sufficiently low.

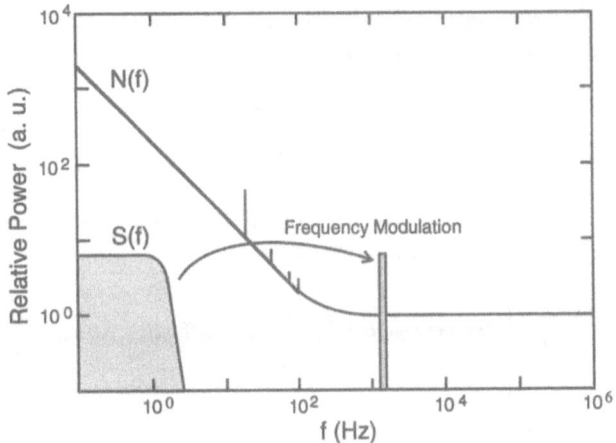

Figure 3.34. The power spectral density of the signal $S(f)$ and noise $N(f)$.

Since the bandwidth of the signal required for most NOM systems is between 0 and 1 kHz, the time constant of the temporal variation of the detected light may be tens of milliseconds or even longer. That is, the signal spectral bandwidth is 0–100 Hz.

The intensity of a slowly varying signal $s(t)$ being sinusoidally modulated with the modulation frequency f_m is represented by $s(t) \cos 2\pi f_m t$. Phase sensitive detection (PSD) has been employed to demodulate the signal $s(t)$ from the sinusoidally modulated one. The method of signal amplification with a high S/N using PSD is called *lock-in detection*.

The principle of operation of the lock-in amplifier used for lock-in detection is schematically explained in Fig. 3.35. The PSD circuit is composed of a multiplication circuit followed by a low-pass filter, i.e., an integrator. The modulated signal $s(t) \cos 2\pi f_m t$ and noise $n(t)$ are amplified by a tuned amplifier and fed into the PSD circuit. The output signal from the multiplication circuit is the product of the input

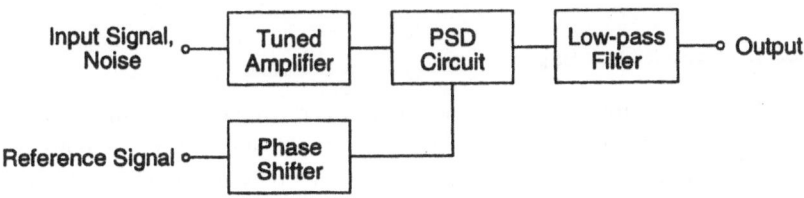

Figure 3.35. Block diagram of a lock-in amplifier. PSD, Phase-sensitive detection.

signal $s(t) \cos 2\pi f_m t + n(t)$ and the reference signal $K \cos(2\pi f_m t + \phi)$, which is synchronized to the modulation signal $\cos 2\pi f_m t$:

$$K \cos(2\pi f_m t + \phi) [s(t) \cos 2\pi f_m t + n(t)]$$

$$= Ks(t)[\cos \phi + \cos(4\pi f_m t + \phi)] + Kn(t) \cos(2\pi f_m t + \phi) \qquad (3.3.13)$$

where ϕ is the phase of the reference signal at $t = 0$. This is then integrated by a low-pass filter to produce the output signal of

$$\frac{K}{T} \int_{-T/2}^{T/2} s(t) \cos \phi \, dt + \frac{K}{T} \int_{-T/2}^{T/2} s(t) \cos(4\pi f_m t + \phi) \, dt$$

$$+ \frac{K}{T} \int_{-T/2}^{T/2} n(t) \cos(2\pi f_m t + \phi) \, dt \qquad (3.3.14)$$

where T is the time constant of the low-pass filter. The first term can be reduced to $Ks(t) \cos \phi$ if the amplitude of the reference signal K is constant and $s(t)$ is approximated to be constant for the period of T, which means that the output signal from the PSD circuit is proportional to $s(t)$. The second term is the Fourier component of $s(t)$ at the Fourier frequency of $2f_m$, which does not contribute to the signal amplification. The third term is the noise component $n(f_m)$ at the Fourier frequency f_m,

$$n(f_m) \propto \frac{1}{T} \int_{-T/2}^{T/2} n(t) \cos(2\pi f_m t + \phi) \, dt \qquad (3.3.15)$$

The bandwidth B is nearly equal to $1/T$, i.e., the overall bandwidth of the lock-in amplifier is determined not by the bandwidth of the tuned amplifier, but by the time constant of the low-pass filter. Thus, the value of S/N is determined by the noise magnitude which falls within this bandwidth. Therefore, the signal can be amplified effectively with a very high S/N by using a very narrow bandwidth at the high-Fourier-frequency region in which the $1/f$ noise magnitude is low.

3.3.2.4. Photon Counting

When the photoluminescence or fluorescence is detected by the NOM, the detected light power is several picowatts or lower. At an incident light power of 1 pW (10^5 photons/sec), the output signal becomes a pulse train, where each pulse corresponds to a single photon. Under this condition, the detected pulse height becomes independent of the photon flux rate.

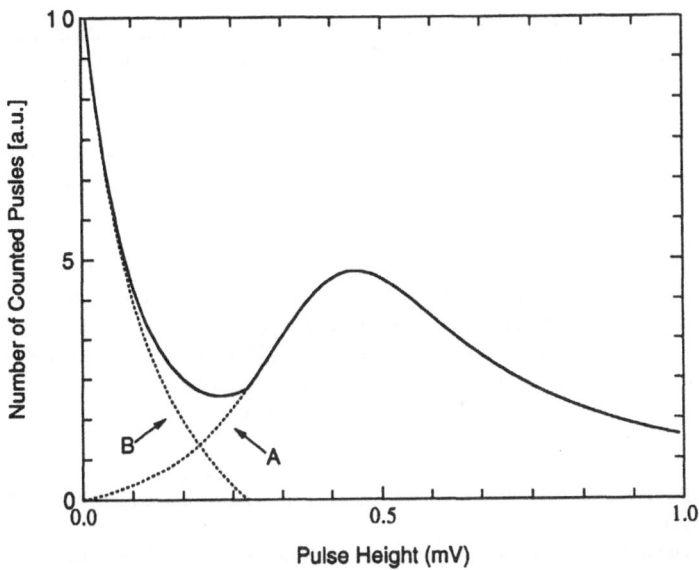

Figure 3.36. Relation between the pulse height and number of counted pulses for the photon counting system. (A) Pulses generated by the incident photons, (B) pulses generated by the dark current.

Photon counting is the method used to count the number of pulses. PMTs are often used for photon counting because of their high gain (a voltage pulse as high as 1 mV is possible for a load impedance as low as 50 Ω). The output voltage is sufficiently high to maintain a high S/N value even for a very weak incident light. The key noise source in photon counting is the dark current pulse, whose pulse height distribution is shown in Fig. 3.36. Since a pulse height discriminator can reject the dark current pulse, a very high S/N is obtained.

3.4. REFERENCES

1. C. J. Chen, *Introduction to Scanning Tunneling Microscopy*, Oxford University Press, Oxford, 1993.
2. B. Hecht, H. Bielefeldt, Y. Inoue, and D. W. Pohl, Facts and artifacts in near-field optical microscopy, *J. Appl. Phys.* **81**: 2492–2498 (1997).
3. M. Stahelin, M. A. Bopp, G. Tarrach, A. J. Meixner, and I. Zschokke-Granacher, Temperature profile of fiber tips used in scanning near-field optical microscopy, *Appl. Phys. Lett.* **68**: 2603–2605 (1996).
4. E. Betzig, P. L. Finn, and J. S. Weiner, Combined shear force and near-field scanning optical microscopy, *Appl. Phys. Lett.* **60**: 2484–2486 (1992).

5. R. Toledo-Crow, P. C. Yang, Y. Chen, and M. Vaez-Iravani, Near-field differential scanning optical microscope with atomic force regulation, *Appl. Phys. Lett.* **60**: 2957–2959 (1992).

6. K. Karrai and R. D. Grober, Piezoelectric tip–sample distance control for near field optical microscopes, *Appl. Phys. Lett.* **66**: 1842–1844 (1995).

7. Y. Toda, M. Kourogi, M. Ohtsu, Y. Nagamune, and Y. Arakawa, Spatially and spectrally resolved imaging of GaAs quantum-dot structures using near-field optical technique, *Appl. Phys. Lett.* **69**: 827–829 (1996).

8. R. D. Grober, T. D. Harris, J. K. Trautman, and E. Betzig, Design and implementation of a low temperature near-field scanning optical microscope, *Rev. Sci. Instrum.* **65**: 626–631 (1994).

9. C.-H. Shin and M. Ohtsu, Stable semiconductor laser with 7 Hz linewidth by an optical-electrical double feedback technique, *Opt. Lett.* **15**: 1455–1457 (1990).

10. A. C. Nillson, E. K. Gustafson, and R. L. Byer, Eigenpolarization theory of monolithic nonplanar ring oscillators, *IEEE J. Quantum Electron.* **25**: 767–790 (1989).

11. C. H. Henry, Theory of the linewidth of semiconductor lasers, *IEEE J. Quantum Electron.* **QE-18**: 259–264 (1982).

12. M. Okai, T. Tsuchiya, K. Uomi, N. Chinone, and T. Harada, Corrugation-pitch modulated MQW-DFB lasers with narrow spectral linewidth, *IEEE J. Quantum Electron.* **27**: 1767–1772 (1991).

13. M. Ohtsu, *Frequency Control of Semiconductor Lasers*, Wiley, New York, 1996.

14. M. Ohtsu, *Coherent Quantum Optics and Technology*, Kluwer, Dordrecht, 1992.

15. M. Ohtsu, *Highly Coherent Semiconductor Lasers*, Artech House, Norwood, 1992.

16. M. Ohtsu and Y. Teramachi, Analyses of mode partition and mode hopping in semiconductor lasers, *IEEE J. Quantum Electron.* **25**: 31–38 (1989).

17. M. Ohtsu, Y. Teramachi, Y. Otsuka, and A. Osaki, Analysis of mode-hopping phenomena in an AlGaAs laser, *IEEE J. Quantum Electron.* **QE-22**: 535–543 (1986).

18. P. Berge, Y. Pomeau, and C. Vidal, *Orders within Chaos*, Wiley, New York, 1984.

19. M. Ohtsu and T. Tako, Coherence in semiconductor lasers, in *Progress in Optics*, Vol. 25, E. Wolf, ed., Elsevier, Amsterdam, 1988, pp. 191–278.

20. D. Welford and A. Mooradian, Observation of linewidth broadening in (GaAl)As diode lasers due to electron number fluctuations, *Appl. Phys. Lett.* **40**: 560–562 (1982).

21. R. L. Fork, C. H. Brito-Crus, P. C. Becker, and C. V. Shank, Compression of optical pulses to six femtoseconds by using cubic phase compression, *Opt. Lett.* **12**: 483–485 (1987).

22. W. Wang and M. Ohtsu, Generation of frequency-tunable light and frequency reference grids using diode lasers for one-petahertz optical frequency sweep generator, *IEEE J. Quantum Electron.* **31**: 456–467 (1995).

23. S. Jiang, K. Nakagawa, and M. Ohtsu, Reflection-resonance-type photon scanning tunneling microscope, *Jpn. J. Appl. Phys.* **33**: L55–L58 (1994).

24. C.-H. Shin and M. Ohtsu, Heterodyne optical phase-locked loop by confocal Fabry–Perot cavity coupled AlGaAs lasers, *IEEE Photon. Technol. Lett.* **2**: 297–300 (1990).

25. E. A. Ash and G. Nicholls, Super-resolution aperture scanning microscope, *Nature* **237**: 510–512 (1992).

26. M. Fee, S. Chu, and T. W. Hänsch, Scanning electromagnetic transmission line microscope with sub-wavelength resolution, *Opt. Commun.* **69**: 219–224 (1989).

27. F. Zenhausern, Y. Martin, and H. K. Wickramasinghe, Scanning interferometric apertureless microscopy: Optical imaging at 10 angstrom resolution, *Science* **269**: 1083–1085 (1995).

FOUR

FABRICATION OF PROBES

As described in Section 3.1, the probe is the most essential component governing the performance of the NOM. Since the size of the probe tip is of the order of nanometers, special advanced fabrication processes have had to be developed. Fabrication methods such as pulling heated glass capillaries [1] and sharpening quartz rods or optical fibers by chemical etching have been used [2, 3]. However, neither sufficiently high reproducibility nor an apex diameter small enough for resolving the nanometric-scale structure of samples has been obtained. Further, tailoring the profile of the probe for higher transmission efficiency is not possible. To solve these problems, a selective etching method widely used in semiconductor very large-scale integrated chip fabrication has been applied to sharpen a single-mode fiber using buffered hydrofluoric (HF) acid as an etching agent [4]. Sections 4.1 and 4.2 describe the processes of sharpening the fiber and coating a metallic film to fabricate a protruded probe, respectively. Section 4.3 reviews the fabrication and performance of other novel fiber probes.

4.1. SHARPENING OF FIBERS BY CHEMICAL ETCHING

A fiber can be sharpened by utilizing the difference in the etching rates between the core and cladding in a selective etching solution of buffered HF acid. Advantages of this method are high reproducibility, mass production, and capability of tailoring the shape of the resulting fiber by controlling the etching conditions. This method is thus compatible with industrial operation.

Buffered HF acid with composition 40% (weight %) NH_4F, 50% HF, and H_2O is used to sharpen a fiber with pure silica cladding and GeO_2-doped silica core. The sharpening process of such a fiber can be explained by the following chemical reactions. The chemical reaction of SiO_2 with HF in water can be expressed as [5]

$$SiO_2 + 4HF \longrightarrow SiF_4 + 2H_2O \qquad (4.1.1a)$$

$$SiF_4 + 3H_2O \longrightarrow H_2SiO_3 + 4HF \qquad (4.1.1b)$$

$$SiF_4 + 2HF \longrightarrow H_2SiF_6 \qquad (4.1.1c)$$

Similarly, the reaction for GeO_2 is expressed as

$$GeO_2 + 4HF \longrightarrow GeF_4 + 2H_2O \qquad (4.1.2a)$$

$$GeF_4 + 3H_2O \longrightarrow H_2GeO_3 + 4HF \qquad (4.1.2b)$$

$$GeF_4 + 2HF \longrightarrow H_2GeF_6 \qquad (4.1.2c)$$

It has been found that the dissolution rate of the GeO_2-doped core is faster than that of the pure silica cladding in pure HF acid, so that the end of the fiber core becomes concave (similar to the shape shown in Fig. 3b in ref. 4). To obtain a lower dissolution rate of the GeO_2-doped core, NH_4F is added to HF as a buffer solution. In this case, the resultant H_2SiF_6 and H_2GeF_6 [Eqs. (4.1.1c) and (4.1.2c)] react with NH_3, ionized from NH_4F, as follows:

$$H_2SiF_6 + 2NH_3 \longrightarrow (NH_4)_2SiF_6 \qquad (4.1.3a)$$

and

$$H_2GeF_6 + 2NH_3 \longrightarrow (NH_4)_2GeF_6 \qquad (4.1.3b)$$

The difference in the solubility of $(NH_4)_2SiF_6$ and $(NH_4)_2GeF_6$ in the etching solution enables one, by adjusting the conditions of the etching (ratio of NH_4F to HF, the doping ratio of GeO_2, etching time), to obtain sharpened fibers with various cone angles and apex diameters.

Although the discussion given above attributed the reason for sharpening the fiber only to the difference in etching rates between the GeO_2-doped silica core and the pure silica cladding, it is still uncertain why a fiber with a uniformly GeO_2-doped core is sharpened and a nanometric apex diameter is realized. One possible reason may be the existence of a binding energy distribution of the Si–O–Ge network along the core radius resulting from residual stress generated by rapid cooling during the fiber fabrication process [6]. During this etching process, polar materials such as H_2O could change the dissolution rate along the core radius. In addition, it is not clear if nanometric silica maintains the characteristics of amorphous silica (see Section 4.1.4). More intensive studies on the properties of nanometric silica are required. The following subsections describe methods of producing different types of sharpened fibers.

4.1.1. A Basic Sharpened Fiber

Since the degree of buffering is controlled by varying the volume ratio X of NH_4F, the composition of the etching solution will be identified by its value of X.

The volume ratio of HF to H_2O is kept constant and is defined as 1:1. The solution is maintained at room temperature. By introducing the fiber into this solution, the core is sharpened. Figure 4.1a shows a typical scanning electron microscope (SEM) micrograph of an etched fiber. Figure 4.1b shows the magnified SEM micrograph of the top of the sharpened core. The apex diameter (defined in Fig. 4.2) can be estimated to be around 3 nm, with a cone angle of 20 deg (cf. Fig.3.2c). This diameter is smaller than even the conventional pyramid tips attached to the cantilevers used for the atomic force microscope (AFM). Although it has been reported that the cone angle θ_1 depends only on the percent doping [7], θ_1 has been found to depend also on X, especially in the case of fibers with highly GeO_2-doped core. This dependence is shown in Fig. 4.3 for several fibers of different doping ratio. Note that the values of θ_1 for curves A, C, and D are larger than 180 deg for $X <$ 1.6. This means that the core is etched at a rate faster than that of the cladding, resulting in a fiber with a concave facet. However, with increasing value of X, θ_1 decreases and for $X > 4$, θ_1 is almost constant and is independent of slight fluctuations in the composition of the solution. In this domain, θ_1 is determined by the doping ratio. The minimum θ_1 (14 deg) is realized for a 22 mol% GeO_2-doped silica core with 2.1 mol% F-doped silica cladding (curve E).

Finally, it is worth noting that identical nanometric tips have been successfully realized by etching a dual-core fiber (two noncoaxial cores in a common cladding), as shown in Fig. 4.4. This success illustrates the high spatial homogeneity and high

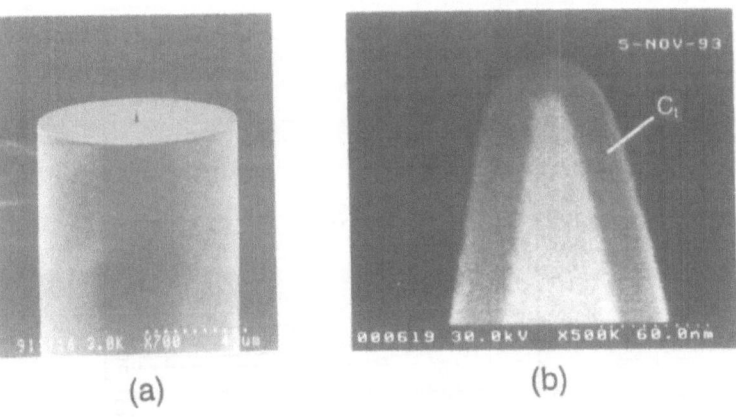

(a)   (b)

Figure 4.1. (a) SEM micrograph of a sharpened fiber with a cladding diameter of 90 μm. A fiber with 23 mol% GeO_2-doped core was etched ($X = 10$). The etching time was 120 mins. (b) Magnified SEM micrograph of the fiber tip. C_t indicates contamination due to bombardment of the electron beam while conducting the SEM observations. The original tip with apex diameter of 3 nm, coated with 7-nm-thick Pt/Pd, is covered by contamination C_t with a thickness of 22 nm.

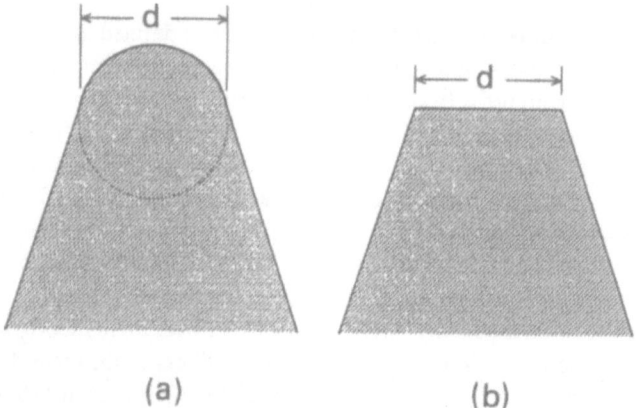

(a) (b)

Figure 4.2. The apex shape and apex diameter d of the sharpened fiber. (a) A rounded apex. The apex diameter is defined as the distance between the intersection points of the tangents drawn along the slope of the sharpened core to the completed circle (indicated by dotted curve) of the rounded apex. (b) A flattened apex.

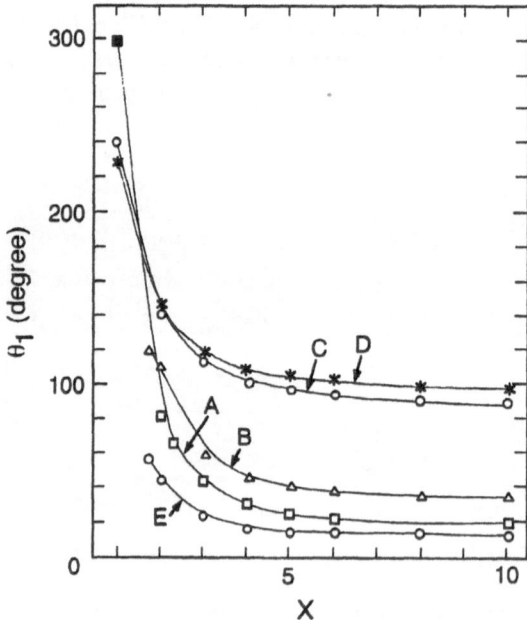

Figure 4.3. Dependence of the cone angle θ_1 of a sharpened fiber on the volume ratio X of NH_4F. (A–D) Results for fibers with a GeO_2-doped silica cores with doping ratio 23, 14, 8.5, and 3.6 mol%, respectively. (E) Result for a fiber with 22 mol% GeO_2-doped silica core and 2.1 mol% F-doped silica cladding.

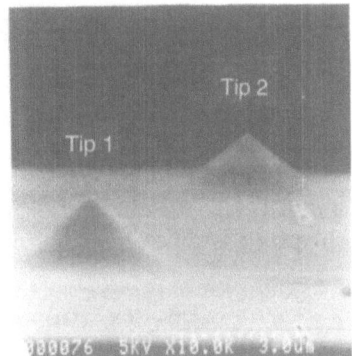

Figure 4.4. SEM micrograph of tips produced on a dual-core fiber with a GeO_2 doping ratio of 3.6 mol%.

reproducibility of the etching process. Based on this selective etching method, a variety of sharpened fibers can be tailored. Four representative examples are shown in Fig. 4.5. These have been used in practical NOMs. Details of the fabrication processes are described in the following subsections.

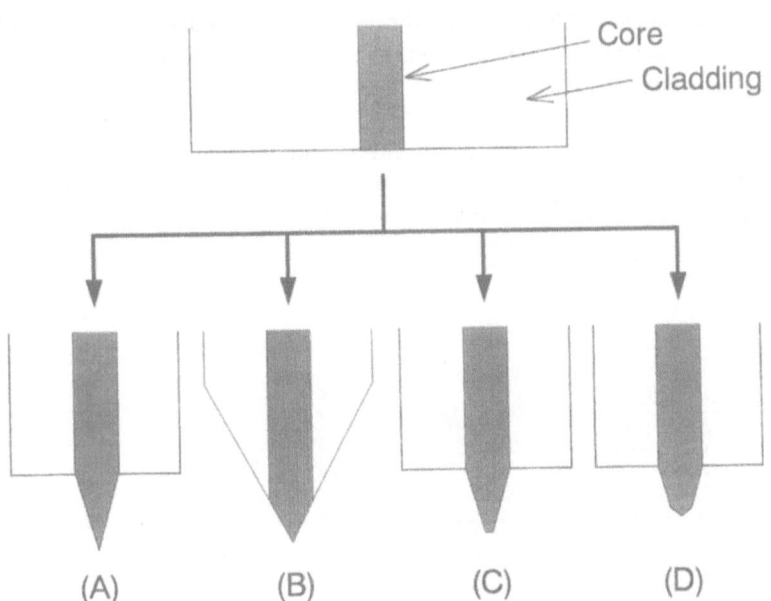

Figure 4.5. Cross-sectional profiles of four representative sharpened fibers which can be fabricated by the selective chemical etching technique. (A) A sharpened fiber with reduced-diameter cladding. (B) A pencil-shaped fiber. (C) A flat-top fiber. (D) A double-tapered fiber.

4.1.2. A Sharpened Fiber with Reduced-Diameter Cladding

As seen from Fig. 4.1a, the cladding face diameter of about 90 μm is much larger than the 10-μm length of the sharpened core. This large difference increases the chances of the sample being damaged after being hit by the cladding face. In order to solve this problem, a two-step etching process has been proposed [8]: In the first step, the cladding diameter is reduced; in the second step, the core is sharpened by the method of Section 4.1.1.

4.1.2.1. Cladding Diameter Reduction

An etching solution with $X = 1.7$ is used so that the cladding and core are etched at equal rates. For a fiber with a 23 mol% GeO_2-doped core, the cladding diameter is reduced at a rate of 0.95 μm/min.

4.1.2.2. Core Sharpening

The core is sharpened by the etching method of Section 4.1.1. The value of X is kept at greater than 3. This step further decreases the cladding diameter by about 20 μm ± 1.5 μm. Figure 4.6 shows the SEM micrograph of a sharpened fiber with a cladding diameter of 8 μm (reduced from an original value of 125 μm). When reducing the diameter of cladding, one should keep in mind that to maintain the optical transmission loss to a minimum, the cladding diameter must be at least twice the core diameter.

In addition to solving the practical scanning problem described above, the additional flexibility of the fiber (due to the reduced cladding diameter) allows its use as a cantilever for AFM measurements. This can be used for position control

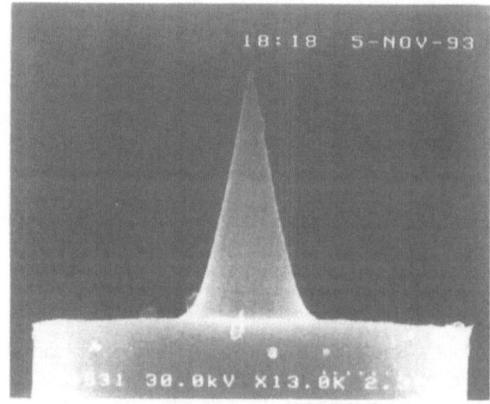

Figure 4.6. SEM micrograph of a sharpened fiber with reduced-diameter cladding. The cladding diameter is decreased to 8 μm.

of the probe (Section 3.1.2.2). Furthermore, this sharpened fiber with a reduced cladding diameter is compatible with the process of three-dimensional nanometric photolithography for realizing a nanometric aperture, as will be described in Section 4.2.2.

4.1.3. A Pencil-Shaped Fiber

The sharpened fibers described in Sections 4.1.1 and 4.1.2 have a discontinuous step at the boundary between the core and the cladding. To fabricate a protruded probe, these fibers have to be coated with metallic film using vacuum evaporation. As the flat end of the cladding is not coated efficiently for these fibers, a protruded probe having a nanometric foot diameter cannot be obtained easily. In order to solve this problem, a four-step etching process has been developed to fabricate a pencil-shaped fiber with a tapered cladding [9].

Since a low-GeO_2-doped fiber could not be hollowed by the second step, a fiber with a high-GeO_2-doped core (25 mol%) is used. All the etching processes are performed at room temperature ($25.0 \pm 0.1°C$).

Step I. The cladding is tapered by etching at an interface of oil and HF acid (denoted "oil/HF acid") as shown in Fig. 4.7[I]. This is because the height of the meniscus formed at the interface around the fiber decreases with decreasing cladding diameter [10]. The HF acid forms a very smoothly surfaced taper as compared with the case where a buffered HF acid solution is used. The floating oil, in addition to slowing down the evaporation of HF acid, also determines the taper angle α of the cladding. If the dissolution rates of the core R_{core} and of the cladding R_{clad} are equal, the fiber is tapered and the meniscus etching is automatically stopped when the meniscus height is zero [11]. However, for a high-GeO_2-doped fiber, the tapered fiber may also be hollowed because of a large difference between R_{core} and R_{clad}. Therefore, etching has to be stopped at the time when the cladding is tapered.

A spindle oil or a silicone oil is floated on the HF acid (about 50 wt%). R_{clad} is about 340 μm/hr at 25°C. When dimethylsilicone oil (0.935 g/cc) is used with a high-GeO_2-doped fiber, a taper of 20 deg is achieved. The required etching time is about 22 min.

Step II. The core is hollowed by HF acid as is shown in Fig. 4.7[II]. The 50 wt% HF acid is used to rapidly hollow the core. From geometrical considerations, $R_{core} > R_{clad}/\sin(\alpha/2)$ is required. To satisfy this condition, a fiber with a highly GeO_2-doped core has to be used since these rates are determined by α and the degree of doping.

Figure 4.8 shows the dependence of the apex diameter d and the cladding diameter D on the etching time of the second step T_2, where the diameters are measured after the four-step etching is completed. This figure shows that an etching time of 2 min realizes an apex diameter d smaller than 10 nm.

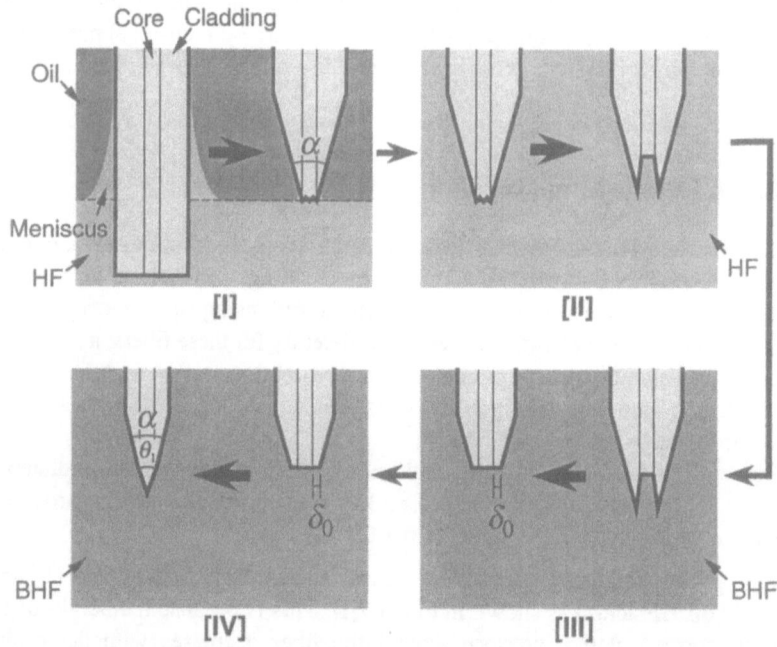

Figure 4.7. The four-step etching process for fabricating a pencil-shaped fiber. Here α is the taper angle of the cladding, δ_0 is the thickness of the flattened cladding, and θ_1 is the cone angle of the sharpened core. BHF, Buffered HF acid.

Step III. The hollowed cladding is flattened by an $X = 1.7$ buffered HF solution as shown in Fig. 4.7[III]. δ_0 represents the thickness of the flattened cladding. The etching time is 90 min.

Step IV. The core is sharpened by buffered HF solution ($X > 1.7$). The etching time is 90 min.

An SEM micrograph of a fabricated pencil-shaped fiber and its magnified image are shown in Figs. 4.9a and 4.9b, respectively. The cone angle of the core θ_1 is 20 deg and the apex diameter d is less than 10 nm including the thickness of the coated Au film. Figure 4.10 shows the dependence of the cone angle θ_1 on X, which agrees with that shown in Fig. 4.3.

The pencil-shaped fibers are fabricated with high reproducibility because the fourth step is performed by selective etching and is independent of the first step. Furthermore, the shape of the pencil-shaped fiber can be modified by controlling the etching time and X of the buffered HF solution used in the fourth step [9].

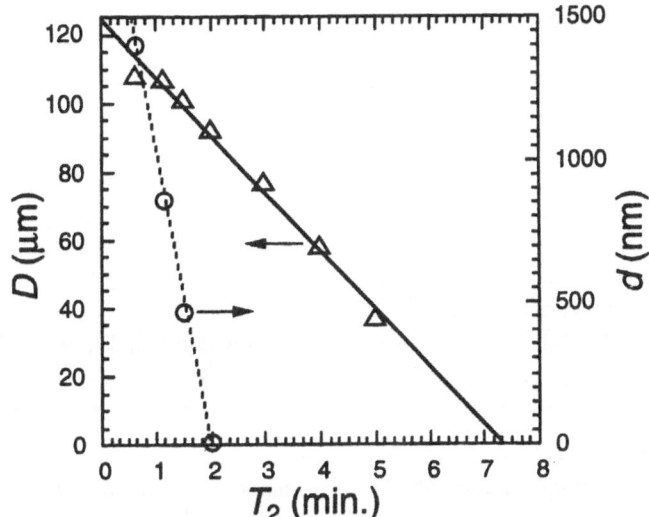

Figure 4.8. Dependence of the cladding diameter D and the apex diameter d of the pencil-shaped fiber on the etching time T_2 of the second step. The values of α and θ_1 are both 20 deg. A fiber with a GeO_2 doping ratio of 25 mol% is used. Dimethylsilicone oil, 0.935 g/cc, is used for step I. The concentration of the HF acid is 50 wt% for steps I and II. The value of X is 10 for steps III and IV.

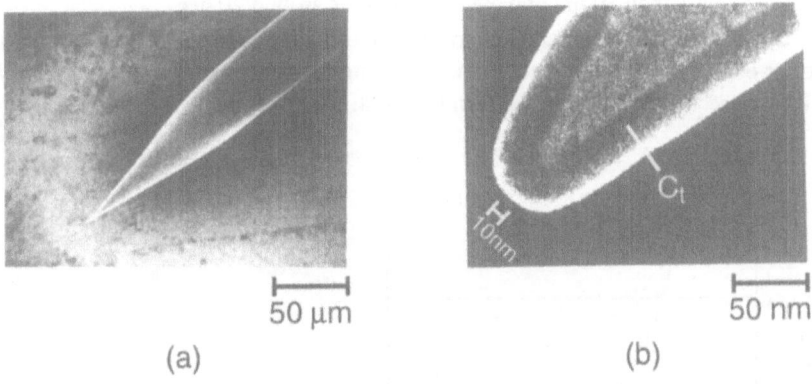

(a)   (b)

Figure 4.9. (a) SEM micrograph of a pencil-shaped fiber. The etching time T_2 for step II is 2 min and the other conditions are the same as for Fig. 4.8. The taper angle of the cladding α is 20 deg. (b) Magnified image of the top of panel (a). C_t indicates a contamination due to bombardment of the electron beam while conducting the SEM observations. The original tip, covered with a 3-nm-thick Au layer is covered by the contamination C_t. The cone angle of the sharpened core θ_1 is 20 deg and the apex diameter d is less than 10 nm including the Au film.

4.1.4. A Flattened-Top Fiber

A sharpened fiber with a nanometric flat top can be fabricated by a three-step etching process. The meaning of "flattened apex diameter d" is defined in Fig. 4.2b. This kind of probe with flattened apex could be advantageously used for aperture fabrication, and especially for fabricating a light-emitting probe (Section 4.3.1). Although this type of fiber has been previously fabricated by the heating and pulling method [12], it has been difficult to produce an apex with dimensions less than 50 nm reproducibly. The advantage of the present chemical etching method lies in its high reproducibility and small size.

A schematic of the three-step etching process is shown in Fig. 4.11 [13]. The first and second steps are the same as for the sharpened probe described in Section 4.1.2. All the etchings are performed at room temperature (22.5 ± 0.5°C).

Step I. A fiber with pure silica cladding and a 27 mol% GeO_2-doped silica core is etched in an etching solution of $X = 1.7$ for an etching time of 90 min to reduce the cladding diameter D to approximately 40 μm. The tip length of the sharpened core L is 10 μm. If $D/2 \leq L$, it is possible to coat a metallic film in a vacuum evaporation unit (Section 4.2.1) on only the sides and on the cladding face. This is due to the fact that no shadow of the cladding is formed on the sharpened core while coating the metallic film at an inclination of around 25 or 30 deg with respect to the evaporation source in a vacuum evaporation unit.

Step II. The fiber is etched in buffered HF acid ($X = 10$) for 120 min to get a sharpened core with rounded apex as shown in Fig. 4.12a. For this case, the apex diameter is less than 10 nm and the cone angle is around 20 deg.

Step III. In order to make a sharpened fiber with a flattened apex, the fiber is further etched in an etching solution containing ten-times diluted solution with $X = 10$. The result of this step is shown in Fig. 4.12b. It can be seen from this figure

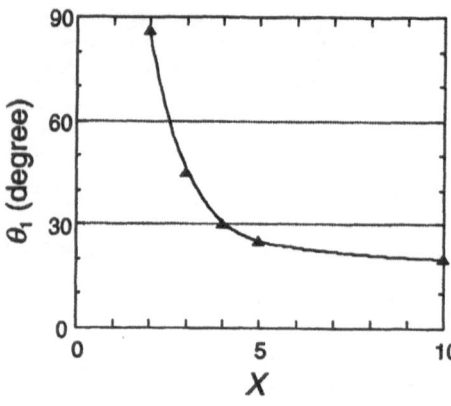

Figure 4.10.   Dependence of the cone angle θ_1 on X of step IV. The GeO_2 doping ratio is 25 mol%.

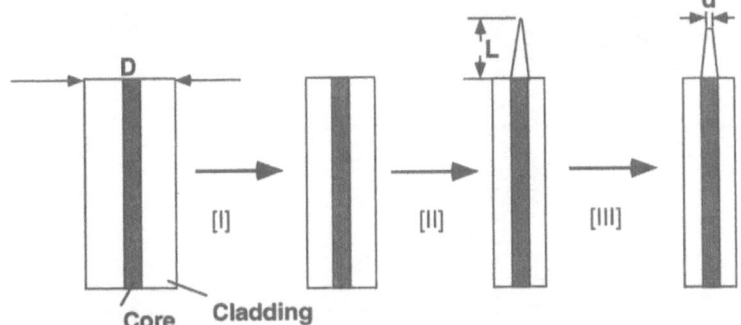

Figure 4.11. The three-step etching process for fabricating a flat-top fiber. Here D is the cladding diameter, L is the length of the sharpened core, and d is the flattened apex diameter defined by Fig. 4.2b.

that the apex of the fiber is changed from a rounded to a flattened shape. The apex diameter d is around 20 nm. By accurate control of the etching time and environment variables such as temperature, a flat apex can be obtained with quite high reproducibility.

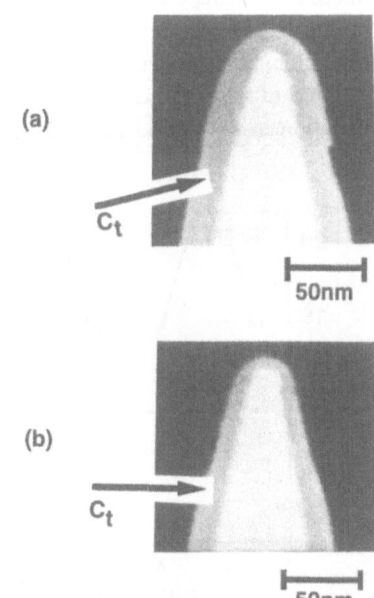

Figure 4.12. SEM micrographs showing the etching results due to (a) step II and (b) step III. C_t indicates contamination due to the bombardment of the electron beam while conducting the SEM observations.

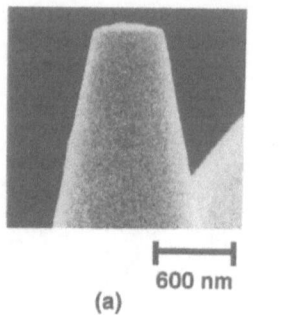
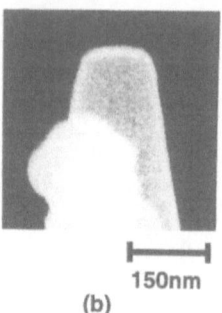

600 nm   150nm

(a)   (b)

Figure 4.13. SEM micrographs showing a flattened apex due to step II for an etching time of (a) 60 min and (b) 90 min.

It should be mentioned that it is possible to fabricate an apex with a flattened shape (as shown in Fig. 4.13) during the second step by controlling the etching time. However, the diameter of the flattened apex d is around 100 nm. It is almost impossible to fabricate an apex with a diameter of less than 50 nm.

Figure 4.14 shows the variation in the shape of the apex as a function of etching time with the vertical axis being the apex diameter d of the third step. Here, the closed circles and squares correspond to an apex with a rounded (Fig. 4.2a) and flattened (Fig. 4.2b) shape, respectively. The closed triangles correspond to an intermediate shape which is difficult to discriminate. The shape corresponding to each data point in this figure is actually the average shape obtained from at least three etching results. As can be seen, the rounded apex at the zero time of the etching changes to a flattened apex at approximately every 2 min. Though there is a little variation in the period of the limit cycle of repeatable shapes, there is almost a

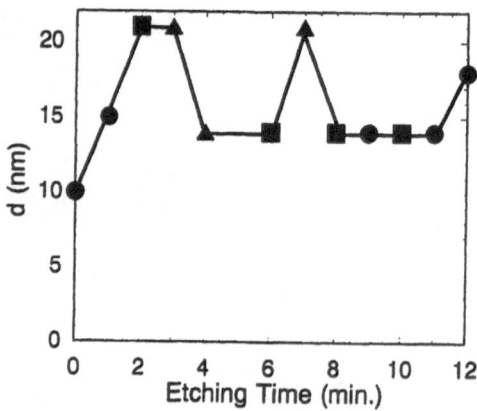

Figure 4.14. Variation of the apex shape and the apex diameter d of the sharpened core as a function of the etching time of step III. The closed circles and squares correspond to an apex with a rounded (Fig. 4.2a) and flattened (Fig. 4.2b), shape, respectively. The closed triangles correspond to an intermediate shape.

certain periodicity occurring almost at every 2 min. This kind of oscillatory change in shape with respect to etching time is fairly reproducible and should provide a clue to understanding the mechanism of the etching process.

As in Fig. 4.14, the diameter of the flattened apex is always greater than that of the rounded apex at the zero time of etching. This can be expected, as the etching process involves only the dissolution of the core with negligible changes in the cladding. Moreover, the average diameter of the flattened apex is around 15 or 20 nm and the largest diameter obtained is around 30 nm when the etching time is as long as 10 min.

In general, when a fiber is introduced into an etching solution [10], a meniscus is formed at the interface. The height of the meniscus is influenced by the fiber radius, glass wettability, and viscosity. The etching proceeds along the meniscus isotropically. The etching process also takes place at the same rate in all directions around the axis of the cylindrical fiber. In Fig. 4.15, the length of the sharpened core L is plotted as a function of the etching time in the third step. As seen from this figure, the length of the sharpened core initially decreases rapidly and then stabilizes at a much slower rate as the etching process proceeds. After around 20 min, there is absolutely no evidence for the previous existence of a tip. This corresponds to an average etching rate of 8.3 nm/sec.

Studies of the etching of oxide-doped silica [14] show that in the etching process of a core containing GeO_2-doped SiO_2, Si–O–Ge and Ge–O–Ge bonds are formed that have to be broken in order to dissolve the material. In the etching of doped silica in a buffered HF solution, the etching process can be seen as a combination of the breaking of Si–O–Se and Ge–O–Si bonds by F^- and HF_2^- species of buffered HF and that of Ge–O–Ge bonds by water. When the doping ratio of GeO_2 is as high as in our case, the Ge–O–Ge bonds become abundant and they are easily attacked by water. Therefore, when the etching solution is dilute,

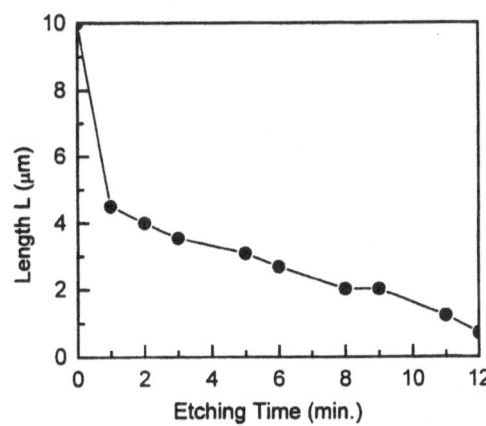

Figure 4.15. Variation of the length of the sharpened core L as a function of the etching time of step III.

with high values of pH (around 7), the etching rate of the core becomes higher and there is a gradual decrease in the length of the sharpened core as in Fig. 4.15.

Even though the length of the sharpened core decreases monotonically as expected, the structure of the sharpened tip does not. Figures 4.16a and 4.16b show the sharpened tips when the etching times of the third step are 1 and 11 min, respectively. As can be seen from Fig. 4.16b, the sharpened core has two structures, in contrast to the sharpened core shown in Fig. 4.16a: The upper part of the sharpened core is an acute-angled cone, while the lower part is oblique. The appearance of two different structures in the third step raises some doubts about the existing explanation of the etching process. If the etching proceeds only in an isotropic fashion with the rate being same in all directions [15–17], then the presence of this double structure cannot be explained. This shows some evidence for the existence of anisotropic dissolution in the vertical direction.

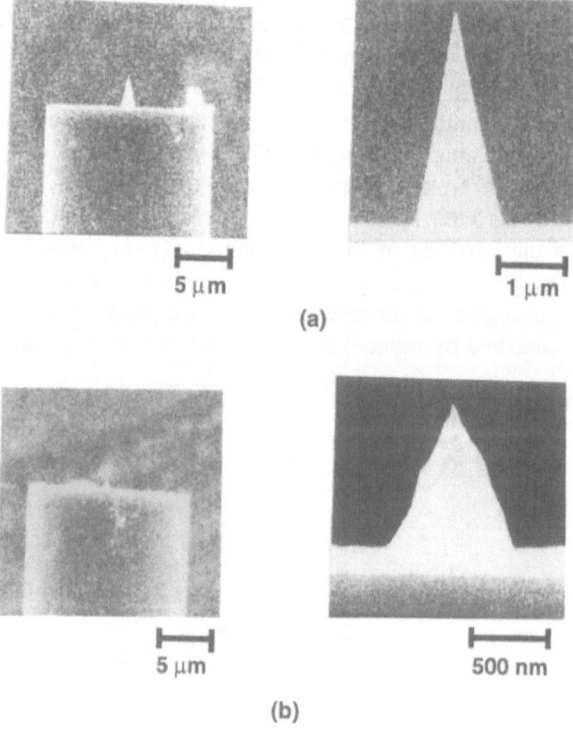

Figure 4.16. Changes in the shape of the sharpened fiber for step III etching time of (a) 1 min and (b) 11 min. The SEM micrograph on the right shows a magnified view of the sharpened core shown on the left.

In order to explain the changes in the morphology of the apex, the etching process must be considered as a discrete process with the dissolution taking place in the form of dissolution of particles or clusters. One possible explanation for the existence of such a discrete nature is that some structural changes are necessary in order for the system to remain in an equilibrium state with minimum free energy. However, more detailed investigations are required to understand why these kinds of changes in the shape of the apex should follow some regularity and happen in a pseudoperiodic fashion.

Further, at the apex of the sharpened core, the dimensions involved are a few or a few tens of nanometers, i.e., a few tens or a few hundreds of atoms are involved. In the case of such nanometric structures of a mesoscopic nature, it appears that the conventional etching theories [17, 18] can no longer be applied. Existing etching theories intended to explain the mechanism of etching of SiO_2 by buffered HF are meant for the planar silica structures usually employed in the fabrication of VLSI devices. They are suited for macroscopic dimensions. The kinds of changes in the morphology on a nanometric scale suggest the presence of some order existing in the macroscopic amorphous glass material [19]. As this kind of phenomenon is reproducible for a slightly different doping level of 25 mol% GeO_2 in the core, the cyclical etching characteristics are not due to any nonuniformity in doping levels.

4.1.5. A Double-Tapered Fiber

When the illumination mode is used for exciting a semiconductor material in order to carry out spatially resolved photoluminescence (PL) spectroscopy, it is advantageous to collect the PL signal by a probe. This is because both the light used for excitation and the PL signal light traverse the same probe. This avoids a deterioration of resolution caused by the carrier diffusion effect. In this case, the most serious problem is that the detected signal is small due to the low excitation and collection efficiency of the tiny probe. To solve this problem, it is necessary to improve the efficiency of transmission to the probe by tailoring the shape of the fiber tip. This subsection describes a novel sharpened fiber with improved transmission efficiency, which is developed by utilizing the high reproducibility and controllability of the selective chemical etching technique [20].

The metal-coated fiber probe, which can be regarded as a metal-cladded tapered waveguide, has a complicated loss mechanism. The existence of a cutoff diameter for even the lowest mode is the most specific feature of the metallic waveguide [21]. Beyond the cutoff diameter, the intensity of the light decreases rapidly. With respect to the transmission efficiency, the distance between the foot of the protrusion and the place where the fiber diameter is greater than the cutoff diameter is a key parameter. The dependence of the transmission power on the cone angle may be due to the evanescent propagation of light in the metallic-cladded

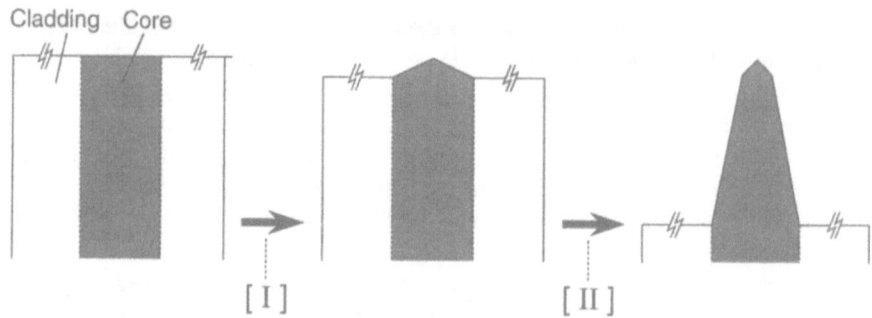

Figure 4.17. The two-step etching process for fabricating a double-tapered fiber.

waveguide where the cladding diameter is less than the cutoff. In order to increase the transmission efficiency, it is reasonable to shorten the distance between the foot of the protrusion and the place where the fiber diameter is greater than the cutoff diameter [22]. This is nearly equivalent to increasing the cone angle.

However, in the case of a short tip (i.e., a sharpened core with a large cone angle for high efficiency), contact between the cladding and the sample surface is a real concern, especially when the sample surface is bumpy. To guard against contact, it is necessary to lengthen the tip while maintaining the large transmission efficiency. For this purpose, a double-tapered fiber has been fabricated with high reproducibility. A two-step etching process is employed as shown in Fig. 4.17 (essentially the same process discussed in Section 4.1.2).

In the first step, using a buffered HF solution of $X = 1.8$, a short tip with a cone angle as large as 150 deg is fabricated. In the second step, a long tapered region which delivers light to the cutoff diameter is obtained using a solution with $X = 10$. As confirmed by the SEM micrograph shown in Fig. 4.18, the resultant top cone

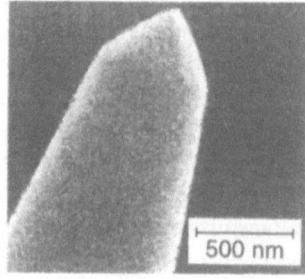

Figure 4.18. SEM micrograph of a fabricated double-tapered fiber.

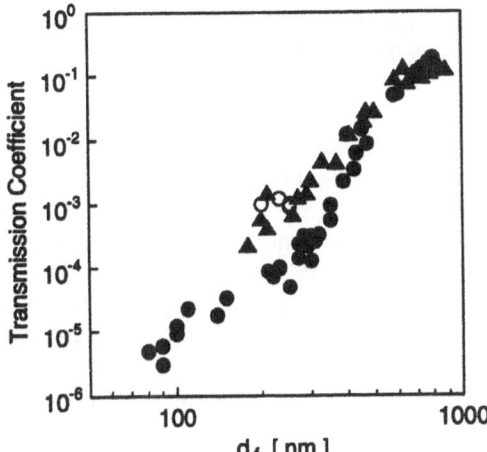

Figure 4.19. Transmission coefficient as a function of the foot diameter d_f. Closed circles: Single-tapered fiber with a 20 deg cone angle (tip 1). Closed triangles: Single-tapered fiber with a 50 deg cone angle (tip 2). Open circles: Double-tapered fiber.

angle is about 90 deg. A metallic film can then be coated on this fiber by the technique described in Section 4.2.1.[*]

To demonstrate the high efficiency of the tip, Fig. 4.19 compares the transmission coefficient of the double-tapered fiber with that of two single-tapered fibers. The transmission coefficient, defined as the ratio of output to input power, is plotted as a function of foot diameter. The results for the double-tapered fiber are indicated by open circles, and those for the single-tapered fibers are given by closed circles (tip 1. cone angle $\theta_1 = 20$ deg, length of conical core $L = 6$ μm) and triangles (tip 2. $\theta_1 = 50$ deg, $L = 2.5$ μm). The double-tapered fiber was coated by 200 nm of Au and had a foot diameter of 200 nm. For the purpose of evaluating the transmission coefficients of the various fibers, 130-μW light from an He–Ne laser ($\lambda = 633$ nm) was coupled into the fibers with 60% efficiency and the light emitted from the probe was collected in the far-field with a 0.4-numerical aperture objective lens.

As can be seen from the figure, the shorter tips (irrespective of whether they were double or single tapered) have an order of magnitude higher transmission in the region where the foot diameter is smaller than the cutoff diameter ($d_c = 400$ nm for $\lambda = 633$ nm in glass). In the region where $d_f > d_c$, the transmission efficiency is relatively independent of the tip length. This confirms that the most important factor in determining the transmission efficiency is the length of the cutoff region of the metallic waveguide; a short tip length is necessary for maximum efficiency.

The double-tapered fiber, while maintaining the advantages (high transmission, low thermal stress) of a short tip length below cutoff, is not as susceptible to damage by contact with sharp features on the sample.

---

[*]Due to the large cone angle, the apex of the double-tapered fiber appears to be coplanar with the metallic film (cf. Fig. 6.7b). In this sense the foot diameter defined by Fig. 3.2c is equivalent to the aperture diameter of the apertured probe formed by pulling a heated fiber.

4.2. METAL COATING AND FABRICATION OF A PROTRUDED PROBE

A nanometric foot diameter must be realized in order to suppress scattering (C-mode) or generation (I-mode) of low-spatial-Fourier frequency components of the evanescent light (for the definition of the C- and I-modes, see Section 3.1.1). This section describes two methods of fabricating such a protruded probe.

The initial common step is that of coating the sharpened fiber with a metallic film of sufficient thickness to block unwanted light. Both Al and Au have been popularly used as a coating metal. Their films have to be thicker than about 100 nm because the skin depth of the light is several tens of nanometers, as shown in Fig. 4.20.

The advantage of Al as compared with Au is its smaller skin depth for a wide range of wavelengths. The disadvantage is its chemical instability. Its surface is easily covered in air with an oxide film of variable thickness. The reproducibility in the thickness of the Al film is also low when a nanometric protruded probe is fabricated by the chemical etching technique. Furthermore, a clear SEM micrograph of the apex of an Al-coated fiber tip cannot be obtained because of the oxide film. In addition, the W boat used for vacuum evaporation gets corroded easily by Al.

Although Au has a larger skin depth at a short wavelength range (especially at 500 nm wavelength), its high chemical stability is an outstanding advantage

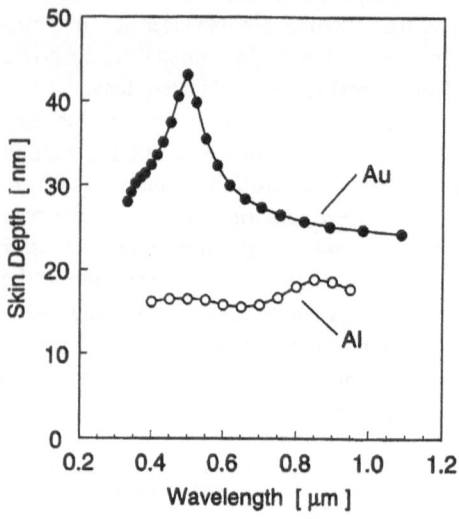

Figure 4.20. Skin depth of Al and Au as a function of the optical wavelength, calculated by using the absorption index (imaginary part of the complex refractive index).

For stronger adhesion, it is necessary to coat the fiber with a Cr film of several nanometers thickness prior to coating with Au or Al. The minimum foot diameter fabricated by either of the methods described in this section is determined by the clustering of the atoms in the metallic film at the stage of coating. In the case of Au, the diameters of the clustered grains are about 10 and 30 nm for the cases of a conventional vacuum evaporation unit and sputtering, respectively. The grain size can be decreased to several nanometers if more advanced vapor deposition techniques are used. It should be noted that using SEM to evaluate the foot diameter (see footnote in Section 4.1.5) is only an auxiliary method to estimate the near-field optically effective foot diameter. For accurate estimation of the effective foot diameter, one should measure the spatial distribution of the evanescent light power generated on the surface of the probe.

After the metallic film is coated on the fiber, the second step is to remove it from the apex part of the sharpened core. In the case of the flattened-top fiber described in Section 4.1.4, preparation of the tip is straightforward. For coating in a vacuum evaporation unit, the fiber is mounted at an inclination of about 25 deg with respect to the evaporation source so as to coat the cladding face and the sides of the sharpened core, and rotated slowly. Figure 4.21a shows an SEM micrograph of a fiber coated with a 150-nm-thick Au film. The encircled tip region is shown magnified in Fig. 4.21b. Figure 4.21c explains schematically the complicated image of Fig. 4.21b which is observed in SEM with the electron beam impinging normal to the fiber. Here, the central circle corresponds to the flat apex of the fiber probe and the hatched region corresponds to the Au film. The dark region corresponds to the rim of the sharpened fiber. The foot diameter is less than approximately 30 nm (see footnote in Section 4.1.5).

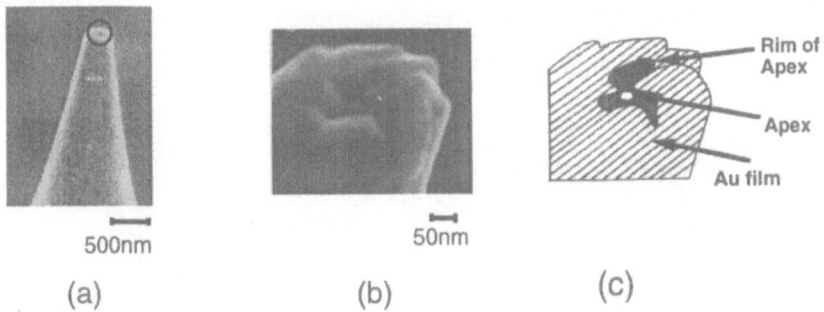

Figure 4.21. (a) SEM micrograph of a flat-top fiber probe coated with a 150-nm-thick Au film, under low magnification. (b) Magnified picture of the encircled tip region of panel (a). (c) A schematic explanation of panel (b) with the hatched region corresponding to the Au coating covering the sides of the sharpened fiber and the central circle corresponding to the flat apex. The dark region around the flat apex corresponds to the rim of the apex.

In the case of sharpened fibers described in Sections 4.1.1–4.1.3 and 4.1.5, the fabrication process is more complicated. Two methods will be discussed in the following subsections.

4.2.1. Removal of Metallic Film by Selective Resin Coating

In order to realize a protruded probe as in Fig. 3.2c, a simple etching technique called selective resin coating (SRC) has been employed [23]. Figure 4.22 shows the SRC method, which consists of four steps: (I) coating metallic film, (II) coating resin, (III) preferential etching of metallic film from the apex, and (IV) removal of resin. Figure 4.23 shows a scanning electron micrograph of two fabricated protruded probes. In Fig. 4.23a, sputtering was used for coating. The protruded probe has an apex diameter $d < 10$ nm, foot diameter $d_f < 30$ nm, and cone angle $\theta_1 = 20$ deg with the 120-nm-thick Au film. In Fig. 4.23b, a vacuum evaporation unit was used to reduce the grain size of the Au film. The values of d, d_f, and θ_1 are almost the same as those of Fig. 4.23a.

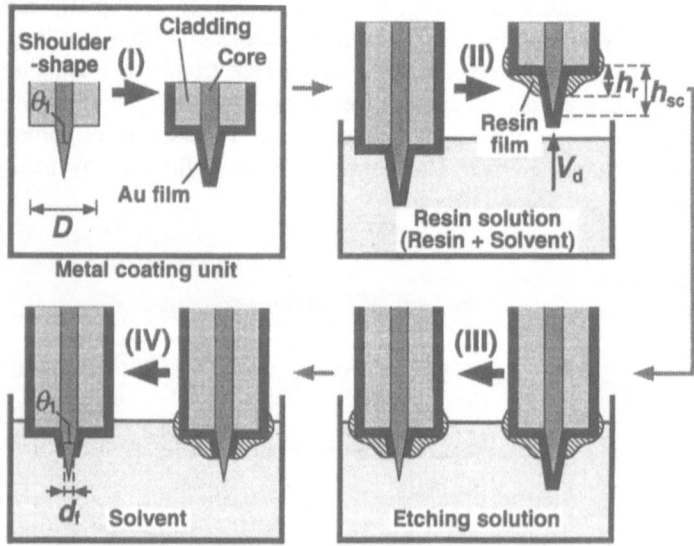

Figure 4.22. The selective resin coating method: (I) coating the Au film, (II) coating the resin by dipping the fiber into the resin solution and pulling up into air with a drawing speed V_d, (III) etching in the KI–I$_2$ solution, and (IV) removal of resin film. Here D is the diameter of the cladding d_f is the foot diameter of the protrusion, θ_1 is the cone angle, h_{sc} is the length of the sharpened core measured from the end of the cladding, and h_r is the length of the resin-coated core measured from the end of the cladding.

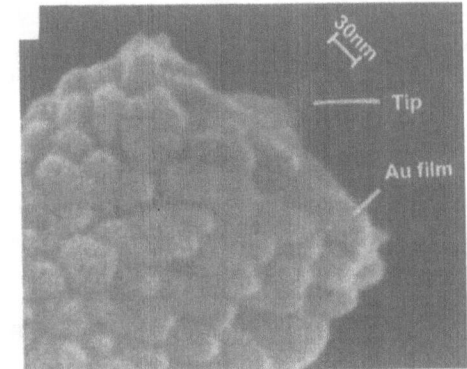

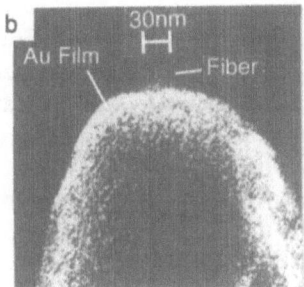

Figure 4.23. Scanning electron micrographs of the protruded probes fabricated by (a) the SRC method and (b) a modified SRC method. Note that the grain size has been made small in panel (b) by using a vacuum evaporation unit for Au film coating. For the case of panel (a), the diameter of the cladding D, the foot diameter d_f, the cone angle θ_1, and the apex diameter d are 45 μm, 30 nm, 20 deg, and 10 nm, respectively. The values of d, d_f, and θ_1 for panel (b) are almost the same as those of panel (a).

In step I, an Au film is coated on a sharpened fiber with a reduced-diameter cladding (Section 4.1.2) by using a magnetron sputtering unit. In step II, the fiber is dipped into an acrylic resin solution and drawn up with a speed V_d of about 5 cm/sec. A commercial acrylic resin solution designed as a primer for painting metal is employed. It has a viscosity of 11 cP at 22°C. Although other resin solutions can be used, they must be both corrosion-resistant against the etching solution in step III and easily removable by solvents in step IV. On revolving the fiber from the acrylic resin solution into air, it is covered with a dried acrylic resin film of submicrometer thickness.

In order to fabricate a protruded tip (Fig. 3.2c), the end of the sharpened core has to be free from the resin film. Since the apex diameter d is negligibly small, one can relate the foot diameter d_f of the protruded probe to the length of the sharpened core h_{sc} and the length of the resin-coated core h_r (Fig. 4.22):

$$d_f = 2\,(h_{sc} - h_r)\,\tan(\theta_1/2) \qquad (4.2.1)$$

where θ_1 is the cone angle. This equation shows that d_f is determined by h_r for fixed values of h_{sc} and θ_1. In turn h_r depends on the viscosity of the resin solution, the drawing speed V_d, and the cladding diameter D.

In step III, the sharpened core with Au and resin films is etched for 2 min in a 50-times-diluted KI–I$_2$ solution of weight ratio KI:I$_2$:H$_2$O = 20:1:400. The etching time has to be controlled according to the thickness of the Au film. For a typical Au film thickness of 100–150 nm, the etching time varies from 2 to 3 min. In step IV, the acrylic resin film is removed by dipping into a solvent such as acetone.

The foot diameter d_f of the protrusion can be varied by controlling the cladding diameter D and the drawing speed V_d (for which the two-step etching method of Section 4.1.2 is necessary). Figure 4.24a shows the dependence of d_f on D for $V_d = 5$ cm/sec. It can be seen that d_f decreases with increasing D and approaches a constant value of 28 nm at $D > 38$ μm. Fabricating 14 probes with $D = 33$ μm gave the average and standard deviation of d_f as 45 and 12 nm, respectively. Figure 4.24b shows the variation of the foot diameter d_f of the protrusion as a function of the drawing speed V_d for $D = 40$ μm. As is seen, d_f decreases with increasing V_d and approaches a constant value of 28 nm at $V_d > 2$ cm/sec.

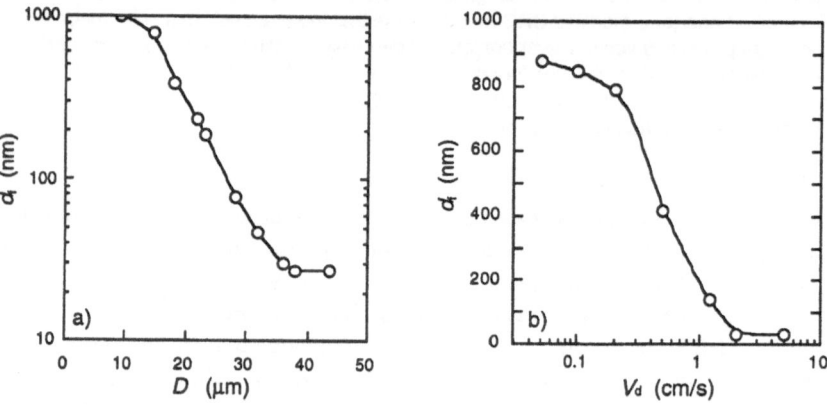

Figure 4.24. Dependence of the foot diameter d_f on (a) the cladding diameter D for a drawing speed V_d of 5 cm/sec, and on (b) V_d for $D = 40$ μm.

4.2.2. Removal of Metallic Film by Nanometric Photolithography

This technique is based on a photolithographic method which has been popular for fabricating semiconductor devices. This photolithographic method differs from the conventional diffraction-limited photolithography used for planar semiconductor devices in that it must be applied to a nanometric and three-dimensional surface. The fabrication process consists of the following six steps as shown in Fig. 4.25 [24]: (I) sharpening the fiber, (II) coating an Au film, (III) coating a photoresist, (IV) exposing the fiber tip and development, (V) removing the Au film from the top of the sharpened core, and (VI) removing the photoresist.

The novelty of this method lies in the method of exposure at step IV: The fiber tip is exposed by the evanescent field generated on a planar surface as shown in Fig. 4.26. Due to the rapid decrease of the evanescent field intensity as a function of fiber–surface separation, the nanometric area at the fiber tip is selectively exposed.

The two-step etching process of Section 4.1.2 is employed for step I in order to control the cladding diameter so that the optimum thickness of the coated photoresist is maintained in step III. In step II, an Au film of about 100 nm thickness is coated by a conventional vacuum evaporation unit. The third and fourth steps are described below.

Step III. Coating a Photoresist. A positive photoresist is coated on an Au-film-coated sharpened fiber. The essential feature of the photoresist used here is its small dissolution speed (<10 nm/min). Furthermore, for high reproducibility of the exposure in step IV, the following method of coating is employed to keep the thickness of the photoresist δ (see Fig. 4.26) constant at the fiber tip: After a droplet

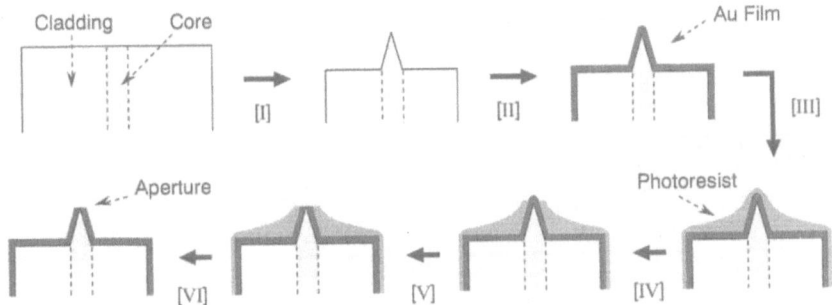

Figure 4.25. Three-dimensional nanometric photolithography for fabricating a protruded probe: (I) Sharpening the fiber. (II) Coating an Au film. (III) Coating a photoresist. (IV) Exposing the fiber tip and development. (V) Removing the Au film from the top of the sharpened core. (VI) Removing the photoresist.

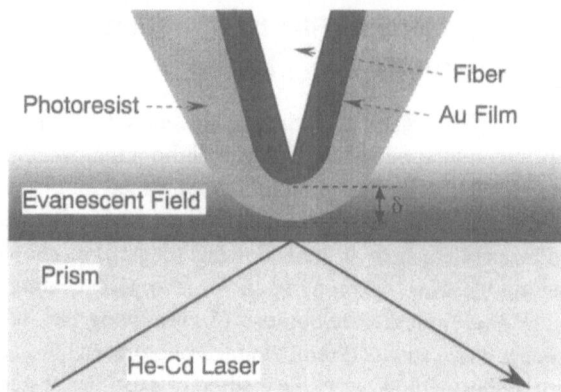

Figure 4.26. Exposing the fiber by using the evanescent field generated on a planar surface. δ is the thickness of the photoresist.

of photoresist forms on the top of the needle of the syringe, it is made to touch to the top of the fiber tip, and then the droplet is pulled up from the fiber.

A photoresist coated by this method becomes thinner with decreasing cladding diameter, as shown by Fig. 4.27. Here, the thickness of the photoresist at the fiber

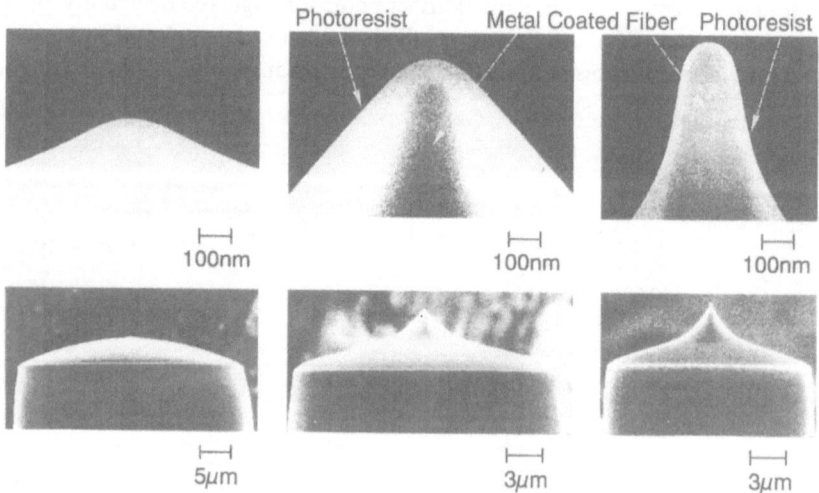

Figure 4.27. SEM micrographs of fibers coated with photoresist for cladding diameters of (a) 40 μm, (b) 23 μm, and (c) 18 μm. The viscosity of the cladding is 60 mPa sec. Upper parts of this figure show the magnified images of the lower parts.

tip is about 100, 70, and 20 nm for cladding diameter 40, 23, and 18 μm, respectively. This dependence on the cladding diameter is highly reproducible as long as the viscosity of the photoresist and the thickness of the Au film are constant. Here, the cladding diameter is empirically adjusted to be 23 μm since the penetration depth of the evanescent field is insufficient for a thicker photoresist, so the Au film cannot protrude out of the photoresist. When the photoresist is thinner, on the other hand, the photoresist on the rim of the exposed area is easily peeled off in the process of step V. Thus, the two-step etching method of Section 4.1.2 is necessary, as was the case for Section 4.2.1, to obtain the optimum cladding diameter for the present photolithography method. The temperature and the prebaking time are 80°C and 20 min, respectively.

Step IV. Exposing the Fiber Tip and Development. The exposure system is shown in Fig. 4.28. The fiber is exposed to the evanescent field generated on the prism surface into which the He–Cd laser light (422 nm wavelength) is incident at the angle of total reflection. The fiber–prism separation is controlled to be 5 nm by using the monitored shear force for negative feedback, as described in Section 3.1.2.2. To realize an axially symmetric exposure, a mirror (M2 in Fig. 4.28) is used so that the fiber is exposed to two mutually "counterpropagating" evanescent fields. Generation of interference fringes between the two evanescent fields is avoided by a slight misalignment of the incident and reflected laser beams.

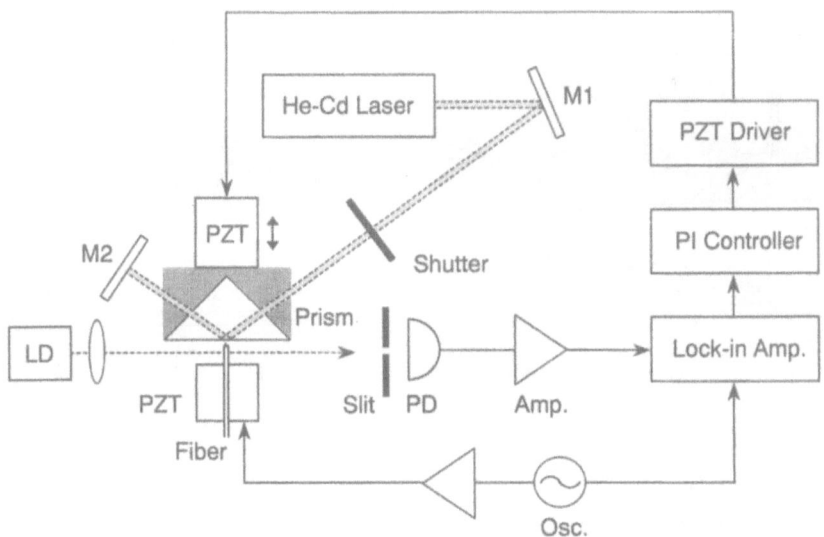

Figure 4.28. Experimental apparatus for exposure of the fiber. LD, Semiconductor laser of 690 nm wavelength; PD, photodetector.

Au Coated Fiber Photoresist

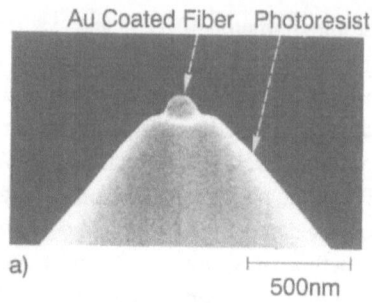

a)

500nm

b) Exposed Area

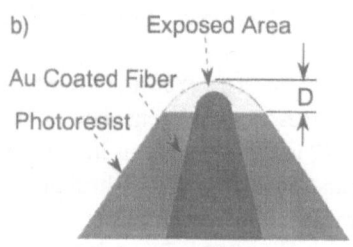

Au Coated Fiber

Photoresist

D

Figure 4.29. (a) SEM micrograph of a fiber exposed by using two mutually counterpropagating evanescent fields. (b) Schematic explanation of panel (a). D is the depth of the exposed photoresist.

Figures 4.29a and 4.29b show, respectively, the SEM micrograph and a cross-sectional schematic of the symmetrically exposed fiber tip. Here the optical

Au Film   Fiber

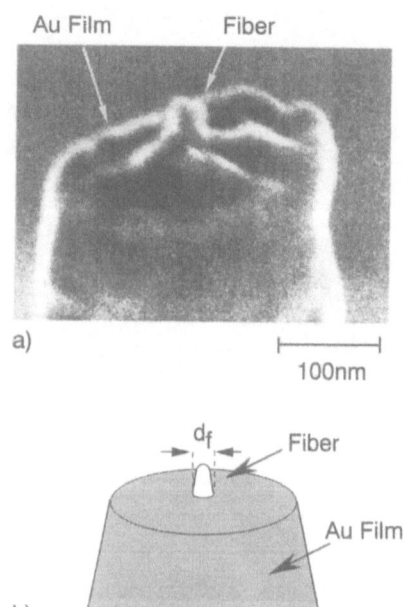

a)

100nm

d_f Fiber

Au Film

b)

Figure 4.30. (a) SEM micrograph of the protruded fiber. (b) Schematic explanation of panel (a). d_f is the foot diameter.

Figures 4.29a and 4.29b show, respectively, the SEM micrograph and a cross-sectional schematic of the symmetrically exposed fiber tip. Here the optical energy for exposure is about 0.12 mJ and the exposed depth D is 150 nm. After the exposure, the fiber is dipped into the developer to remove the exposed photoresist and then baked at 120°C for 30 min.

The fiber is then dipped into a conventional remover of photoresist. Figure 4.30a shows the SEM micrograph taken after step IV. Figure 4.30b is a schematic explanation. The foot diameter d_f is about 30 nm. The $KI-I_2$ solution described in Section 4.2.1 is used for etching the Au film in step V. The fiber is dipped for about 40 sec. The grain size of the Au film formed by vacuum evaporation (~30 nm) determines the minimum size of the aperture in the present case. Reducing the grain size should allow a smaller aperture to be obtained.

This novel photolithographic approach makes nanometric fabrication possible by inducing a photochemical reaction of a few tens of molecules of photoresist that are confined in a three-dimensional nanometric scale even if the evanescent field is two-dimensional. This method demonstrates the possibility of fabricating optically three-dimensional nanometric materials, and can be used as an advanced fabrication technology for nanometric devices.

4.3. OTHER NOVEL PROBES

4.3.1. Functional Probes

A variety of novel probes have been proposed. The following are examples of so-called "supertips," which can convert an optical signal to other physical signals or vice versa [25].

The cleaved edge of a GaAs crystal was used as a light-emitting probe by forming a Schottky contact consisting of GaAs as the semiconductor and aluminum as the metallic contact on two adjacent cleaved faces [26]. Its advantage is that the optoelectronic conversion can be integrated into the probe itself. A second example is a pH sensor fabricated by incorporating fluoresceinamine into an acrylamide-methylenebis (acrylamide) copolymer which was attached covalently to a silanized fiber tip surface by photoinitiated polymerization [27]. The reduction in size allows a short response time of several milliseconds. Further development in fabrication technology is required, as the former is still two-dimensional, while the latter is as large as about 1 μm. Size reduction is necessary to improve the spatial resolution.

Another example of such a supertip is the use of a sharpened Er^+-doped fiber as a functional probe with optical gain. The etching process of Section 3.1 can be applied successfully, as is shown in Fig. 4.31. Figure 4.32 shows another example of such a supertip. A fluorescence-emitting probe is made by fixing dye molecules

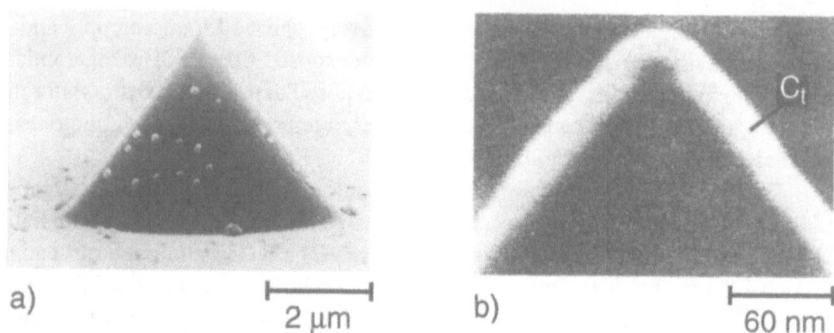

a)   2 μm   b)   60 nm

Figure 4.31. (a) SEM micrograph of a fiber with an Er^+-doped core sharpened by selective chemical etching. (b) Magnified image of the top of panel (a). C_t indicates contamination due to bombardment of the electron beam while conducting the SEM observations. The apex diameter including Au film is smaller than 15 nm.

(Rhodamine 6G) on the flat-top fiber of Section 4.1.4. Here, the flattened apex diameter (d in Fig. 4.2b) of the fiber core is 300 nm, and the number of fixed dye molecules is estimated to be between 1×10^6 and 1×10^7. The process of fixing the dye molecules is as follows [28]. First, an ethanol solution containing dye with a density of 0.06 mol/L is filled into a glass capillary of 500 nm inner diameter. Second, the flat-top fiber is inserted into the capillary to a depth of about 300 nm, and then pulled out so that a nanometric droplet of dye solution is attached on top of the fiber. The dye molecules are left over after the ethanol is evaporated.

Advantageous properties of this probe are as follows: (1) By detecting selectively the fluorescence intensity, the performance of imaging experiments, especially for the case of the C-mode, is independent of the size of the foot of the sharpened fiber core, its cone angle, and the aperture diameter. It depends only on

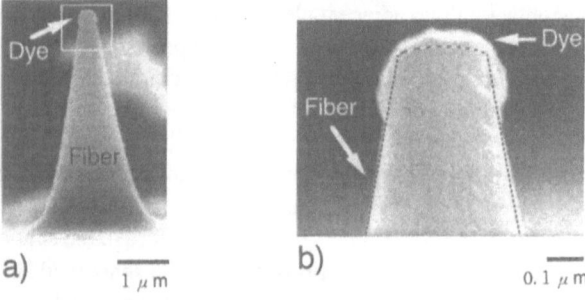

a)   1 μm   b)   0.1 μm

Figure 4.32. (a) SEM micrograph of Rhodamine 6G molecules fixed on a flat-top fiber. (b) Magnified image of the top of panel (a).

the total size of the dye molecules, which makes the analysis of the obtained image simpler. Thus, it is equivalent to using a floating nanometric particle as a probe. (2) It can be used as a nanometric light source to study cavity quantum electrodynamic phenomena [29]. (3) It can be applied to develop nanometric sensors for several physical and/or chemical quantities, e.g., a pH sensor.

On the other hand, high chemical stability may not be assured as the dye molecules are exposed to the air. Although it has been confirmed by imaging experiments (Section 5.4.1) that they are stable at least for 20 min, stability and lifetime need to be improved.

The attachment of other functional materials, such as semiconductor and inorganic oxide nonlinear optical materials, might extend the field of applications.

4.3.2. Optically Trapped Probes

In another novel probe, a small particle suspended in a solution trapped by a focused laser beam can act as a probe [30]. It can be advantageously used to observe, e.g., biological samples in water without damaging their surfaces because the optical force pressing the particle to the sample surface is even smaller than the atomic force. However, an essential disadvantage is that control of the particle position becomes more difficult with decreasing size of the particle because it is trapped by using conventional far-field optics, which is not compatible with near-field optical microscopy.

4.4. REFERENCES

1. K. Liberman, S. Harush, A. Lewis, and R. Kopelman, A light source smaller than the optical wavelength, *Science*, **247**: 59–61 (1990).
2. D. W. Pohl, W. Denk, and M. Lanz, Optical stethoscopy: Image recording with resolution $\lambda/20$, *Appl. Phys. Lett.* **44**: 651–653 (1984).
3. H. Pagnia, J. Radojewski, and N. Sotnik, Operation conditions of an optical STM, *Optik* **86**: 87–90 (1990).
4. T. Pangaribuan, K. Yamada, S. Jiang, H. Ohsawa, and M. Ohtsu, Reproducible fabrication technique of nanometric tip diameter fiber probe for photon scanning tunneling microscope, *Jpn. J. Appl. Phys.* **31**: L1302–L1304 (1992).
5. D. T. Burns, A. Townshend, and A. G. Catchpole, *Inorganic Reaction Chemistry: Systematic Chemical Separation*, Ellis Horwood, West Sussex, England, 1980.
6. F. L. Galeener, Planar rings in vitreous silica, *J. Non-Crystalline Solids* **49**: 53–62 (1982).
7. M. Kawachi, T. Edahiro, and H. Toba, Microlens formation on VAD single-mode fiber ends, *Electron. Lett.* **17**: 71–72 (1982).
8. T. Pangaribuan, S. Jiang, and M. Ohtsu, Two-step etching method for fabrication of fibre probe for photon scanning tunneling microscope, *Electron. Lett.* **29**: 1978–1979 (1993).
9. S. Mononobe and M. Ohtsu, Fabrication of a pencil-shaped fiber probe for near-filed optics by selective chemical etching, *J. Lightwave Technol.* **14**: 2231–2235 (1996).

10. K. M. Takahashi, Meniscus shapes on small diameter fibers, *J. Colloid Interface Sci.* **134:** 181–187 (1990).

11. P. Tomanek, Fiber tips for reflection scanning near-field optical microscopy, in *Near Field Optics,* D. W. Pohl and D. Courjon, eds., Kluwer, Dordrecht, 1993, pp. 295–302.

12. E. Betzig, A. Lewis, A. Harootunian, M. Issacson, and E. Kratschmer, Near-field scanning optical microscopy(NSOM); development and biological applications, *Biophys. J.* **49:** 269–279 (1986).

13. R. Uma Maheswari, S. Mononobe, and M. Ohtsu, Control of apex shape of the fiber probe employed in photon scanning tunneling microscope by a multi-step etching method, *J. Lightwave Technol.* **13:** 2308–2313 (1995).

14. A. S. Tenney and M. Ghezzo, Etch rates of doped oxides in solutions of buffered HF, *J. Electrochem. Sci.* **120:** 1091–1095 (1973).

15. J. S. Judge, A study of the dissolution of SiO_2 in acidic fluoride solutions, *J. Electrochem. Soc.* **118:** 1772–1775 (1971).

16. G. A. C. M. Spiering, Wet chemical etching of silicate glasses in hydrofluoric acid based solutions, *J. Mat. Sci.* **28:** 6261–6273 (1993).

17. H. Kikuyama, M. Waki, M. Miyashita, T. Yabune, N. Miki, J. Takano, and T. Ohmi, A study of the dissolution state and the SiO_2 etching reaction for HF solutions of extremely low concentration, *J. Electrochem. Soc.* **141:** 366–374 (1994).

18. S. Verhaverbeke, I. Teerlinck, C. Vinckier, G. Steven, R. Cartuyvels, and H. M. Heyns, The etching mechanisms of SiO_2 in hydrofluoric acid, *J. Electrochem. Soc.* **141:** 2852–2857 (1994).

19. S. Iijima and T. Ichihashi, Structural instability of ultrafine particles of metals, *Phys. Rev. Lett.* **56:** 616–619 (1986).

20. T. Saiki, S. Mononobe, and M. Ohtsu, Tailoring of a high–transmission fiber probe for photon scanning tunneling microscope, *Appl. Phys. Lett.* **68:** 2612–2614 (1996).

21. E. L. Buckland, P. J. Moyer, and M. A. Paesler, Resolution in collection-mode scanning optical microscopy, *J. Appl. Phys.* **73:** 1018–1028 (1993).

22. G. A. Valaskovic, M. Holton, and G. H. Morrison, Parameter control, characterization, and optimization in the fabrication of optical fiber near-field optics, *Appl. Opt.* **34:** 1215–1228 (1995).

23. S. Mononobe, M. Naya, T. Saiki, and M. Ohtsu, Reproducible fabrication of a fiber probe with a nanometric protrusion for near-field optics, *Appl. Opt.* **36:** 1496–1500 (1997).

24. T. Matsumoto and M. Ohtsu, Fabrication of a fiber probe with a nanometric protrusion for near-field optical microscopy by a novel technique of three-dimensional nanophotolithography, *J. Lightwave Technol.* **14:** 2224–2230 (1996).

25. R. Kopelman, W. Tan, Z.-Y. Shi, and D. Birnbaum, Near field optical and excition imaging, spectroscopy and chemical sensors, in *Near Field Optics,* D. W. Pohl and D. Courjon, eds., Kluwer, Dordrecht, 1993, pp. 17–24.

26. H.-U. Danzebrink and U. C. Fisher, The concept of an optoelectronic probe for near field microscopy, in *Near Field Optics,* D. W. Pohl and D. Courjon, eds., Kluwer, Dordrecht, 1993, pp. 303–308.

27. W. Tan, Z.-Y. Shi, and R. Kopelman, Development of submicron chemical fiber optic sensors, *Anal. Chem.* **64:** 2985–2990 (1992).

28. K. Kurihara, K. Watanabe, and M. Ohtsu, Photon scanning tunneling microscopy with light-emitting probes, in *Conference Proceedings of the Eleventh International Conference on Optical Fiber Sensors, Sapporo, Japan, May 1996,* pp. 694–697.

29. D. Meschede, W. Jhe, and E. A. Hinds, Radiative properties of atoms near a conducting plane: An old problem in a new light, *Phys. Rev. A* **41:** 1587–1596 (1990).

30. S. Kawata, Y. Inoue, and T. Sugiura, Near-field scanning optical microscope with a laser trapped probe, *Jpn. J. Appl. Phys.* **33:** L1724–L1727 (1994).

FIVE

IMAGING EXPERIMENTS

5.1. BASIC FEATURES OF THE LOCALIZED EVANESCENT FIELD

5.1.1. Size-Dependent Decay Length of the Field Intensity

With regard to superresolution of the NOM beyond the diffraction limit, it has been pointed out in Section 2.4.3 that the confinement of a three-dimensional evanescent field on a subwavelength object strongly depends on its size and topography. The approach of a sharpened probe tip to a single object with precise positioning enables one to obtain direct information about the evanescent field distribution with the sample–probe separation as a parameter. This subsection describes measurements of the characteristic decay length of the evanescent field from a subwavelength sphere [1].

Figure 5.1 illustrates schematically the measurement of an evanescent field with a sharpened fiber probe. The decay length can be measured by observing the dependence of the evanescent signal intensity on the sample–probe separation. When one observes the scattered field from a subwavelength-size object irradiated by excitation light as shown in Fig. 5.1a, the planar evanescent wave from the substrate due to the incident light is also detected superimposed on the exponential decay curve of the desired signal intensity. This complexity of the signal structure makes it difficult to determine the decay length of the evanescent field explicitly. In order to overcome this difficulty, a wavelength conversion technique is used. That is, the fluorescence light from dye molecules doped in a sphere is detected (see Fig. 5.1b) and the background signal is rejected by optical filters so that the evanescent field on the sphere is selectively detected.

An experimental setup of fluorescence detection is shown in Fig. 5.2, which employs a C-mode NOM. As a sample, rhodamine-doped polystyrene spheres are dispersed on a silica substrate using an atomizer. The density of the dispersed spheres is lower than 0.2 particle/μm^2, which is sufficiently low to suppress the background fluorescence from the adjacent spheres. Dye molecules in the sphere are excited by light from a p-polarized Ar^+ laser ($\lambda = 488$ nm wavelength) which is incident under the condition of total internal reflection. A sharpened fiber probe

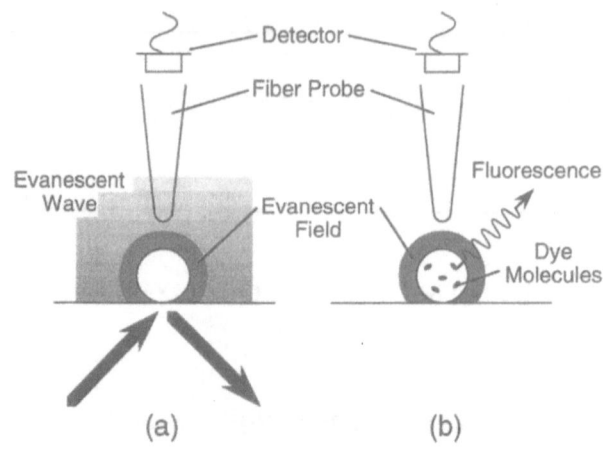

Figure 5.1. Experimental configuration for measuring the decay length of an evanescent field. (a) Conventional C-mode NOM. (b) C-mode NOM with wavelength conversion by detecting selectively the fluorescence from dye molecules doped in a subwavelength-size sphere. Excitation light is omitted for clarity.

(fabricated by the technique described in Section 4.1.2) with a 20-nm apex diameter (d of Fig. 3.2c) is used as a probe. The shear force between the probe and sphere (cf. Section 3.1.2.2) is monitored and used to control the sample–probe separation. The fluorescence ($\lambda = 600$ nm) from a single sphere is picked up by the probe and is coupled to the guided mode of the fiber. Rejecting the excitation light with optical filters, the fluorescence signal is selectively measured by the photon counting method.

Note that a sphere of radius $a = 50$ nm is used as a sample in this study for the following technical reasons: (1) Measurement for spheres of $a < 50$ nm is rather difficult due to the low collection efficiency of present C-mode NOMs (direct detection of the decay length is not possible with the I-mode NOM despite its higher

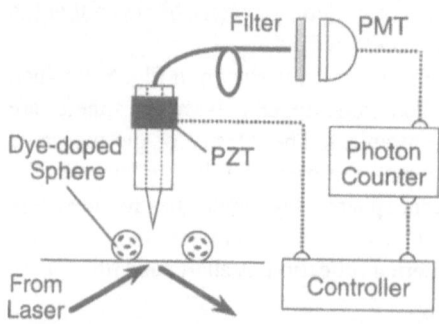

Figure 5.2. Experimental setup of a C-mode NOM for fluorescence detection. PZT, Piezoelectric transducer; PMT, photomultiplier.

collection efficiency). (2) The decay length of the planar evanescent field intensity at the present incident angle of 50 deg is about 70 nm, which imposes a maximum size constraint on the sphere.

For precise evaluation of the decay length of the evanescent field, the dependence of detected fluorescence intensity on sample–probe separation is measured by positioning the probe just above the sphere (see Fig. 5.3a). A rapid change in signal intensity is observed for $z < 80$ nm, where z represents the sample–probe separation (curve A of Fig. 5.3b). The same measurement was performed 500 nm away from the sphere. The result is given by curve B of Fig. 5.3b). Its intensity is constant in the region $z < 80$ nm. The baseline of the signal from the single sphere lies between the horizontal axis and the curve B. From curve A, the decay length is estimated to be 15 nm, where the decay length is defined as the value of z at which the signal intensity is half that at $z = 0$. The enhancement factor η, which is defined as the ratio of the signal intensity at $z = 0$ to that at $z > a$, is estimated to be in the range $3 < \eta < 4$.

These experimental results are compared with a theoretical calculation based on Mie scattering theory given in Section 2.4.3 [2]. Here we will briefly review this calculation methodology as applied to this case. The conventional Mie scattering theory concerns the far field produced by the scattering of an incident homogeneous plane wave by a dielectric sphere. However, by considering a general incident plane wave (inhomogeneous and homogeneous [3]) and focusing on the scattering region just near the sphere, one can obtain the intensity distribution of the evanescent field produced due to the scattering of an s- or p-polarized planar evanescent wave. Since

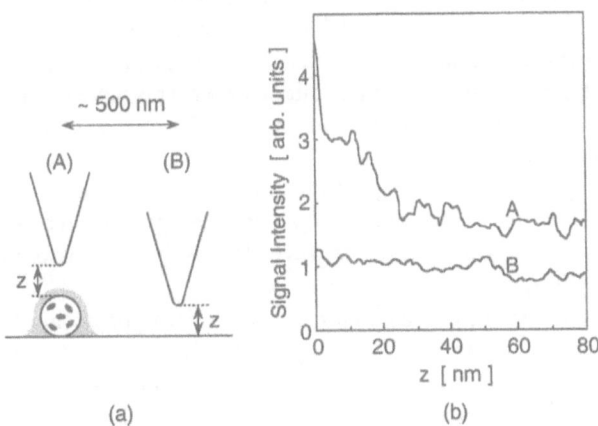

Figure 5.3. (a) Schematic explanation of the relative positioning of the probe and the sphere. The probe is set just above (A) or 500 nm away (B) from the sphere. (b) Dependence of fluorescence detection intensity on the sample–probe separation. Curves A and B are taken for the experimental configuration given by A and B in panel (a), respectively.

the fluorescent dye molecules are uniformly distributed inside the sphere, one can approximate the object as a macroscopic fluorescent dielectric sphere. By averaging the evanescent field intensity distribution over the possible polarization directions of the induced dipoles, one obtains the theoretical curve shown in Fig. 2.6. Here the size parameter ka is assumed to be 0.5, which corresponds to the experimental condition of $a = 50$ nm and $\lambda = 600$ nm. Note that coherent and interference effects between the radiated fluorescences from the dipolar molecules are neglected due to the random molecular orientations inside the dielectric sphere.

The curve in Fig. 2.6 shows a decay length of $0.3a$ (= 15 nm) in good agreement with the experimental result. Theoretical calculation using other values of ka is expected to show that the evanescent field is strongly confined around the sphere within a fraction of its diameter and is independent of the wavelength of light used for excitation.

5.1.2. Manifestation of the Short-Range Electromagnetic Interaction

The emission properties of fluorescent materials in near-field optics are both very interesting and important. Results obtained by some groups seem inconsistent [4–6]. To improve understanding, independent experimental results are needed. The scattering of the evanescent field of a fluorescent material by a subwavelength-size probe modifies its radiation properties due to the short-range electromagnetic interaction. For example, Xie $et. al.$ observed a change in the radiative lifetime of a single molecule due to the presence of a probe in the near-field region [4, 5]. They reported that the lifetime is reduced from 2.7 nsec in the far-field to 1.6 nsec when the sample–probe separation is reduced to 5 nm. The shortened lifetime (enhanced radiation rate) results in increased radiation power. This increase can be associated with the complementary enhancement factor η ($3 < \eta < 4$) given in the previous section. Moreover, it is claimed that these larger values of η can be attributed to another cavity effect: the change of the emission property of each dye molecule doped in the spherical dielectric cavity [7]. For example, for a radiating dipole near a surface, its emission rate is enhanced several times, irrespective of the polarization of emission [7–9].

This enhancement can be confirmed by directly measuring the increase in fluorescence detection sensitivity. The approach of a probe to a fluorescent material in the near-field region largely modifies its radiation mode. This modification leads to a change of its radiation pattern and of its lifetime, which offers the possibility of a highly effective collection of fluorescence. To investigate this coupling, an experiment was carried out using the experimental setup of Fig. 5.2 to measure the increase in detection sensitivity [10]. The protruded probe had a 20-nm apex diameter d and 400-nm foot diameter d_f (see Fig. 3.2c for definition of d and d_f).

Figure 5.4 is a near-field fluorescence image of microspheres. The probe is scanned while maintaining the sample–probe separation at a constant value of 10–20 nm. The maximum photon counting rate is 1000 counts/sec and the image acquisition time is about 200 min. A high-contrast image is obtained. From a cross-sectional view of the image, the size (full-width at half-maximum) of a single sphere is estimated as 180 nm. Since the wavelength of the fluorescence is about 600 nm, the spheres are observed with a resolution beyond the diffraction limit. It should be emphasized that in this measurement the small apex diameter of 20 nm contributes to the highly resolved imaging.

To confirm the increase in fluorescence detection sensitivity, the pickup efficiency Γ is defined as the ratio between the fluorescence intensity coupled to the guided mode of the fiber and the total fluorescence intensity. Γ is experimentally estimated to be larger than 0.1, by using the values of the excitation intensity of the Ar^+ laser (20 W/cm^2), the absorption cross section of a single sphere (1.7×10^{-12} cm^{-2}), the quantum efficiency of the rhodamine molecule (0.8), and the total

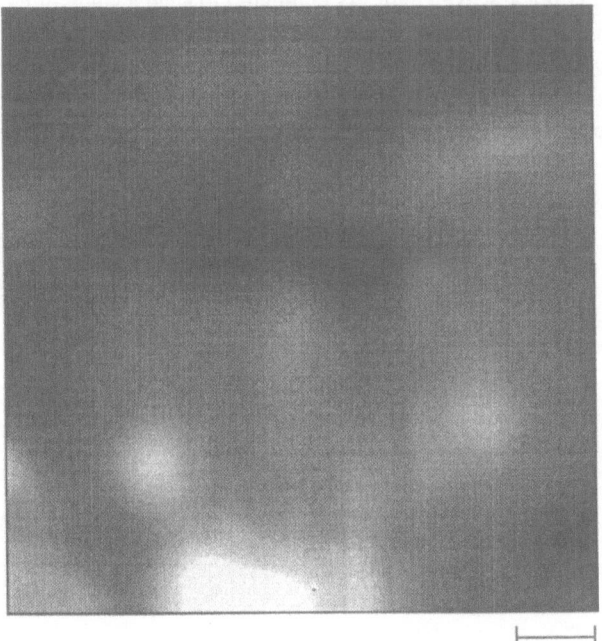

100 nm

Figure 5.4. Near-field fluorescence image of dye-doped microspheres. The scan area is 800 × 800 nm^2.

efficiency of the detection system (7×10^{-3}; filter transmission and photomultiplier quantum efficiency).

On the other hand, Γ can be expressed as $(\Omega/4\pi)\gamma$ if the radiation pattern of the dye molecules is assumed to be isotropic even if the probe is within a distance of 10–20 nm from the sphere. Here Ω (= 0.1) is the solid angle at the foot of the protruded probe with respect to the sample, and γ is the coupling coefficient to the guided mode (see Fig. 5.5a). Due to the small foot diameter of the probe and absorption by the Au film, γ is measured to be less than 0.01. (This value is deduced from the experimental transmission coefficient of a probe with 400 nm foot diameter. Details of this experiment are equivalent to the one discussed in Fig. 4.19.) Thus, Γ is estimated to be smaller than 1×10^{-4}. The experimental value of Γ in the near-field region is much greater than the estimate neglecting near-field optical effects (see Fig. 5.5b). Taking into account the imaging result of Fig. 5.4, it is concluded that the subwavelength resolution and enhancement of sensitivity strongly depend on the spatial profile of the evanescent field intensity for small spheres.

Modification of the spontaneous emission probability, similar to the observed radiation pattern modification, has been observed in a closed optical cavity with the size of an optical wavelength. As compared to such a study of cavity quantum electrodynamics (QED), the present study corresponds to the observation of a quantum optical phenomenon with subwavelength dimension. These experimental and theoretical results clearly show that the application of NOM will provide valuable and novel insight into quantum optics including cavity QED and nanooptics.

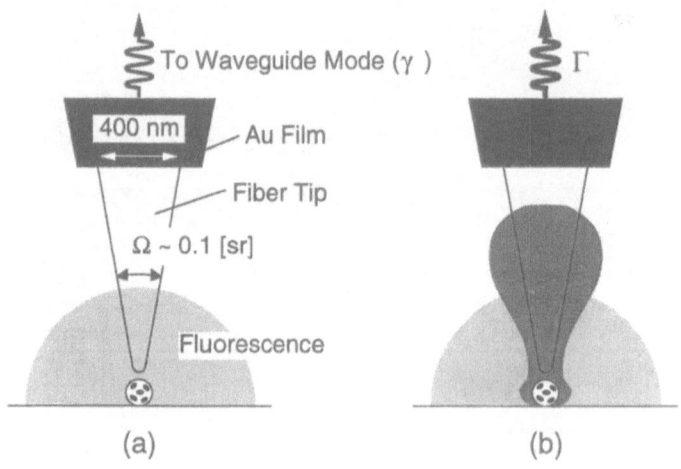

Figure 5.5. (a) Pickup efficiency determined by the solid angle at the aperture plane from the sample (Ω) and coupling coefficient to the guided mode (γ). (b) Enhanced pickup efficiency Γ due to the short-range electromagnetic interaction between the probe tip and the fluorescent material.

5.1.3. High Discrimination Sensitivity of the Evanescent Field Intensity Normal to the Surface

Curve A of Fig. 3.13 shows that the discrimination sensitivity of the intensity variation of the evanescent field normal to the material surface can be higher than 1 nm (subnanometer normal resolution). This section reviews experimental results with the C-mode NOM which demonstrate high discrimination sensitivity.

This demonstration requires a nanometric probe tip, a nanometric reference sample, and an accurate servocontrol system for scanning the probe. The protruded probe with nanometer-size apex diameter has already been discussed (Chapter 4). The next necessary factor is a nanometric reference sample homogeneously fixed on the substrate. The topography must be calibrated by other complementary and reliable methods. An atomic-level step on an ultraflat, semitransparent sapphire (α-Al_2O_3) plate is a promising candidate. It has been recently fabricated using an advanced surface processing technology for oxide materials [11]. Theoretical analysis based on molecular dynamics [12] shows that Al and O atoms on the stabilized top surface form a step of a monoatomic layer with a height of 0.2 nm, as explained schematically by Fig. 5.6. A multiatomic layer step can be also fabricated with a height of up to 10 nm. The width of the step can be varied from 20 to 1000 nm. Figure 5.7a shows a commercial AFM image of a multiatomic step layer. Figure 5.7b shows the magnified AFM image of the 2.5-nm step. As the step width is as large as 150 nm, this sample cannot be used for evaluating lateral resolution.

This step was measured by C-mode NOM. As described in Section 3.1.2.1, this measurement utilizes the outstanding advantage of the C-mode NOM that an auxiliary method such as shear-force monitoring does not have to be used for controlling the probe position, making the analysis of the image characteristics simpler. The sample–probe separation must be of the order of several nanometers

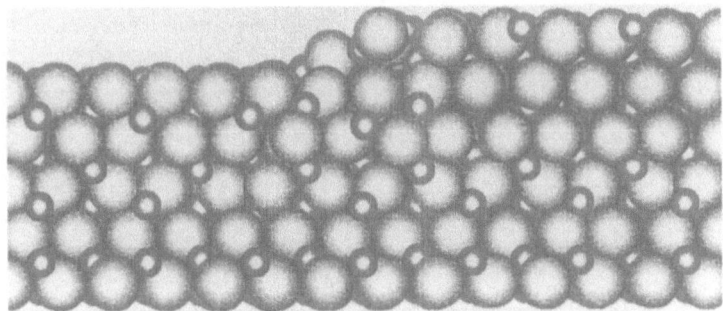

Figure 5.6. Schematic explanation of sapphire with a step of a mono- atomic layer (side view). The big and small spheres represent oxygen and aluminum atoms, respectively.

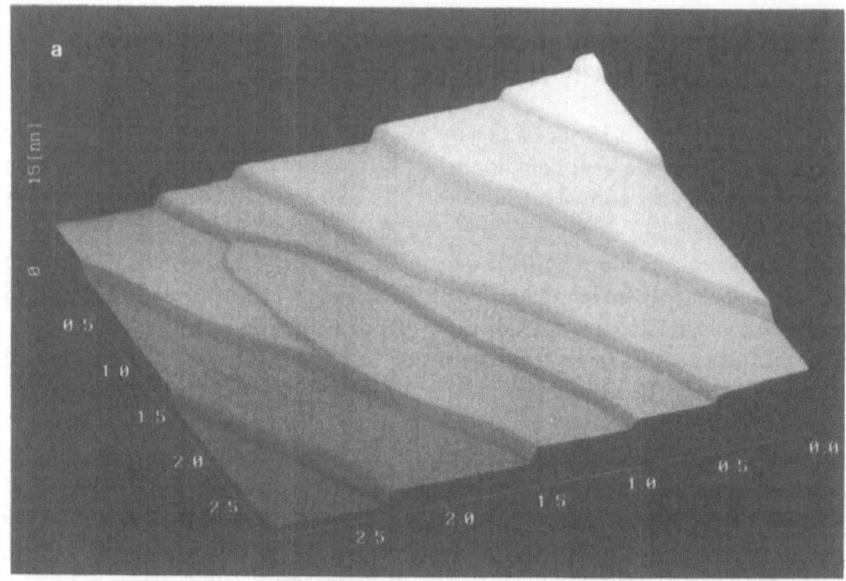

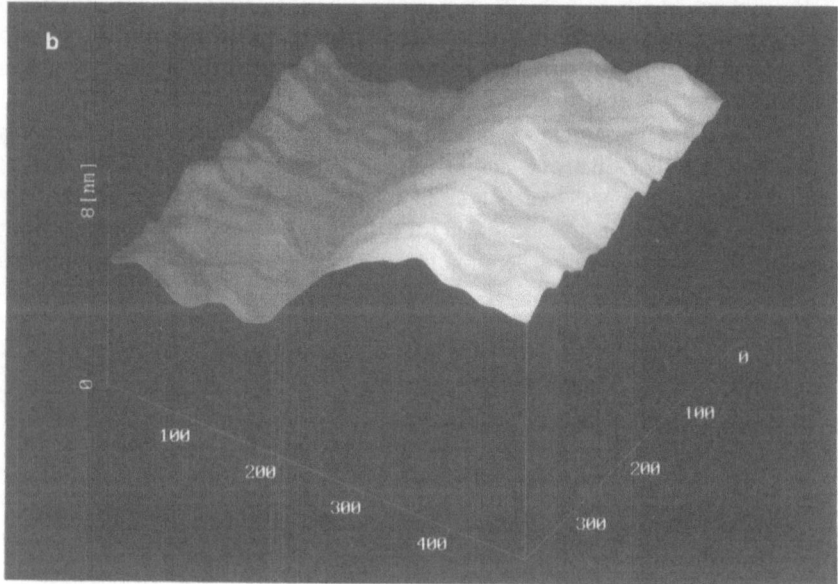

Figure 5.7. (a) Image of a sapphire surface with a multiatomic layer, taken by a commercial atomic force microscope (3×3 μm^2). (b) Magnified image of panel (a), from which the step height is found to be 3 nm, and the lateral span is 150 nm. Scan area is 490×390 nm^2.

to pick up the high-spatial-Fourier-frequency components of the evanescent field, which has a nanometric decay length depending on the step size. For this purpose, the light detection sensitivity and loop gain for controlling the probe position have to be sufficiently high.

Figure 5.8 shows the NOM image of the step taken by carefully maximizing the performances of detecting devices and optimizing the control parameters [13]. The probe used here is equivalent to that of Fig. 4.23. The light source is a diode laser (λ = 685 nm). One mW of power is incident to the optically flat rear surface of the sapphire. The s-polarized laser beam is incident normal to the step (i.e., the electric field vector of the incident light is parallel to the direction of the step). A clear image of the 2.5-nm step can be seen in Fig. 5.8, demonstrating that the present C-mode NOM has subnanometric discrimination sensitivity to the intensity variation of the evanescent field normal to the surface. Although fully quantitative estimation of the spatial resolution is not possible from this result, comparison with Fig. 5.7b shows that the image is as clear as the AFM image. This implies that the resolution as high as that of the commercial AFM.

The quality of this image decreases rapidly with increasing sample–probe separation. The image disappears when the separation is larger than 10 nm. This clear dependence of image visibility on sample–probe separation originates from the size-dependent decay length of the evanescent field, as described in Section 5.1.1. Section 5.2.1 will describe a more quantitative evaluation of image characteristics by using a biological reference sample.

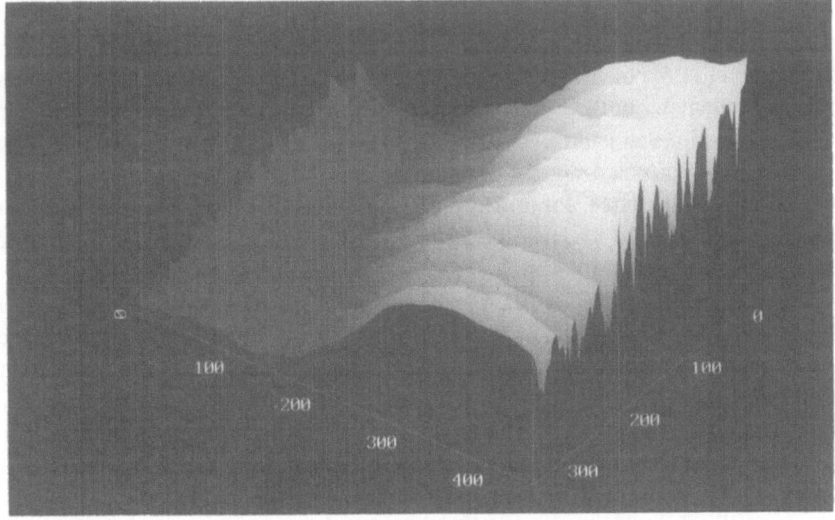

Figure 5.8. NOM image of a multiatomic layer step. Scan area is 460 × 350 nm^2.

5.2. IMAGING BIOLOGICAL SAMPLES

Optical microscopes have long been used in biology. Among the most significant biological samples, the thickness of cell walls, the sizes of functional polymers such as ion channels, the sizes of intercellular connecting elements, and the diameters of actin filaments are of the order of nanometers. These have been primarily observed by using the differential contrast video microscope, SEM or TEM, and AFM. Since observations by optical methods can provide unique information about these samples, it is of interest to obtain the images of nanometric biological samples by NOM. For imaging biological samples, NOM should be capable of high resolution, operation in liquids, and specific or local feature detection. The latter can be done through fluorescence techniques. This section demonstrates these capabilities for the C- and I-modes.

5.2.1. Imaging by the C-Mode

5.2.1.1. Basic Characteristics of an Image Taken in Air

Imaging experiments of stretched straight flagellar filaments of *Salmonella* (FFS) are reviewed here as an example of the biological application of C-mode NOM, and simultaneously to demonstrate the interpretation of image characteristics [14]. FFS are representative examples of a molecular motor [15], and thus their size and structure have been well calibrated. FFS are fixed on a hydrophilized glass plate along random directions. Figure 5.9 shows a transmission electron micrograph (TEM) of FFS which are fixed along random directions on the glass plate. The diameter of the FFS can be measured to be 25 nm from this figure.

The experimental setup for the C-mode NOM (cf. Fig. 3.4) uses a half-wavelength plate to control the polarization state of the incident light. An Ar^+ laser ($\lambda = 488$ nm) is used as a light source. Protruded probes are used (the apex diameter and foot diameter are 3 and 30 nm, respectively). The rapidly decreasing evanescent field intensity is used to control sample–probe separation during scanning.

In the images of Fig. 5.10a, the bright streaks correspond to the FFS. The incident light is s-polarized (i.e., its electric field vector lies in the substrate surface plane (refer to the gray arrow in the figure). The direction of the incident light propagation (i.e., the wavevector) is from upper left to lower right, as identified by the black arrow. The pixel size is 20 nm $\times$ 20 nm. The sample–probe separation is estimated to be 15 nm in the following way: After finishing the scan, the probe is pushed toward the sample by applying a voltage corresponding to a 15-nm excursion to the PZT. At this position, when the probe is scanned repeatedly, streaks in the image are observed due to the probe scratching the sample surface.

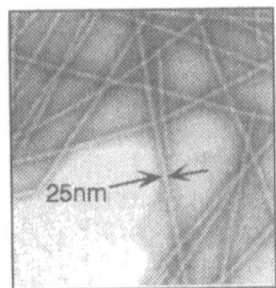

Figure 5.9. TEM micrograph of straight-type flagellar filaments of *Salmonella* (FFS) (provided by Prof. S.-I. Aizawa, Teikyo University).

Figure 5.10a shows that the FFS fixed perpendicular to the direction of propagation of the incident light beam are clearly imaged, while those parallel to the incident light beam are not. Although this feature is quite reproducible, it does not agree with the conventional theoretical result [16]. A more accurate theory which considers the complete sample–probe system is required for describing the experimental results.

Figure 5.10b shows an image taken under the same conditions as that of Fig. 5.10a except that the sample–probe separation is increased to 65 nm. Compared with Fig. 5.10a, almost the same image is obtained except for a lowering of

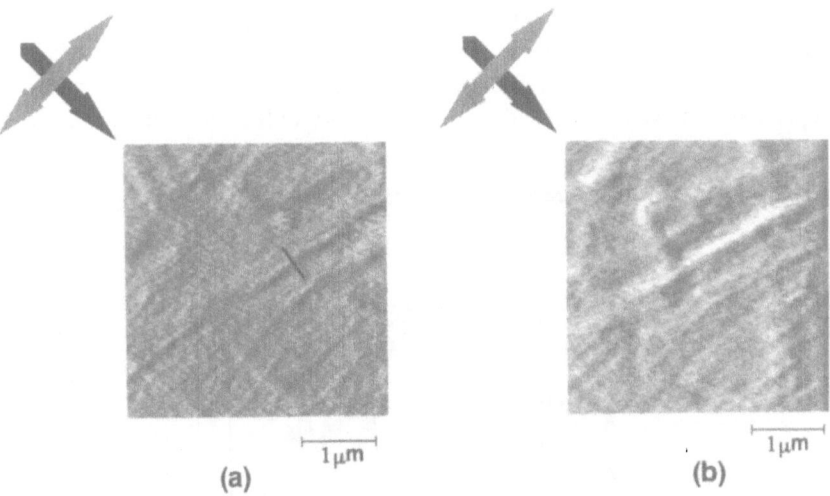

(a)   (b)

Figure 5.10. NOM image of FFS by *s*-polarized incident light. Black and gray arrows represent the directions of the wavevector and the electric field vector of the incident light, respectively. A protruded probe is used, as shown in Fig. 4.23. The apex diameter d and foot diameter of the Au film d_f are 3 and 30 nm, respectively. The sample–probe separation is fixed to be (a) 15 nm or (b) 65 nm.

resolution and contrast. The dependence of resolution and contrast on the sample–probe separation originates from the intrinsic nature of the evanescent field. In order to investigate this dependence, the spatial power spectral density of the detected light intensity is calculated from all the pixels of these figures. Figure 5.11 shows the calculated values, which are plotted as a function of spatial Fourier frequency. Comparison between the two curves in this figure shows that the magnitude of high-spatial-Fourier-frequency components decreases with increasing sample–probe separation. This is the origin of the lower resolution and contrast of the image of Fig. 5.10b. This decrease in the high-spatial-Fourier-frequency components originates from the size-dependent confinement feature of the evanescent field (see Figs. 2.6 and 5.3) and the resonance phenomenon with respect to the relative size of sample and probe (see Fig. 2.7). In summary, detection efficiency decreases rapidly with increasing sample–probe separation, since the decay length of the high-spatial-Fourier-frequency components of the evanescent field intensity distribution is short.

Figure 5.12 shows the image of FFS obtained by using a protruded probe with a foot diameter as large as 100 nm. The sample–probe separation is fixed to be 15 nm. It should be noted that the scan area is slightly different from that of Fig. 5.10a. However, comparison of this figure with Fig. 5.10a easily confirms the lowering of resolution and contrast. This is because the detection efficiency of lower spatial-Fourier-frequency components increases with increasing foot diameter. This is also due to size-dependent confinement of the evanescent field and resonance with respect to the sizes of the sample and the probe.

In order to study the dependence of the image characteristics on the vectorial property of the evanescent field, the image of the FFS is taken under the same

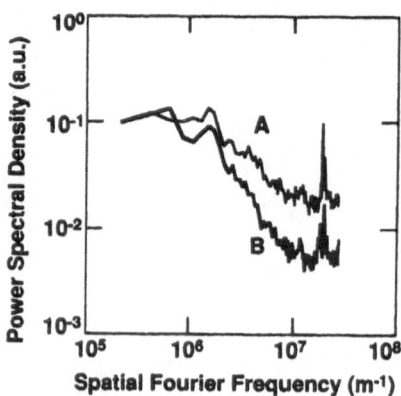

Figure 5.11. The spatial power spectral density of the detected light intensity, calculated from the image of (A) Figure 5.10a and Figure 5.10b.

Figure 5.12. NOM image of FFS by s-polarized incident light. The sample–probe separation is 15 nm. The foot diameter d_f is 100 nm. The black and gray arrows represent the directions of the wavevector and electric field vector of the incident light, respectively.

1 μm

conditions as in Fig. 5.10a, except that the incident light is p-polarized (the direction of its electric field vector is normal to the substrate surface). Figure 5.13 shows the result. Comparing this figure with Fig. 5.10a, it can be seen that the FFS seen as bright streaks in Fig. 5.10a appear dark with bright edges on either side. (Compare, for example, the cross-sectional profile of the FFS identified by a black bar in these figures.) This difference can be explained by considering the distributions and directions of the dipoles on the sample surface and the probe tip, which are induced by the evanescent field mediating the short-range electromagnetic interaction between the sample and probe.

Figure 5.14 explains the origin of this difference. It depicts a cross-sectional view of the induced dipoles on the sample surface and probe tip, and the electric lines of force of the evanescent field. In the case of s-polarization (Fig. 5.14a) the direction of the induced dipole in the sample is parallel to the substrate surface plane. Thus, one lobe of the electric line of force of the evanescent field is generated above the sample. Since the dipole induced in the probe tip by this evanescent field aligns with the lines of electric force of this lobe, the dipole is parallel to the substrate surface plane only when the probe tip is just above the sample. The wavevector of the scattered light generated by this dipole is thus parallel to the probe axis and efficiently couples to the guided mode of the fiber. As the probe moves away from the sample, the coupling efficiency is reduced. A single peak in the cross-sectional intensity distribution is obtained as seen along the black bar in Fig. 5.10a.

In the case of p-polarized incident light (Fig. 5.14b), the direction of the induced dipole on the sample surface is perpendicular to the substrate surface. Lobes of the generated evanescent field are on both sides of the sample. Thus, when

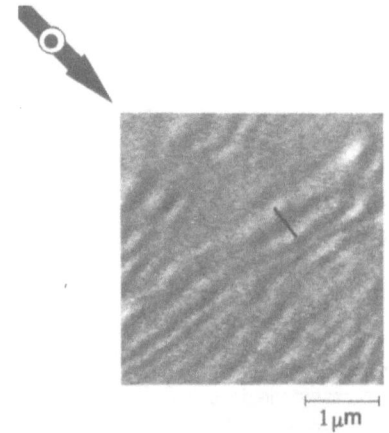

1 μm

Figure 5.13. NOM image of FFS by p-polarized incident light. The experimental parameters are the same as those of Fig. 5.10.

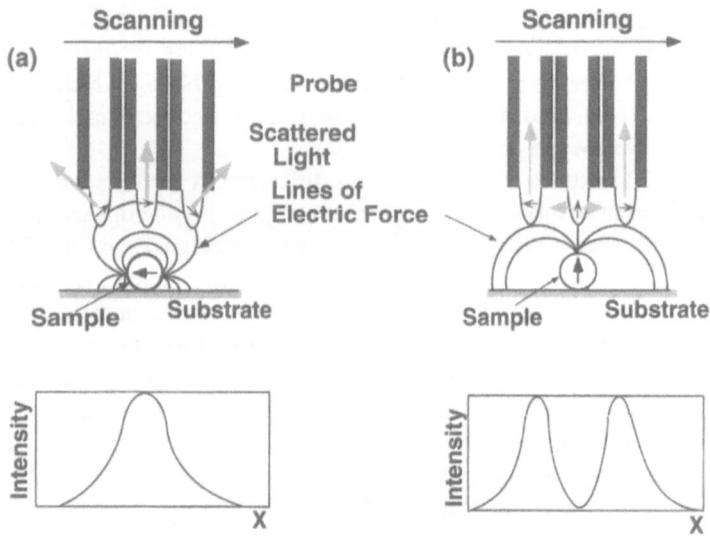

Figure 5.14. Schematic explanation of the dependence of the image characteristics on the incident light polarization. (a) s-polarized incident light corresponding to Fig. 5.10a. (b) p-polarized incident light corresponding to Fig. 5.13. Lower parts represent the cross-sectional profile of the detected light intensity for the FFS.

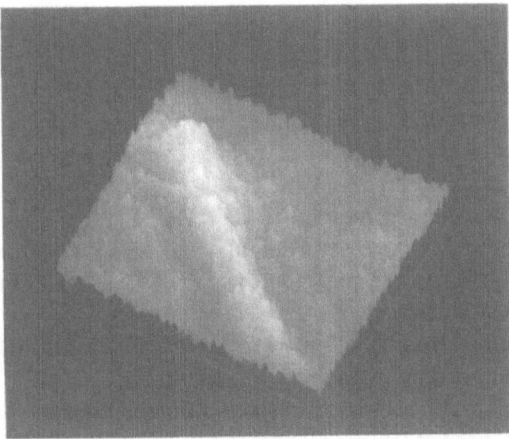

Figure 5.15. Magnified image of part of an FFS taken under the same condition as for Fig. 5.10a. Scan area is 480×480 nm^2.

the probe tip is just above the sample, the direction of its induced dipole is perpendicular to the substrate surface, parallel to the probe axis. The wavevector of the scattered light is then perpendicular to the probe axis and does not couple into the fiber. As the probe moves away from the sample and into the lobes, the dipole rotates and the light is coupled into the fiber. Therefore, double peaks are seen on the cross-sectional intensity distribution along the black bar of Fig. 5.13.

There are several other ways of explaining the image dependence on the incident light polarization (the single- or double-peak feature). One is in terms of induced image dipoles [17]. A second is by the concept of symmetry breaking. That is, in the case of p-polarization (Fig. 5.14b), the system composed of the electric field vector of the incident light, induced dipoles, and distributions of lines of electric force of the evanescent field is axially symmetric when the probe is just above the sample. In such a situation of axial symmetry, the output signal from the system is the lowest, which results in a double peak. On the other hand, since the system is axially asymmetric for s-polarization (Fig. 5.14a), this situation of broken symmetry gives the highest output signal and thus a single-peak feature is realized.

Figure 5.15 shows the magnified image of part of the FFS taken under the same condition as for Fig. 5.10a. The full-width at half-maximum of the cross-sectional intensity distribution is 30–50 nm, depending on the position along the FFS axis. The increment during scanning (pixel size) is 20 nm. Estimating the difference between the width and the increment, the spatial resolution of this imaging experiment can be regarded to be as high as that of the TEM observation given by Fig. 5.9. The origin of such a high resolution could be the contribution of high-spatial-

Fourier-frequency components of the evanescent field, which are scattered preferably at the apex region of the protruded probe.

Although Fig. 5.15 demonstrates the qualitatively high-resolution capability of the present C-mode NOM system, the resolution needs to be estimated with the help of the high cutoff frequency of the transfer function or 0-dB frequency of the spatial power spectral density of the image. For reference, it should be pointed out that the noise magnitudes of present C-mode NOM are far lower than those of curves A and B in Fig. 5.11, which means that the resolution estimated by the 0-dB frequency of the spatial power spectral density (i.e., f_c of Fig. 3.3b) is much higher than that limited by the pixel size.

The discussions given above demonstrate the capability of high-resolution imaging by a C-mode NOM using a protruded probe. Similar high-resolution imaging for a biological specimen has also been demonstrated for bacteriophage T4 with a spherical head of 100 nm diameter and a cylindrical tail of 10 nm diameter [18].

5.2.1.2. Imaging in Water

For a real understanding of biological samples, it is important to conduct *in vivo* observations in their natural environment (i.e., water or liquids). The C-mode NOM can be advantageously used for these observations because the rapid variation of the evanescent field power normal to the sample surface can be used as a control signal to regulate precisely the sample–probe separation.*

During imaging, the evanescent field intensity on the sample surface is monitored to detect the sample–probe separation. Scanning is performed so as to maintain the monitored intensity constant. Moreover, this monitoring technique is almost free from the strong viscous drag of water and hence from injuring the surface of the soft sample.

This Section demonstrates the results of high-resolution observation of straight-type flagellar filaments of *Salmonella* (FFS) in water [19]. Figure 5.16 shows a simplified schematic view of the experimental setup of the C-mode NOM, which is nearly the same as that of Fig. 3.4 used for the imaging experiment of FFS in air described in the previous Section. In the present experiment, a diode laser ($\lambda = 685$ nm) with wideband modulation capability is used as a light source. A

---

*In an I-mode NOM, an auxiliary control signal is needed for this regulation. Shear force is commonly used for this purpose (Section 3.1.2.2). However, the use of shear force in water has several complications: (1) the strong viscous drag of liquids arising from dithering the probe reduces the detection sensitivity of the sample–probe separation. (2) Dithering may scratch and injure soft biological samples. (3) With shear force or atomic force as feedback control signal, the equipotential contour on the sample surface is mapped. This is not necessarily always equal to the equienergy contour of the evanescent field, making the interpretation of near-field optical image rather complex.

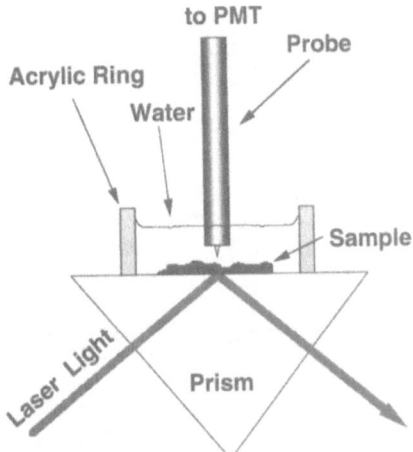

Figure 5.16. Simplified schematic explanation of the experimental setup of C-mode NOM used for imaging biological samples in water. The acrylic ring makes the surface tension of the water uniform allowing the probe tip to immerse into the water without bending.

half-wavelength plate is used to control the polarization of the incident light, which illuminates the glass prism at an incident angle of 63 deg to satisfy the condition of total internal reflection. A thin glass plate is used as the sample substrate and is fixed on the prism with an index-matching oil sandwiched in between.

The probe used here is the same as that employed in the previous section. The cone angle of the sharpened core (θ_1) is 14 deg. The apex diameter d is smaller than 10 nm and it makes the main contribution to the realization of high resolution by picking up the high-spatial-frequency components of the evanescent field. The foot diameter d_f of 30 nm determines the cutoff of lower spatial-frequency components (cf. Fig. 3.3a). The sample is fixed on a tube-type PZT actuator for three-dimensional scanning. A photomultiplier and a lock-in amplifier are used for phase-sensitive detection. The injection current of the diode laser is modulated.

The FFS is used as a reference sample because its diameter of 25 nm has been calibrated by TEM observation (see Fig. 5.9). In the present experiment, after the FFS are fixed on a hydrophilized glass plate, an acrylic ring is attached to allow observation in water. The ring makes the surface tension of the water uniform allowing the probe tip to be immersed into the water without bending. The acrylic ring (10 mm inner diameter, 2 mm height) is filled with water during the scanning.

Figure 5.17 shows the images of FFS obtained in water. The scan area is 600×600 nm^2 and the pixel size is 10×10 nm^2. During this scan, the sample–probe separation is fixed to be 30 nm as estimated from the rapid variation of the evanescent field intensity as a function of the sample–probe separation. The decay length of the evanescent field (the position at which the field power is e^{-1}

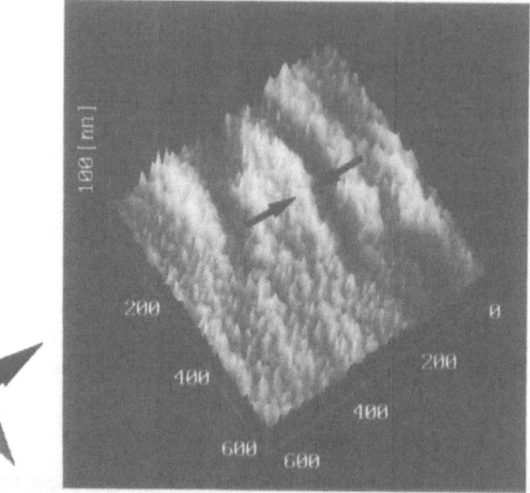

Figure 5.17. Images of FFS obtained in water. Scan area is 600×600 nm^2 with a pixel size of 10×10 nm^2. Sample–probe separation is fixed to be 30 nm. Black and gray arrows represent the directions of the wavevector and the electric field vector of the incident light, respectively. Small arrows are used to estimate the diameter of a fragment of FFS.

times that of the maximum) in water is estimated to be 284 nm, while that in air is around 60 nm. The probe was scanned with sufficiently low speed (0.3 nm/sec) to reduce the effect of viscous drag of water. Black and gray arrows in this figure represent the directions of the wavevector of the incident light and electric field vector, respectively) i.e., the incident light is s-polarized, corresponding to that of Fig. 5.15.

Bright segments in this figure represent fragments of FFS. Five fragments of FFS lying in different orientations (but still nearly perpendicular to the direction of the wavevector) can be seen. The diameter of a single segment of FFS (indicated by the arrows) is 50 nm as estimated from the full-width at half-maximum of the intensity. Comparing this with its nominal value of 25 nm, evaluated from TEM observation in vacuum, the difference is only around 25 nm, which includes the pixel size of 10 nm.

The difference in these values can be attributed to the enlargement or swelling of FFS in water. It is worth pointing out that the diameter was estimated to be 30–50 nm by imaging in air (see Fig. 5.15), which is comparable to the present result. Further, during actual scans, Brownian motion, capillary forces, etc., which arise due to the water do not affect high resolution of the C-mode NOM. Although C-mode NOM can be applied only when the substrate is not strongly absorbing, substrates for most biological nanometric samples meet this requirement. This means that present monitoring technique can become a general method for imaging biological samples.

5.2.2. Imaging by the I-Mode

A neuron consists of a cell body and a neural process (Fig. 5.18). Inside this branching neural process are many microtubules made up of a protein called tubulin [20]. They are responsible for the transport of proteins and intracellular vesicles (packets of protein substances) generated in the cell body to other parts. The end of the neural process is called a growth cone and has a flat, palmlike structure known as a lamellipodium. Inside this growth cone the shaded region in Fig. 5.18 corresponds to many tubulin and actin filaments* forming a networklike structure. The growth cone is believed to stretch to make connections with other cells.

Since understanding neuron structure is very important in discovering how these interconnections are formed, neurons have been studied extensively by electron microscope and other varieties of optical microscope such as the differential interference contrast microscope (DIC) and the confocal microscope. While the electron microscope offers higher resolution than conventional optical microscopes (resolution limited to the order of the wavelength), a high vacuum is required, nondestructive observation is impossible, and metallic plating on the sample surface is necessary.

From the point of view of neurobiologists, surface information about the growth cones, which stretch to develop neural interconnections such as synaptic contacts, is important to understand the mechanism of the formation of neural interconnections. Although it is possible to perform such studies on growth cone with an optical microscope (DIC or confocal microscopes) [21], resolution is limited to the order of the wavelength. With the contact-mode atomic force microscope (AFM), the presence of strong forces makes it difficult to investigate the soft surface membrane.

NOM has the advantage of attaining high resolution while being nondestructive and eliminating the need for a high vacuum. In addition, it has the capability of spectroscopic labeling. By making use of these features, this section demonstrates that neurons can be imaged nondestructively in air with a resolution comparable to that of an electron microscope [22]. The I-mode NOM is employed instead of the C-mode for the following two reasons: (1) Neuron samples are relatively thick and difficult to image by the C-mode NOM. (2) When samples are stained with absorption dyes, large absorption leads to very weak signals in the case of the C-mode.

An Au-coated flat-top fiber is used as a probe. Its fabrication process has been described in Section 3.1.4. Figure 4.21 is an SEM micrograph of the probe used for the present imaging experiment. A flat-top fiber is used to avoid scratching the

---

*Actins are the basic protein elements involved in carrying out the functions of neurons and are responsible for the contraction of muscles.

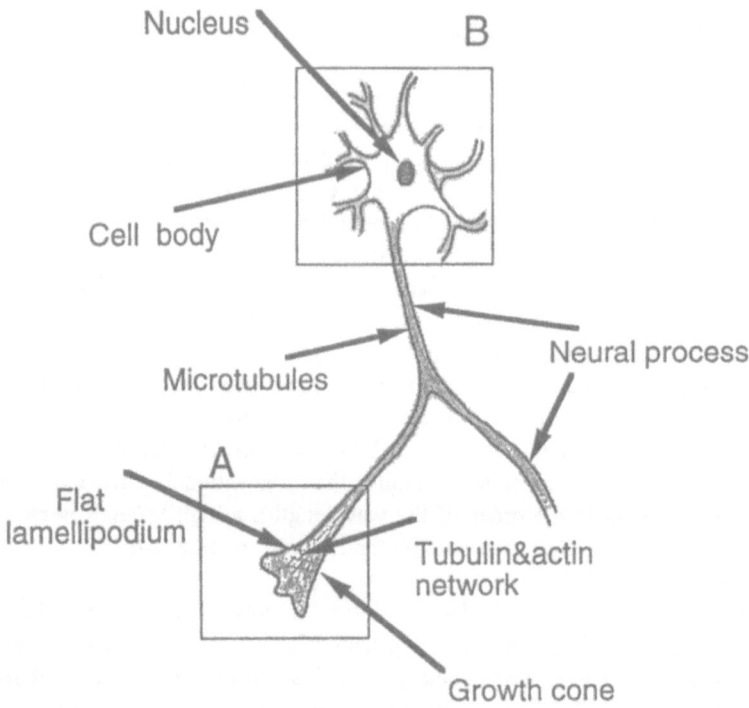

Figure 5.18. Schematic explanation of a neuron with its cell body and neural processes. The end of a neural process has a fingerlike structure called a growth cone. In the growth cone over the flat lamellipodium there are many actin and tubulin filaments forming a network. Inside the neural process are many tubes called microtubules taking part in chemical transport. Areas A and B are imaged and shown in Figs. 5.19 and 5.20, respectively.

fragile biological sample while controlling the probe position by monitoring the shear force. The experimental setup is shown in Fig. 3.6. The light from an Ar^+ laser ($\lambda = 488$ nm) is coupled to the fiber probe to illuminate the neuron sample. The scattered light is collected by an objective lens fixed on the glass substrate. To maintain constant sample–probe separation, the shear force between the sample and probe is monitored and used as a feedback signal, as described in Section 3.1.2.2 (cf. Fig. 3.8).

The neurons are taken from tissues of the hippocampal region of the brain of a Wistar rat. The functions of this region are related to learning and memory [20]. The neuron tissues are cultured for 30 days, and are then fixed with paraformaldehyde (4%) and glutaraldehyde phosphate (2%)-buffered saline solution and rinsed with distilled water. Finally they are dried on a microscope cover-glass substrate. Two kinds of neuron samples are prepared, one labeled with toluidine blue (a dye with absorption band centered at $\lambda = 600$ nm) and the other unlabeled. It is expected that the dye will enhance the contrast of the overall image.

5.2.2.1. Imaging Neurons without Dye Labeling

Figure 5.19 shows the simultaneously obtained optical intensity (a, c) and shear force (b, d) over a neuron. In Figs. 5.19a and 5.19b the sample–probe separation as calibrated by shear force measurements was maintained constant at less than 40

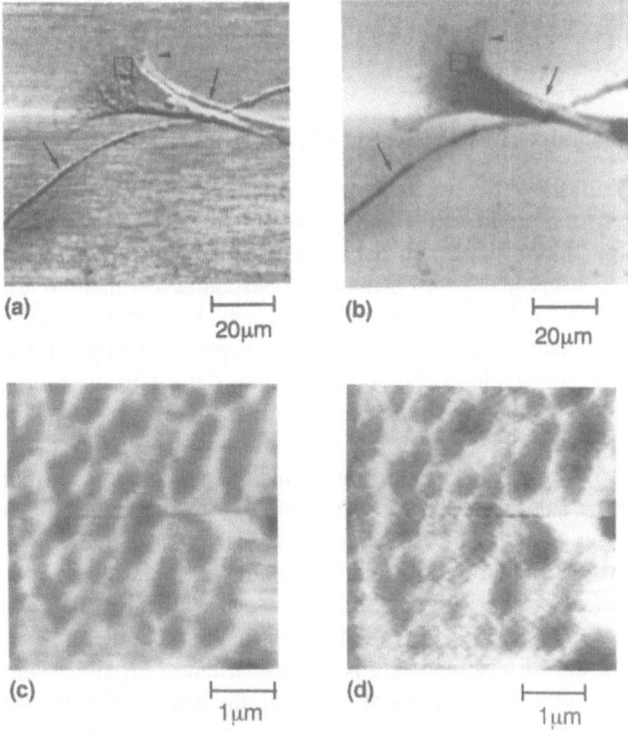

Figure 5.19. NOM images (a, c) and shear-force topographic images (b, d) obtained simultaneously (corresponding to region A in Fig. 5.18). Scan area for panel (a) and (b) is 93 $\times$ 89 μm^2, while that for panels (c) and d is 4.4 $\times$ 4.5 um^2, which show a magnified scan over the flat lamellipodium indicated by the square in panels (a) and (b). In panels (a) and (b) the arrows and arrowheads indicate the neural process and growth cone, respectively. In the shear-force topographic image, bright and dark regions correspond to valleys and hills, respectively. The networklike structure consisting of tubulin and actin filaments is believed to have been formed during the drying of the sample.

nm across the $93 \times 89 \ \mu m^2$ scan area. In both the optical and shear-force topographic images, arrows indicate the neural processes, and arrowheads indicate the growth cones (area A of Fig. 5.18). This growth cone structure actually stretches to communicate with other cells [21]. Thus, understanding its structure is important in the study of cell–cell communication. Comparison between these figures confirms that the optical image shows the edges of the neural process with better contrast.

Figures 5.19c and 5.19d focus on part of the flat lamellipodium (indicated by a square in Figs. 5.18a and 5.18b). Here the sample–probe separation was maintained constant at less than 20 nm over the $4.5 \times 4.5 \ \mu m^2$ scan area. The honeycomblike network structure seen in these figures is believed to be the network of the cytoskeleton consisting of actin and tubulin filaments formed during the drying of the sample. As seen in Fig. 5.19c, the bright filaments are seen with better contrast than in Fig. 5.19d.

5.2.2.2. Imaging Neurons Labeled with Toluidine Blue

Figure 5.20 shows the simultaneously obtained optical intensity (a, c) and shear force (b, d) over a section of neuron (area B of Fig. 5.18) labeled with toluidine blue. In Figs. 5.20a and 5.20b the sample–probe separation as calibrated by shear-force measurements was maintained constant at less than 50 nm across the $44 \times 45 \ \mu m^2$ scan area. In both figures, the arrows indicate neuron processes and arrowheads indicate cell bodies. In this case, in spite of the large thickness of the sample and large sample–probe separation, the optical image is obtained with enhanced contrast compared to the shear-force topographic image. This is due to the presence of strong absorption variations.

Figures 5.20c and 5.20d focus on a small part of the image (indicated by a square in Figs. 5.20a and 5.20b. Here the sample–probe separation was maintained constant at less than 10 nm over the $860 \times 920 \ nm^2$ scan area. In the optical measurement (Fig. 5.20c) a fringelike structure can be seen, in contrast to the shear-force measurement (Fig. 5.20d), where one sees only a constant variation. Since shear force is sensed by variations on the surface, the presence of a cell membrane on the surface renders this interior structure invisible. One concludes that NOM excludes the necessity of removing the cell membrane with detergent (damaging ultracellular structures present underneath the membrane) which is necessary when making observations by an electron microscope or an atomic force microscope. In other words, NOM can resolve structures which lie below the cell membrane.

It has been known from earlier research reports that microtubules are the main constituent of the cytoskeletal element of the neural process [23]; the fringelike structures seen in Fig. 5.20c can be identified as microtubules. Furthermore, presence of microtubules has been confirmed by the observation of immunohisto-

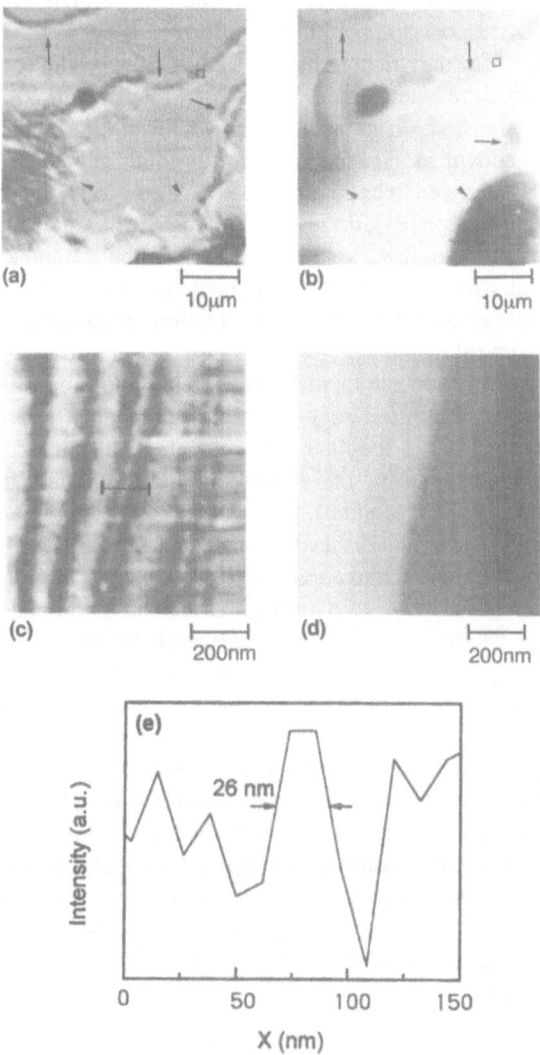

Figure 5.20. NOM images (a, c) and shear-force topographic images (b, d) obtained simultaneously (corresponding to region B in Fig. 5.18) labeled with toluidine blue. Scan area for panels (a) and (b) is $44 \times 45 \ \mu m^2$, while that for panels (c) and (d) is $860 \times 920 \ nm^2$, which show magnified scans over the neural process indicated by the square in panels (a) and (b). In the shear-force topographic image, bright and dark regions correspond to valleys and hills, respectively. In panels (a) and (b), arrows and arrowheads indicate neural processes and cell bodies, respectively. The fringelike structures seen in panel (c) are microtubules, and panel (e) shows the cross-sectional intensity variation across the tube indicated by the line in panel (c).

chemical-stained* tubulin with a confocal microscope. The bright regions of different width correspond to microtubules observed either in the form of bundles or as a single tube. The dark regions having a still smaller width seen on either side of the bright regions are empty spaces between the microtubules. A cross-sectional view of the intensity variation across the narrowest bright region indicated by a line in Fig. 5.20c is shown in Fig. 5.20e. The full-width at half-maximum of this narrowest tube is estimated to be 26 nm. Based on the observations of microtubules by an electron microscope [24], the nominal diameter of a single microtubule is 25 nm. Comparison between these values suggests that the resolution of the present observation by the I-mode NOM is comparable to that of the electron microscope. To the best of the authors' knowledge, observations at this resolution are at the leading edge of current results.

Such a high resolution can be realized by using a flat-top fiber with an apex diameter as large as 30 nm (or less) because of the edge effect of the boundary between the flattened top of the fiber and the coated Au film [25] (the very high spatial Fourier components of the evanescent field confined at the boundary contribute to improve the resolution).

One can clearly see a remarkable difference in the overall contrast between the image of neurons with and without toluidine blue labeling. In order to investigate this difference quantitatively, a contrast parameter C is defined as $C = I_{max} - I_{min})/(I_{max} + I_{min})$, where I_{max} and I_{min} are respectively the maximum and minimum values of the intensity of the two-dimensional array of the image data. Contrast C was calculated for a number of both dyed and undyed neuron samples. Scanned images contained 256×256 pixels each. The average value of C for undyed samples was 0.17, while it was 0.35 for dyed samples. This shows an increase in the contrast of the overall image by a factor of almost two by absorption labeling with dye.

This subsection has thus demonstrated that with the I-mode NOM, it is possible to observe internal ultrafine structures such as microtubules lying underneath the cell membrane. These results show that NOM eliminates the necessity of removing the cell membrane (as is done to make observations by AFM and electron micros-copy), demonstrating the immense potential of NOM in conducting observations on biological samples.

5.2.2.3. Imaging Neurons under Optical Feedback

In I-mode NOM, a shear-force technique has often been employed as the feedback signal for controlling sample–probe separation. However, this technique is not without problems, as represented by characteristics 6–9 in Section 3.1.1.2.

---

*Neurons are stained by monoclonal antibodies of mouse to tublin and then labeled with rhodamine-conjugated anti-mouse immunoglobulin.

In order to solve these problems, section 3.1.2.2 proposed an optical feedback technique that uses the rapidly decaying signal of the evanescent field intensity to control sample–probe separation.

The experimental system is shown in Fig. 3.10. Light from an Ar^+ laser is coupled into a protruded probe fabricated by the techniques given in Sections 4.1.2 and 4.2.1 with apex and foot diameters less than 10 and 30 nm, respectively, to illuminate the sample. In order to generate a rapidly decaying evanescent field for feedback control, a small right-angle prism ($1 \times 1 \times 2$ mm^3) is bonded to the parallel plate holding the sample, and light from a semiconductor laser ($\lambda = 685$ nm) is made to be incident at the angle of total internal reflection. The sample substrate with neurons fixed on the surface is mounted on the parallel plate with a sandwich of index-matching oil. The evanescent field generated on the sample surface is picked up by the same probe and detected by a photomultiplier tube. Phase-sensitive detection is employed for both the detection of the evanescent field and the Ar^+ laser light. Proper filters are installed prior to the detectors in order to reduce the background signal. To compare the images obtained under optical feedback, a shear-force detection scheme is also included in the system.

Figure 5.21 shows the variation of the evanescent field intensity and the amplitude of the laterally dithered probe as a function of sample–probe separation. It can be seen that the rise of the evanescent field intensity is as steep as the decrease

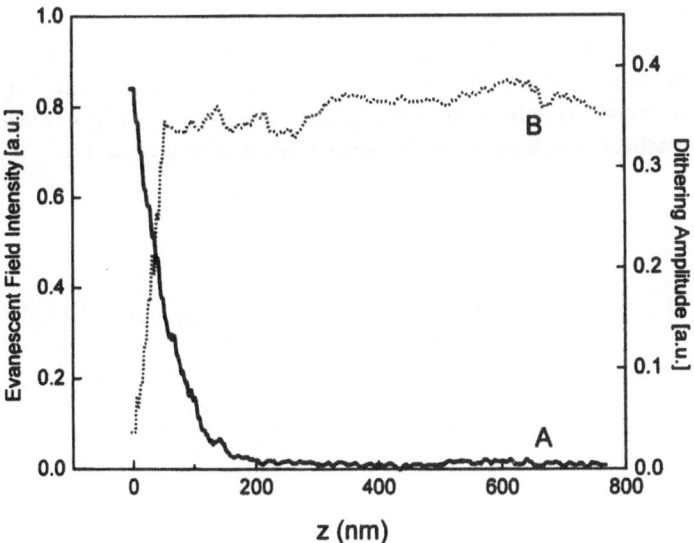

Figure 5.21. Variation of the evanescent field intensity (A) and the dithering amplitude (B) of the probe as a function of the sample–probe separation.

5μm

Figure 5.22. I-mode NOM image of the section of a neural process obtained under shear-force feedback.

of the dithering amplitude. Hence the evanescent field intensity signal can be used for optical feedback to control the sample–probe separation with high sensitivity.

Figure 5.22 shows the I-mode NOM image of a section of a neural process (see Fig. 5.18) under shear-force feedback. Inside the neural process, bundles of tubulin filaments as indicated by the arrow can be seen. Figure 5.23 is a magnified view of the area indicated by the square in Fig. 5.22 under (a) shear-force and (b) optical feedback control. On the right corners, as indicated by arrows, dark, fringelike structures can be seen which are believed to be bundles of tubulin. These results show that under optical feedback, it is possible to obtain topographic variations almost the same as under shear-force control. Further, by comparing Figs. 5.23a and 5.23b, it can be confirmed that the spatial resolution capability of the image by optical feedback is as high as that by shear-force feedback control.

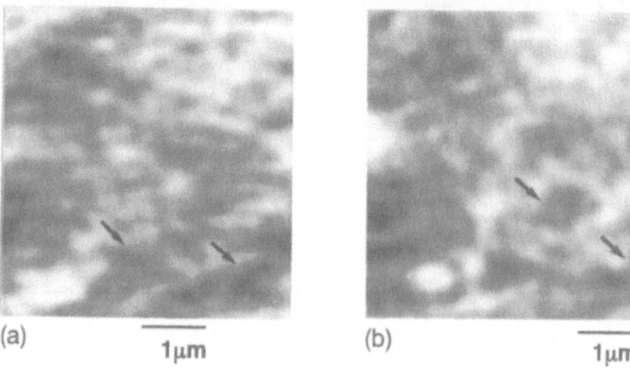

(a) 1μm (b) 1μm

Figure 5.23. Magnified view of the area indicated by the square in Figure 5.22 obtained under (a) shear-force feedback and (b) optical feedback. Arrows indicate tubulin bundles.

However, besides the existence of similar features in Figs. 5.23a and 5.23b, there also exist some differences. This indicates that under optical feedback, one obtains more information regarding the sample and new insight into the physics behind near-field optical imaging phenomena.

5.2.2.4. Fluorescence Imaging

One outstanding advantage of the NOM is that fluorescence can be imaged and spectroscopy carried out. This allows specific labeling of biological samples by fluorescence dye, which is impossible with atomic force or electron microscopes. This section describes the I-mode NOM detection of fluorescence from fibroblasts taken from the cell line of a normal rat kidney.

The fibroblasts [26] were obtained from the subcloned cell line of a normal rat kidney. The cell line was grown in Eagle's minimum essential medium, washed with concentrated H_2SO_4, warmed and dried, and then transferred to a coverglass plate. Twenty-four hours later the cell line was fixed with a 3% paraformaldehyde and phosphate-buffered saline (PBS) solution at $25°C$ for 30 min. After washing with 0.01% Triton X-100 detergent, a hole was made in the surface lipid layer. The sample was placed in 0.2 M glycine for around 15 min to quench the fluorescence from the paraformaldehyde. Next, the cell line was made to react with fluorescein (a dye emitting fluorescence at around $\lambda = 550$ nm) attached to a goat anti-rabbit antibody. During this step, in order to avoid nonspecific attachment of the antibodies, blocking is done using a 10% goat serum/phosphate-buffered saline. Finally, the cell line with specific labeling done for actin filaments is put in distilled water and incubated in 10 mm dithiothreitol and rinsed for short periods in alcohol and acetic acid consecutively.

Figure 5.24 shows a schematic view of a fibroblast cell as used for this experiment. There is a nucleus at the center with actin filaments radially extending over the cell. The square indicates the scanned region. Fluorescence was detected using the photon counting method. The fluorescent dye-labeled fibroblast sample was mounted on a PZT scanner and excited through the probe (cf. Fig. 4.21). Shear

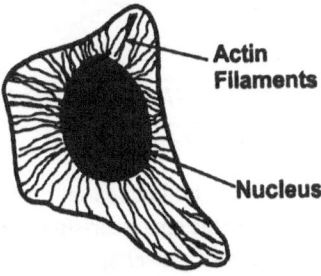

Actin Filaments

Nucleus

Figure 5.24. Schematic view of a fibroblast cell. The square indicates the region scanned by the I-mode NOM.

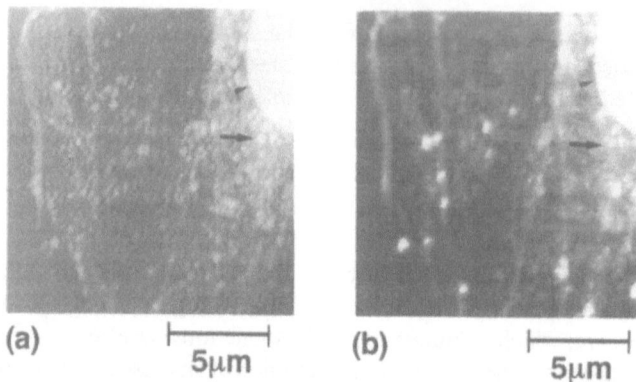

(a) $5\mu m$ (b) $5\mu m$

Figure 5.25. Shear-force topographic image (a) and fluorescent intensity distribution (b) obtained simultaneously over a scan area of $14.3 \times 14.5 \ \mu m^2$. Open arrows: Actin bundles. Filled arrows: Places where clusters of dye attached to the actin bundles. Arrowheads: A section of the nucleus.

force was used to regulate the sample–probe separation. Light from an Ar^+ laser ($\lambda = 488$ nm) was used for excitation. The fluorescent light emitted from the sample was collected by a microscope objective lens, and detected by an avalanche photodiode. A bandpass filter ($\lambda = 550$ nm) and a notch filter (488 nm cutoff filter, with a factor of 1×10^6) are used for selective detection of fluorescence while rejecting excitation light.

Figure 5.25 shows (a) the shear-force topographic image and (b) fluorescence intensity distribution obtained simultaneously over a scan area of $14.3 \times 14.5 \ \mu m^2$. The bright streaks indicated by an open arrow in Fig. 5.25b correspond to actin bundles. The filled arrows in Fig. 5.25 show the places where clusters of dye attached to the actin bundles. Arrowheads indicate a section of the nucleus.

Although the scan area of this figure is large, for preliminary demonstration, spatial resolution is expected to improve since the signal-to-noise ratio of the present photon counting system is still very high. Improvements of the system will allow *in vivo* observation of the cell providing more detailed information on cell dynamics.

5.3. SPATIAL POWER SPECTRAL ANALYSIS OF THE NOM IMAGE

The method of spatial power spectral analysis was employed in Section 5.2.1 to evaluate the characteristics of near-field optical images. This section will describe the details of power spectral analysis, and an estimate of the transfer function of the I-mode NOM will be presented. (The C-mode was discussed in Section 5.2.1.) Randomly distributed gold particles are employed as a reference sample because the diameter of a single particle has been accurately measured by a scanning electron microscope to be 20 ($\pm$ 0.5)nm.

A drop of solution containing colloidal gold particles was dispersed onto a glass substrate and allowed to dry at room temperature. The experimental setup and the fiber probe used here are essentially the same as those shown in Figs. 3.6 and 4.21, respectively.

The NOM image in Fig. 5.26 was obtained at a sample–probe separation of less than 5 nm (maintained constant by a shear-force detection technique). The feedback gain of the shear-force control is maintained very low in order to avoid the problem of cross-talk (see characteristic 9 given in Section 3.1.1.2). Hence it can be considered that scanning is done under a nearly free-running condition. The scan area of this figure is $1.8 \times 1.8 \ \mu m^2$ with one pixel corresponding to approximately $7 \times 7 \ nm^2$. In Fig. 5.26a, the dark region A corresponds to gold particles formed as large clusters. Figure 5.26b shows a magnified view of region B in Fig. 5.26a away from the large cluster region A. The arrow indicates a single gold particle.

By analyzing the image of Fig. 5.26, one can obtain information related to the size of a single gold particle and the degree of randomness of the particle distribution. In turn, with this information, the transfer function of the I-mode NOM system can be estimated. On the basis of the transfer function, the resolution and contrast (as defined in Section 3.1; see Fig. 3.3) of the image can be obtained.

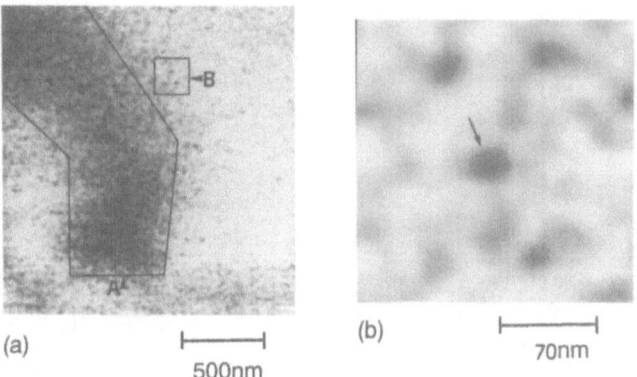

Figure 5.26. (a) Near-field image of 20-nm-diameter gold particles over a scan area of 1.8 $\times$ 1.8 um^2 with one pixel corresponding to an area of 7×7 nm^2. Regions A and B correspond to gold particles formed as large clusters and a section in the isolated gold particle region, respectively. (b) Magnified view of region B, corresponding to a section away from the large clustered region. The arrow indicates an isolated gold particle.

The method of spatial power spectral analysis was used to estimate the transfer function. The two-dimensional spatial power spectral density (PSD) $F(\nu_x, \nu_y)$ of the image shown in Fig. 5.26 was obtained by Fourier transformation. The circular symmetry of the two-dimensional PSD allows the spectral variation to be averaged over angular direction θ. This PSD, averaged with respect to θ, is called the averaged power spectral density (APSD). The variation of the APSD as a function of spatial frequency f is shown in Fig. 5.27. A log–log scale is used. The spectral variation shows clearly a minimum at $f_0 = 5.6 \times 10^7$ m^{-1}. It should be pointed out that the noise level of the I-mode NOM is well below the lowest level of this figure. The minimum f_0 in the APSD contains valuable information regarding the resolving ability of the NOM.

A simple model consisting of dots distributed randomly in space [27] is considered in order to explain the presence of a minimum in the APSD in Fig. 5.27. Such a random distribution can be represented as

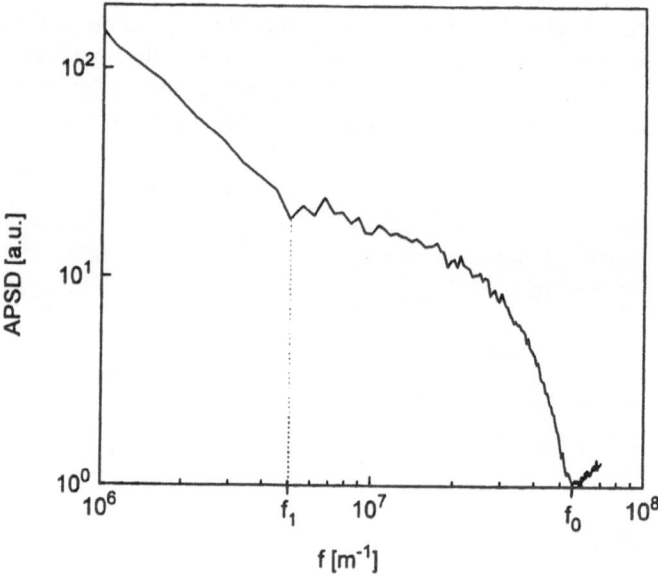

Figure 5.27.   Averaged power spectral density (APSD) as a function of the spatial frequency f given on a log–log scale. Here f_1 is the frequency below which the APSD is dominated by effects due to the random arrangement and clustering of the gold particles. For $f > f_1$, the APSD resembles that due to a single gold particle and is used in the calculation of the transfer function. f_0 is the frequency at which the APSD is minimum.

$$u(x, y) = \sum_{n=0}^{N-1} \sum_{m=0}^{N-1} g(x - nd - \xi_{nm}, y - md - \eta_{nm}) \tag{5.3.1}$$

where $g(x,y)$ is the amplitude variation due to a single dot, n and m are the integers specifying each dot, d is the mean spacing between the dots, and ξ_{nm} and ξ_{nm} are the random displacements given to the center of the (n,m)th dot along the x and y directions, respectively. From the Fourier transform $U(\nu, \mu)$ of this random distribution $u(x, y)$, the PSD $I(\nu, \mu) = |U(\nu, \mu)|^2$ is calculated.

The ensemble average of this PSD distribution is given by

$$\langle I(\nu, \mu) \rangle = |G(\nu, \mu)|^2 \sum_{n} \sum_{n'} \sum_{m} \sum_{m'} \exp\{-i2\pi[(n - n')\nu d + (m - m')\mu d]\}$$

$$\times \langle \exp\{-i2\pi[(\xi_{nm} - \xi_{n'm'})\nu + (\eta_{nm} - \eta_{n'm'})\mu]\} \rangle \tag{5.3.2}$$

where $G(\nu, \mu)$ is the Fourier transform of the amplitude $g(x, y)$ of a single dot, and the $\langle \rangle$ stands for the ensemble average over the random variables ξ and η. (It is assumed that the random displacements ξ and η are independent of each other and are stationary Gaussian random variables with zero mean value.) Further utilizing the property that the linear combination of Gaussian random variables is again a Gaussian random variable, the term $\langle \rangle$ in Eq. (5.3.2) can be rewritten as

$$\langle \exp\{-i2\pi[(\xi_{nm} - \xi_{n'm'})\nu + (\eta_{nm} - \eta_{n'm'})\mu]\} \rangle$$

$$= \exp[-4\pi^2\sigma^2(\nu^2 + \mu^2)(1 - \delta_{nn'}\delta_{mm'})] \tag{5.3.3}$$

Here, σ^2 represents the variance of random displacements. $\delta_{mm'}$ and $\delta_{nn'}$ are the Kroneckers deltas, which take a value of unity for $m = m'$ and $n = n'$ and become zero otherwise.

Substituting Eq. (5.3.3) into Eq. (5.3.2) and decomposing the multiple sum, the ensemble-averaged PSD can be evaluated as (for details of this transformation, see ref. 28)

$$\langle (I\nu, \mu) \rangle = N^2|G(\nu, \mu)|^2 \{1 + \exp[-4\pi^2\sigma^2(\nu^2 + \mu^2)]$$

$$+ [S^2(\nu, d)S^2(\mu, d)/N^2] \exp[-4\pi^2\sigma^2(\nu^2 + \mu^2)]\} \tag{5.3.4}$$

where

$$S^2(\nu, d) = [\sin^2(N\pi\nu d)/\sin^2(\pi\nu d)] \tag{5.3.5}$$

Equation (5.3.4) represents the PSD due to a random distribution of dots. It is composed of two terms. The first is the PSD $|G(\nu, \mu)|^2$ of a single dot with the line spectral components given by Eq. (5.3.5), which results from the regular arrangement of dots of mean spacing d. The other is the exponential term which represents the effect of a random arrangement of the dots. When the variance of the fluctuations in the position of the dots becomes large, e.g., $\sigma^2 \to \infty$, Eq. (5.3.4) reduces to

$$\langle I(\nu, \mu) \rangle \cong N^2 |G(\nu, \mu)|^2 \tag{5.3.6}$$

This implies that under large fluctuations, the ensemble-averaged PSD reduces to that due to a single dot. For a circular dot of diameter ρ, the amplitude variation due to a single dot can be expressed as

$$g(r) = \begin{cases} 1, & |r| \le \rho/2 \\ 0, & \text{otherwise} \end{cases} \tag{5.3.7}$$

where $r = \sqrt{x^2 + y^2}$, and the APSD is the well-known Airy pattern, which is expressed in terms of a first-order Bessel function:

$$|G(\alpha)|^2 = [2J_1(2\pi\alpha\rho)/(2\pi\alpha\rho)]^2 \tag{5.3.8}$$

where

$$\alpha = \sqrt{\nu^2 + \mu^2} \tag{5.3.9}$$

The first zero occurs at the spatial frequency

$$\alpha_0 = 1.22/\rho \tag{5.3.10}$$

The following two conclusions can be drawn by comparing the calculated results with the experimental power spectral density:

1. The experimental APSD shows a monotonic variation for spatial frequencies below $f_1 = 5.0 \times 10^6$ m^{-1}. This is due to the random arrangement of the gold particles given by the exponential terms of Eq. (5.3.4). The monotonic variation is also due to the presence of clusters of gold particles of various sizes.

2. In the spatial frequency region beyond f_1, the APSD is largely determined by the PSD due to a single gold particle given by Eq. (5.3.8).

In the following discussion, based on the two observations given above, we consider only the spectrum beyond the frequency f_1 in determining the effect of a single particle. In the frequency range beyond f_1, Fig. 5.27 shows that the APSD is minimum at the frequency $f_0 = 5.6 \times 10^7$ m^{-1}. Substituting this value of frequency into Eq. (5.3.10), we can estimate the diameter of a single gold particle to be 22 nm. This estimated value is very close to the SEM measured size of 20 ($\pm$ 0.5) nm for a single gold particle.

Furthermore, based on the power spectral analysis, the transfer function of the I-mode NOM system can be estimated by considering the image as a convolution of the system function with a function describing the sample properties. Its transfer function is given by

$$|H(\alpha)| \equiv \frac{I(\alpha)}{U(\alpha)} \tag{5.3.11}$$

where α is defined in Eq. (5.3.9), $I(\alpha)$ is the experimentally obtained APSD (considering the region beyond f_1), and $U(\alpha)$ is the theoretical power spectral density describing the sample properties. For imaging randomly distributed gold particles, $U(\alpha)$ can be approximated by the ensemble-averaged power spectral density given by Eqs. (5.3.4) and (5.3.8).

The transfer function can be estimated using Eq. (5.3.11) by comparing the experimental APSD of Fig. 5.27 with the theoretical APSD given by Eq. (5.3.8) with $\rho = 20$ nm. The estimated transfer function shows almost a linear variation on a log–log scale for the range of $f > f_1$, and it can be expressed as

$$H(f) = 0.81 \left(\frac{f}{f_0} \right)^{-0.16} \tag{5.3.12}$$

where H is normalized to unity at $f = f_1$. From this equation, the –3-dB cutoff frequency, giving a measure of resolution, can be determined to be 1.2×10^9 m^{-1}, which corresponds to a size of 0.8 nm. This is much higher than the cutoff frequency determined by the apex diameter of the probe. Here, it should be remembered that the sample–probe separation can be much smaller than 5 nm. This is experimentally confirmed because, on repeating the scan over the same area where the image in Fig. 5.26 is obtained, only scratches can be found, without any trace of gold particles.

The high resolution of the present NOM can be attributed to the boundary effect mentioned in Section 5.2.2.2 (see Figs. 5.20c and 5.20e). This is related to the size-dependent intensity distribution of the evanescent field on a subwavelength-size particle. That is, very high spatial frequency evanescent fields are generated at the metal–glass boundary around the foot of the probe. These are detected after being converted into propagating fields by the sample.

To conclude this section, results of a computer simulation of the APSD are presented for comparison with experimental (Fig. 5.27) and calculated [Eqs. (5.3.4) and (5.3.8)] results. In this simulation, the PSD of a two-dimensional random dot array based on Eq. (5.3.1) is calculated using the fast Fourier transform algorithm. The simulation is done starting from a regular array of 32×32 dots (dot size $\rho = 9/512$; spacing $d = 16/512$; total pixel size $= 512 \times 512$). The square root of the variance of the fluctuations σ^2 in the spacing is $2d$. Figures 5.28a and 5.28b show the resulting amplitude profile of the two-dimensional random distribution and its APSD, respectively. The profile of the simulated APSD of Fig. 5.28b agrees well with that of Fig. 5.27, i.e., the monotonic variation in the lower frequency region due to the random arrangement of the dots, and the Airy pattern in the higher frequency region due to a single dot with the first minimum occurring at the value of 67, are almost the same as determined from Eq. (5.3.10).

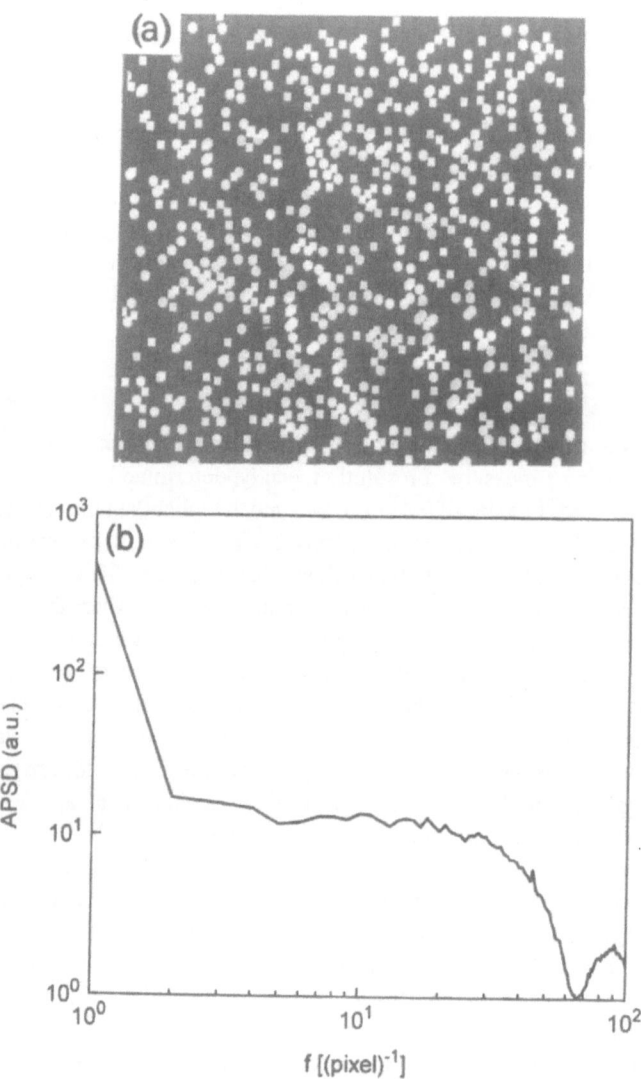

Figure 5.28. (a) Computer-simulated two-dimensional distribution of randomly distributed circular dots and (b) its APSD. Simulation parameters are $\rho = 9/512$, $d = 16/512$, and $\sigma = 2d$.

5.4. REFERENCES

1. T. Saiki, M. Ohtsu, K. Jang, and W. Jhe, Direct observation of size-dependent feature of optical near field on a subwavelength spherical surface, *Opt. Lett.* **21**: 674–676 (1996).

2. K. Jang and W. Jhe, Nonglobal model for a near-field scanning optical microscope using diffraction of the optical near field, *Opt. Lett.* **21**: 236–238 (1996).

3. H. Chew, D. S. Wang, and M. Kerker, Elastic scattering of evanescent electromagnetic waves, *Appl. Opt.* **18**: 2679–2687 (1979).

4. X. S. Xie and R. C. Dunn, Probing single molecule dynamics, *Science* **265**: 361–364 (1994).

5. R. X. Bian, R. C. Dunn, X. S. Xie, and P. T. Leung, Single molecule emission characteristics near-field microscopy, *Phys. Rev. Lett.* **75**: 4772–4775 (1995).

6. J. K. Trautman, J. J. Macklin, L. E. Brus, and E. Betzig, Near-field spectroscopy of single molecules at room temperature, *Nature* **369**: 40–42 (1994).

7. W. Jhe and K. Jang, Cavity quantum electrodynamics inside a hollow spherical cavity, *Phys. Rev. A* **53**: 1126–1129, (1995).

8. J. M. Wylie and J. E. Sipe, Quantum electrodynamics near an interface, *Phys. Rev.* **A30**: 1185–1193 (1984).

9. D. Meschede, W. Jhe, and E. A. Hinds, Radiative properties of atoms near a conducting plane: An old problem in a new light, *Phys. Rev.* **A30**: 1587–1596 (1990).

10. T. Saiki, S. Mononobe, and M. Ohtsu, Nanometric integrating tip: Enhanced sensitivity of fluorescence detection in photon STM, in *Technical Digest, Quantum Electronics and Laser Science Conference*, Baltimore, 1995, pp. 84–85.

11. M. Yoshimoto, T. Maeda, T. Ohnishi, H. Koinuma, O. Ishiyama, M. Shinohara, M. Kubo, R. Miura, and A. Miyamoto, Atomic-scale formation of ultrasmooth surfaces on sapphire substrates for high-quality thin-film fabrication, *Appl. Phys. Lett.* **67**: 2615–2617 (1995).

12. A. Miyamoto, K. Takeuchi, T. Hattori, M. Kubo, and T. Inui, Mechanism of layer-by-layer homoepitaxial growth of $SrTiO_3(100)$ as investigated by molecular dynamics and computer graphics, *Jpn. J. Appl. Phys.* **31**: 4463–4464 (1992).

13. R. Micheletto, S. Mononobe, M. Ohtsu, M. Yoshimoto, T. Maeda, T. Ohnishi and H. Koinuma, Observation of an atomic sapphire step by a collection mode near-field optical microscope, in *Abstracts, The First Asia-Pacific Workshop on Near Field Optics, Seoul, Korea, August 1996*, pp. 54–55.

14. M. Naya, S. Mononobe, R. Uma Maheswari, T. Saiki, and M. Ohtsu, Imaging of biological samples by a collection-mode photon scanning tunneling microscope with an apertured probe, *Opt. Commun.* **124**: 9–15 (1996).

15. S. Kanto, H. Okino, S.-I. Aizawa, and S. Yamaguchi, Amino acids responsible for flagellar shape are distributed in terminal regions of flagellin, *J. Mol. Biol.* **219**: 471–480 (1991).

16. O. J. F. Martin, C. Girard, and A. Dereux, Generalized field propagator for electromagnetic scattering and light confinement, *Phys. Rev. Lett.* **74**: 526–529 (1995).

17. W. Jhe and K. Jang, Simple-image-dipole method for photon scanning tunneling microscopy, *Ultramicroscopy* **61**: 81–84 (1995).

18. S. Jiang, H. Ohsawa, K. Yamada, T. Pangaribuan, M. Ohtsu, K. Imai, and A. Ikai, Nanometric scale biosample observation using a photon scanning tunneling microscope, *Jpn. J. Appl. Phys.* **31**: 2282–2287 (1992).

19. M. Naya, R. Micheletto, S. Mononobe, R. Uma Maheswari, and M. Ohtsu, Near-field optical imaging of flagellar filaments of salmonella in water with optical feedback control, *Appl. Opt.* **36**: 1681–1683 (1997).

20. J. G. Nicholls, A. R. Martin, and B. G. Wallace, *From Neuron to Brain*, Sinauer, Sunderland, Massachusetts, 1992.

21. H. Tatsumi, H. Sasaki, and Y. Katayama, Elongation of growth cone filopodia observed with video-enhanced differential interference contrast microscopy, *Jpn. J. Physiol.* **43** (Suppl. 1): S221–S223 (1993).

22. R. Uma Maheswari, H. Tatsumi, Y. Katayama, and M. Ohtsu, Observation of subcellular nanostructure of single neurons with an illumination mode photon scanning tunneling microscope, *Opt. Commun.* **120**: 325–334, (1995).

23. B. Alberts, D. Bray, J. Lewis, M. Raff, K. Roberts, and J. D. Watson, *Molecular Biology of the Cell*, Garland, New York, 1983.

24. H. Hartwig, Mechanisms of actin rearrangements mediating platelet activation, *J. Cell Biol.* **118**: 1421–1442 (1992).

25. M. Specht, J. D. Pedaring, W. M. Heckl, and T. W. Hänsch, Scanning plasmon near-field microscope, *Phys. Rev. Lett.* **68**: 476–479 (1992).

26. E. R. Solomon, L. R. Berg, D. W. Martin, and C. Villee, *Biology* Saunders, Sunderland, Massachusetts, 1993, Chapter 37.

27. R. Uma Maheswari, N. Takai, and T. Asakura, Power spectral distributions of dot arrays with Gaussian random spatial fluctuations, *J. Opt. Soc. Am. A* **9**: 1391–1397 (1992).

28. R. Uma Maheswari, H. Kadono, and M. Ohtsu, Power spectral analysis for evaluating optical near-field images of 20 nm gold particles, *Opt. Commun.* **131**: 133–142 (1996).

DIAGNOSTICS AND SPECTROSCOPY OF PHOTONIC DEVICES AND MATERIALS

Conventional photonic devices are too large for use in the NOM because they are larger than the optical wavelength, although the NOM is useful for their study. In contrast, the NOM is indispensable for the study of semiconductor quantum structures. Take, for example, the case of a quantum dot. Using conventional optical techniques it is impossible to study the light output from a single quantum dot: What one sees is the superposition spectrum of an ensemble of quantum dots, each shifted slightly in frequency due to its unique individual surroundings. The NOM allows one to optically pump and collect the light from a single quantum dot. This chapter will describe the application of the NOM to the study of both passive and active photonic devices with subwavelength resolution.

6.1. DIAGNOSING A DIELECTRIC OPTICAL WAVEGUIDE

An optical waveguide is a representative passive photonic system. With the development of optical communications and optical information processing, designs of functional optical waveguides with smaller sizes are required for a variety of applications such as local area networks and optical computing. Mass spectroscopy using secondary ion-mass spectroscopy (SIMS) has been popular for materials analysis. To evaluate propagation loss, a cutback method, a scattering method, an interferometric method, and a method using a TV camera have been conventionally employed [1]. The problem with these methods is that the resolution is diffraction limited. Furthermore, diagnostics using SIMS and the cutback method are destructive. For future devices with higher integration, high-resolution nondestructive methods will become indispensable. The NOM is such a method.

The following features of importance for the design and fabrication of a waveguide can be characterized by the NOM method:

1. Scattering caused by surface deformation, defects, and compositional inhomogeneity.
2. The intensity profile of a guided mode.

3. The profile of the refractive index.

This section demonstrates the feasibility of the NOM for these characterizations [2].

In the case of a waveguide whose upper cladding layer is air, it is possible to monitor directly the evanescent field of the guided mode generated on the upper boundary layer, i.e., the guided mode profile can be monitored through the short-range electromagnetic interaction between the waveguide and the probe via the evanescent field.

The waveguide discussed in this section is a proton-exchanged Z-cut $LiTaO_3$ waveguide with a normalized cross-sectional area of $3.8 \times 2.7 \ \mu m^2$ operating at a single mode at $\lambda = 633$ nm. The propagation loss measured by the cutback method is 0.7 dB/cm. Only a transverse magnetic (*TM*) mode can be excited in the waveguide.

The experimental setup is shown in Fig. 6.1 and uses a C-mode NOM. The light from a diode laser (LD; $\lambda = 680$ nm) is coupled to the waveguide, and the evanescent field on the guided layer is sensed by a fiber probe fabricated with the method of Section 4.1.2; it has an apex diameter d of 10 nm. Since the waveguide is made of a ferroelectric material, electric charges are induced on the surface. Thus,

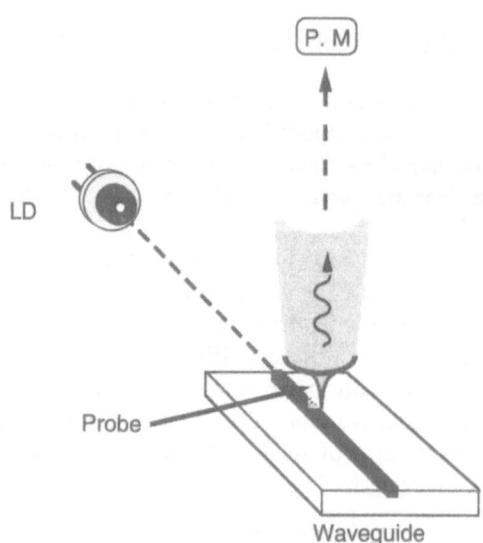

Figure 6.1. Experimental setup for diagnosing a dielectric optical waveguide by using a C-mode NOM. LD, Diode laser; WG, waveguide; PM, photomultiplier.

a Coulomb force is induced between the waveguide surface and the probe, by which the probe is pulled and deformed. In order to prevent the pulling and deformation, the sample–probe separation is modulated by dithering the fiber probe normal to the substrate. The modulation amplitude and frequency are 50 nm and 800 Hz, respectively. Synchronous detection can demodulate the signal for maintaining constant sample–probe separation.

Guided light can be scattered due to deformations and/or inhomogeneities near the waveguide surface [3]. This scattering offers information about the loss mechanism of the waveguide. Figure 6.2a shows the measured profile of a linear waveguide with a scan area of $9.6 \times 8.2 \ \mu m^2$. Figure 6.2b shown a magnified image corresponding to one of the scattering sources (indicated by the white square in Fig. 6.2a). The scan area is $1.5 \times 1.3 \ \mu m^2$. The full-width at half-maximum of the intensity distribution of the picked-up scattered light in Fig. 6.2b is 500 nm, which is a subwavelength value.

This scattering can be interpreted as mainly due to dust particles in the air and inhomogeneity of the composition since the waveguide surface is sufficiently flat as shown by SEM and AFM measurements. In order to investigate the scattering, the near-field and far-field images are compared; results are shown in Fig. 6.3. For these images, the sample–probe separation is fixed to be (a) less than 50 nm or (b) larger than 500 nm. In regions A and B in Fig. 6.3a there are some bright spots. On one hand, in region A' in Fig. 6.3b, corresponding to region A, some bright spots are also seen due to scattering, while on the other hand, no bright spots are seen in the region in Fig. 6.3b, corresponding to region B in Fig. 6.3a. This difference in the images implies a difference in the light-induced dipole distributions on the waveguide surface. In the case of the near-field image, the short-range electromagnetic interaction with the probe tip modifies the distribution of the dipole on the waveguide surface, while this modification is negligible for the far-field image. In

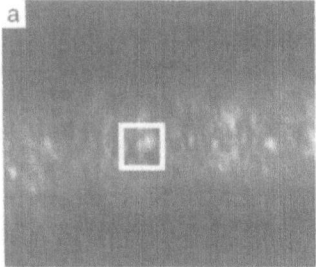 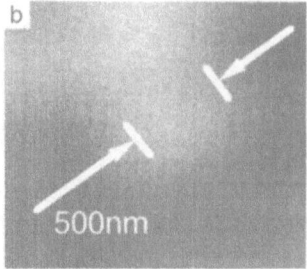

Figure 6.2. (a) Measured profile of an LiTaO$_3$ waveguide with a scan area of $9.6 \times 8.2 \ \mu m^2$. (b) Magnified image of the part indicated by the square in panel (a). Scan area is $1.5 \times 1.3 \ \mu m^2$.

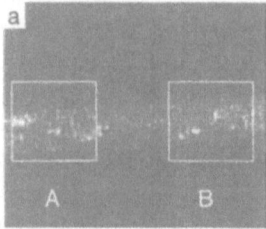
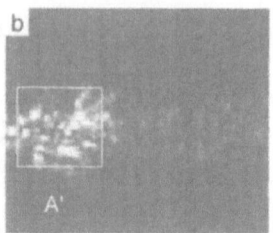

Figure 6.3. Scattered light profile induced by the guided mode of the LiTaO3 waveguide. Scan area is 15×13 μm^2. The image is taken with a sample–probe separation of (a) less than 50 nm and (b) larger than 500 nm.

this context, the NOM is optically destructive even though the waveguide itself is not physically destroyed as in the cutback method [4].

An important parameter for evaluating a waveguide is its effective width, which can be estimated from the transverse spatial profile of the guided mode. The inset in Fig. 6.4 shows the measured two-dimensional profile of the guided mode. The main figure shows its cross-sectional profile, obtained by averaging 256 times at every 60-nm step along the longitudinal direction (x axis in Fig. 6.4). Assuming a parabolic refractive index profile in the guided layer, the evanescent field amplitude of the TM mode can be expressed as a Gaussian, i.e., $\exp(-y^2/2\sigma^2) + A$, where σ is

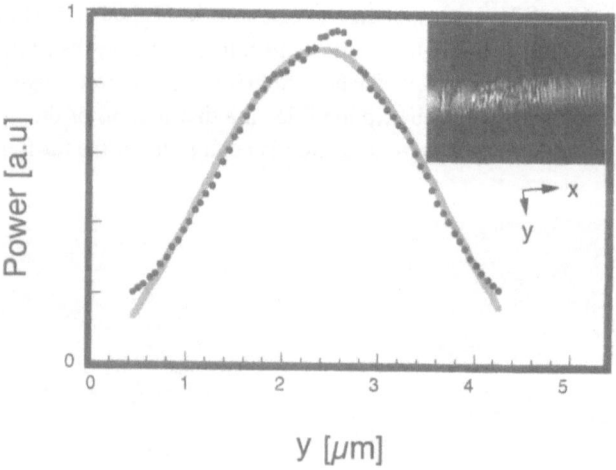

Figure 6.4. Cross-sectional profile of the guided mode obtained by averaging 256 times at every 60-nm step along the x axis. Solid curve represents the least-square-fitted Gaussian profile. Inset: Measured profile of the guided mode of the LiTaO3 waveguide. Scan area is 11×15 μm^2.

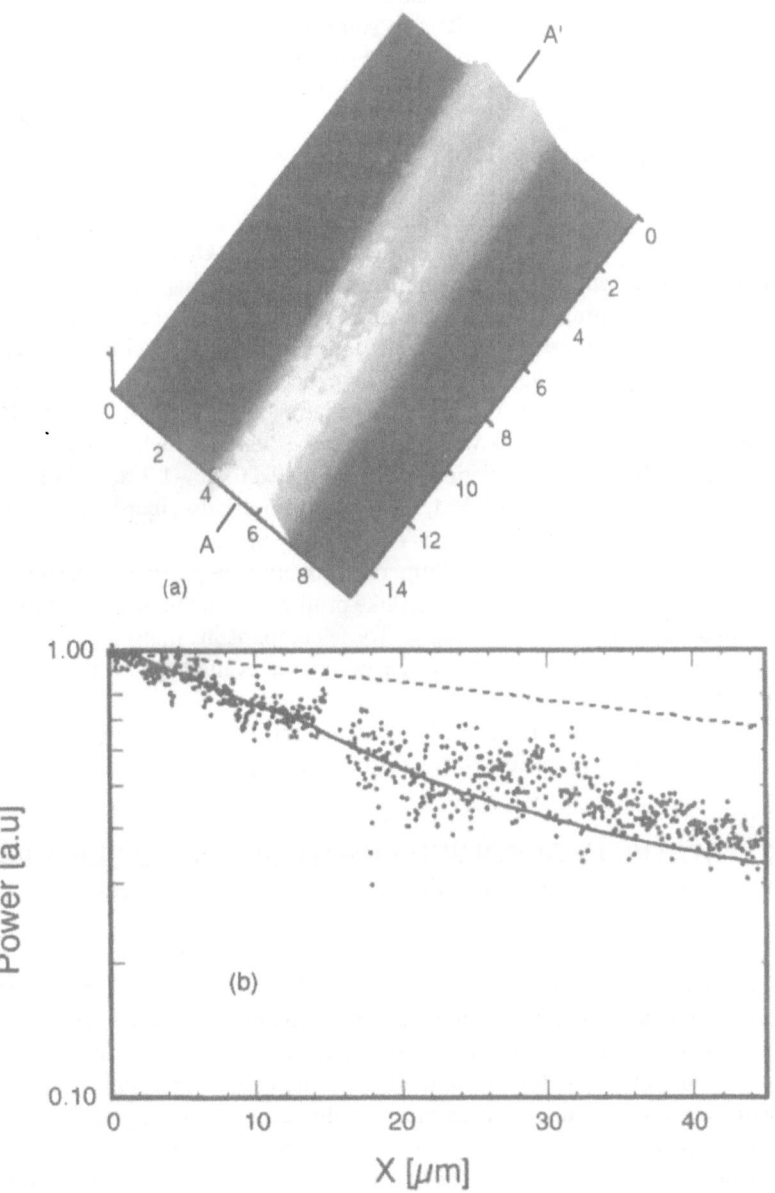

Figure 6.5. (a) Perspective view of the measured image of the guided mode of the Y-branch LiTaO$_3$ waveguide. Scan area is 10×15 μm^2. (b) Intensity variation along the line A–A' of panel (a). The solid and broken curves are the least-square-fitted exponential functions obtained by assuming steplike and Gaussian profiles of the refractive index, respectively.

a measure representing the mode width and A is the noise magnitude introduced at the stage of NOM measurement. The solid curve in Fig. 6.4 is the Gaussian least-square-fitted to the measured value. From this fitting, the effective full-width at half-maximum ($2\sqrt{2}\ \sigma$) is evaluated to be 3.7 μm, in good agreement with the designed value (3.8 μm).

One advantage of using NOM is the capability of two-dimensional imaging of the intensity profile of the guided mode, which is particularly useful for planar integrated optical devices such as a directional coupler or a Y-coupler. To demonstrate this capability, a Y-branch proton-exchanged $LiTaO_3$ waveguide was tested and the results compared with a numerical calculation to map the refractive index profile. The beam propagation method (BPM) was used for the calculation. (The BPM is one of the most powerful methods for modeling and simulating optical waveguide devices [5]). Figure 6.5a shows the measured image of the guided mode of the Y-branch waveguide. The intensity variation along the line A–A′ is plotted in Fig. 6.5b. This measured variation is least-square-fitted by an exponential function, and the intensity variation rate is estimated to be -1.2×10^{-1} dB/μm by this fitting. For calculation by BPM, the refractive index distribution was approximated as (1) steplike or (2) Gaussian. The intensity variation rate was estimated to be -1.2×10^{-1} and -0.30×10^{-1} dB/μm, respectively. Compared with the measured result, the approximation using the steplike profile is found to be more accurate in the Y-branching region of this sample. These comparisons imply that the present diagnostic method using NOM can be useful for evaluating the profile of the refractive index distribution.

The present method is more advantageous for diagnosing integrated passive optical devices with more complicated structures. It expected to find wide use.

6.2. SPATIALLY RESOLVED SPECTROSCOPY OF LATERAL $p–n$ JUNCTIONS IN SILICON-DOPED GALLIUM ARSENIDE

Since the NOM has been used for collecting information on a sample surface, it can be applied for investigating surface light-emitting devices. By detecting submicrometer spatial inhomogeneities formed at the crystal growth stage, useful information on the structure, electronic state, and carrier transport of lateral $p–n$ junctions can be obtained. For example, spatially resolved photoluminescence (PL) spectroscopy [6], electroluminescence (EL) spectroscopy, and photocurrent measurements have been carried out for lateral $p–n$ junctions prepared with Si-doped GaAs. These results are reviewed in this section.

In the investigation of semiconductor optical devices with novel structures (e.g., quantum dots and quantum wires) a spatially resolved spectroscopy technique in strict correspondence with their size and structure is required. The cathode-

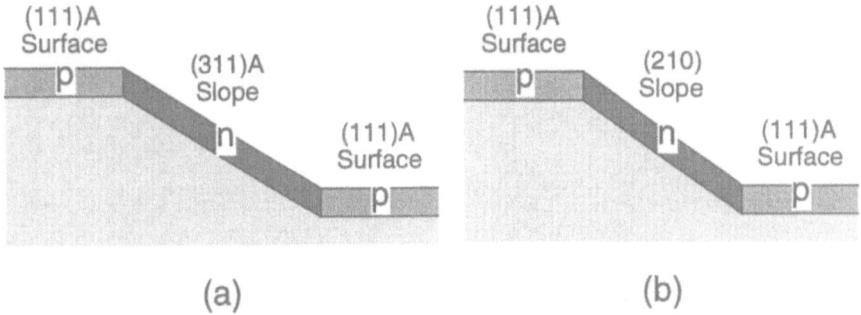

Figure 6.6. Structure of the lateral $p-n$ junctions of samples 1 (a) and 2 (b).

luminescence (CL) technique, whose resolution is limited by the diffraction of the propagating light, has been frequently used for this purpose [7]. NOM, in addition to offering higher resolution than CL, can measure surface structures and optical properties simultaneously. This makes it a powerful tool of choice for the characterization of such kinds of devices.

The lateral $p-n$ junction can be applied to lateral carrier confinement structures, such as surface-emitting laser diodes. The conduction type of GaAs layers with Si dopant depends on the growth conditions and the orientation of the substrate [8, 9]. By using this amphoteric nature of Si, both n- and p-type regions can be grown simultaneously on a patterned substrate, and lateral $p-n$ junctions are formed at the boundary of the two regions [10, 11]. The photoluminescence peak wavelength strongly depends on the conduction type and the carrier concentration in the GaAs layers [12]. By measuring spatially resolved PL spectra, one can examine precisely the carrier distribution in the transition region of the $p-n$ junction.

A semiinsulating GaAs(111)A substrate was etched by a photolithographic technique to obtain a triangular (111)A surface surrounded by three slopes [10, 11]. Two different slopes were prepared, as shown in Fig. 6.6, one a (311)A slope (sample 1) and the other a (210) slope (sample 2). After thermal cleaning, a Si-doped GaAs layer with a thickness of 1 μm is grown on the patterned substrate at 600°C. The Si concentration is estimated as 1×10^{18} cm^{-3}.

6.2.1. Photoluminescence and Electroluminescence Spectroscopy

6.2.1.1. Photoluminescence Spectroscopy

Figure 6.7a shows the experimental setup of an I-mode NOM. The Au-coated, double-tapered fiber probe (Fig. 6.7b) has a 200 nm foot diameter d_f. Taking into

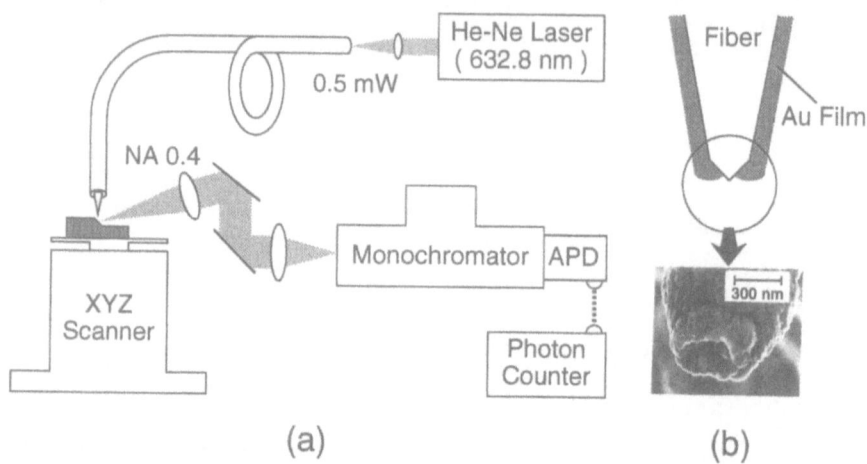

(a)   (b)

Figure 6.7. (a) Experimental setup of an I-mode NOM for PL spectroscopy of the lateral p–n junction. (b) Schematic of the cross-sectional profile of the double-tapered fiber probe and its SEM micrograph.

account both experimental utility and transmission efficiency, the shape of the probe is optimized by the chemical etching method described in Section 4.1.5. As an excitation light source, the 0.5-mW light of a He–Ne laser ($\lambda = 633$ nm) is coupled into the fiber probe. The transmission coefficient is estimated as 1.0×10^{-3} (cf. Fig. 4.19). By fixing the sample–probe separation to be less than 20 nm, the excitation region is restricted to a lateral area of about 100×100 nm^2 and 200 nm in depth, determined by the size of the probe and the absorption coefficient of GaAs. The PL signal from the sample is collected on the same side of the sample surface with an objective lens of 0.4 numerical aperture and transported to a grating monochromator with an avalanche photodiode for photon-counting detection. For shear-force position control, a 1.3-μm diode laser whose photon energy is far below the absorption edge of GaAs is used.

Figure 6.8 shows the shear-force topographic images of two samples in the vicinity of the slopes. The slope width and height of the slope of sample 1 (Fig. 6.8a) are 10 and 6 μm, and those of sample 2 (Fig. 6.8b) are 8 and 5.5 μm, respectively. Near the top surface on the slope some bumps appear (indicated by arrows) which may have been produced in the etching process of the substrates. Comparing the two samples, larger bumps are observed in sample 2, which can be attributed to the difference of etching behavior on the (311)A slope and the (210) slope.

Figure 6.9a shows the normalized PL spectra at some characteristic points on the slope of sample 1. The peak wavelength at the top surface (position A in Fig. 6.9b) and that at the bottom surface (E) show the same value of 870 nm. This peak

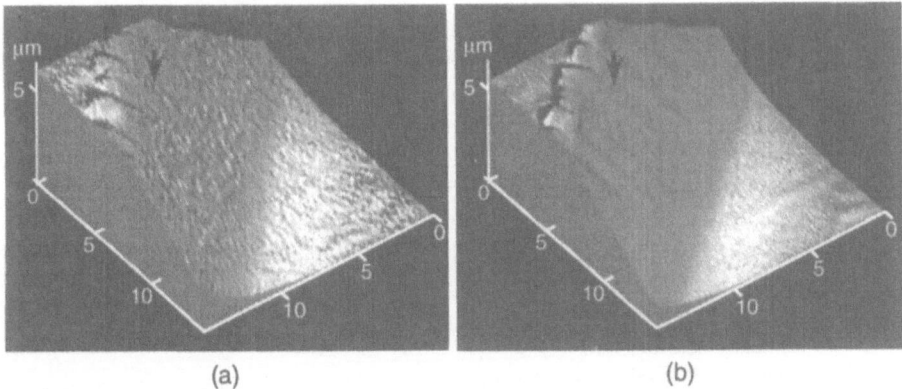

(a) (b)

Figure 6.8. Perspective views of shear-force topographic images of samples 1 (a) and 2 (b). Scan area is $15 \times 15 \ \mu m^2$. Arrows in these figures indicate the positions of bumps.

wavelength corresponds to that of a flat (111)A surface with the same Si concentration. Its conduction type (p-type) has been confirmed by the measurement of capacitance–voltage characteristics [10]. On the (311)A slope (C), the peak wavelength shifts to the higher energy side. Its value of 855 nm is also in agreement with that of an n-type flat (100)A surface investigated previously [10]. In the transition

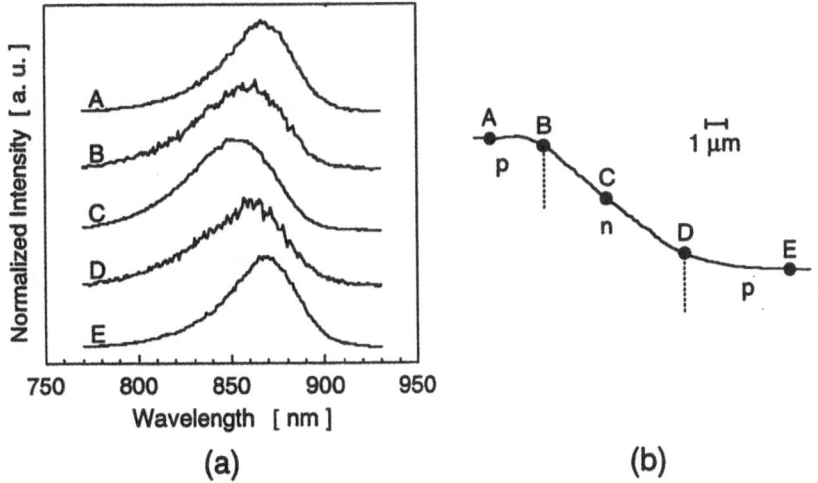

(a) (b)

Figure 6.9. (a) Normalized PL spectra at five points in the vicinity of the slope of sample 1. (b) Cross-sectional view of the p–n junctions indicating five measuring points (A–E).

region (B and D), where the conduction type and carrier concentration gradually change, the peak wavelengths show intermediate values.

Figure 6.10 shows the PL peak wavelength, PL spectral linewidth, and total intensity along the slopes overlaid on the sample structure. A variety of features are observed in the transition regions. At the lower junction, the peak wavelength shows a gradual change with a transition width of 5 μm in both samples. Since conduction type and carrier concentration vary with the tilt angle of the substrate from the (111)A surface [12], the transition width is closely related to the structure of the junction. The gradual change in tilt angle at the lower junction, as is shown in Fig.

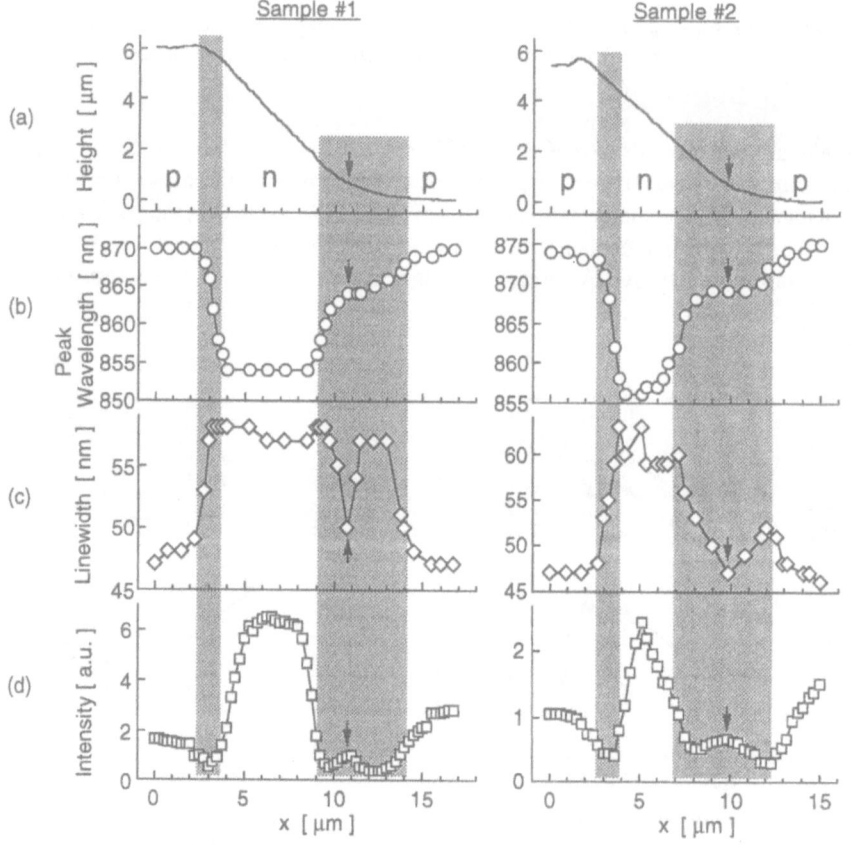

Figure 6.10. (a) Cross-sectional views of the shear-force topographic image of the slopes. (b–d) Plots of PL peak wavelength, PL spectral linewidth, and total PL intensity, respectively, at the slopes of the two samples. The shaded areas indicate the transition regions of the upper and lower junctions.

6.10a, causes a wide transition width of the carrier concentration. At the upper junction, on the other hand, the top surface and the slope make a clear ridge. This abrupt change results in a narrower transition width of 1 μm. From the fast rise of the peak wavelength in the upper junction, it is roughly estimated that a spatial resolution of better than 400 nm has been attained. By comparing the spectral change with the shear-force topographic image carefully, it can be seen that the transition region is formed on the slope, not at the ridge. Previously, to explain the results of capacitance–voltage measurements, it has been inferred that the modulation of carrier concentration occurs on the slope side of the intersection [13]. This was supported by a CL measurement with a resolution of about 1 μm [10]. The present NOM study supports this inference by establishing the position of the boundary with a higher resolution through the accurate correspondence between the surface structure and the optical response.

The total PL intensity in the transition regions is much lower than that from other regions. This is mainly due to the separation and the drift of photoexcited electrons and holes caused by the internal electric field in the transition regions. The width of the depletion layer can be estimated quantitatively from the slow rise of PL intensity on both the p and n sides. In the lower junction, some anomalous optical structures (plateau in the change of the peak wavelength, decrease of spectral linewidth, and increase of PL intensity) are observed, as indicated by arrows in Fig. 6.10. It is important to investigate these local optical properties in the vicinity of the junctions in detail since the distribution of defects and strains and the nonuniformity of dopants in the active region will affect the PL emission efficiency of a device.

To measure the two-dimensional distribution of the PL intensity, the grating monochromator is replaced by a band-pass filter, and the PL signal is collected by the probe. In this mode of detection, the probe both excites and collects the PL signal, allowing a resolution higher than 200 nm to be achieved, independent of the carrier diffusion length, limited only by the probe size. Figure 6.11 shows the PL intensity from sample 1. In this image, the maximum and the minimum counting rate are 2×10^4 and 1×10^4 counts/sec, respectively. In the transition region, as was shown previously, the PL intensity is much lower than in the other regions. Some bright areas, where the PL intensity increases locally, appear in the lower junction. These signals indicate the existence of subsurface inhomogeneities since no corresponding structure is found in the shear-force topographic image (Fig. 6.8a). A nonuniform distribution of Si dopants and of the resultant internal electric field, or that of defects and strains, will affect the local optical responses.

6.2.1.2. Electroluminescence Spectroscopy

In order to diagnose more practical device characteristics, electroluminescence (EL) spectroscopy is carried out by injecting current into the p–n junctions. The

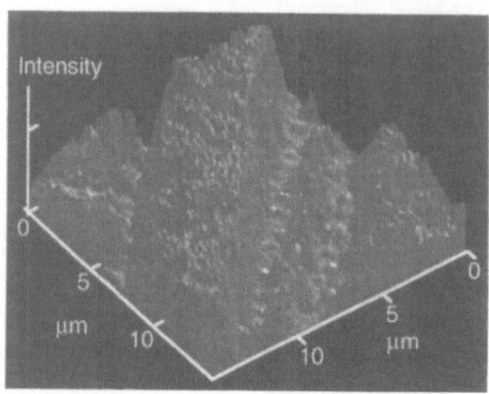

Figure 6.11. Perspective view of PL intensity image, where the PL signal is collected by the probe. The scan area is $15 \times 15 \ \mu m^2$.

samples for this experiment have been fabricated under the same condition as that of sample 1 of Fig. 6.6. Figure 6.12 shows the cross-sectional profile of two samples on which electrodes have been coated so as to detect the EL signal from the p–n junctions at the lower and upper ends of the slope, respectively.

The experimental setup of Fig. 6.7 was modified to work as a C-mode NOM for electroluminescence detection. Figure 6.13 shows the EL intensities from the upper and lower junctions. Shear force topographic images and PL intensity distributions taken for these samples using the experimental setup in the I-mode are the same as those given in Fig. 6.10. By comparing Fig. 6.10 with Fig. 6.13, it is found that the full-width at half-maximum of the spatial distribution of the EL intensity (1.1 μm) is narrower than the width of the transition region of the upper and lower junctions, identified by the shaded areas in these figures.

The reason for the difference is as follows: The spatial distribution of the EL emission given by Fig. 6.13 represents the positions of the p–n junction because only the proximity region of the surface can contribute to the EL emission due to the limited area of the injected current flow in this proximity region. On the other

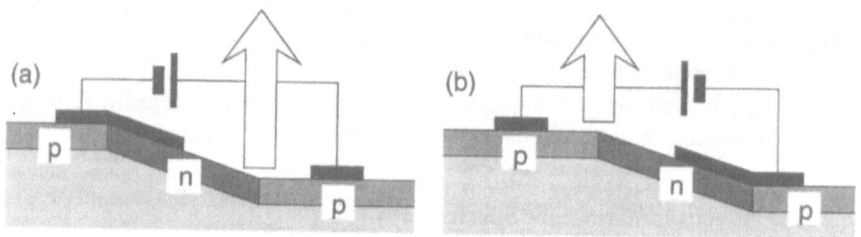

Figure 6.12. Cross-sectional structure of lateral p–n junctions with electrodes for EL from lower (a) and upper (b) ends of the slope. White arrows represent the direction of EL emission.

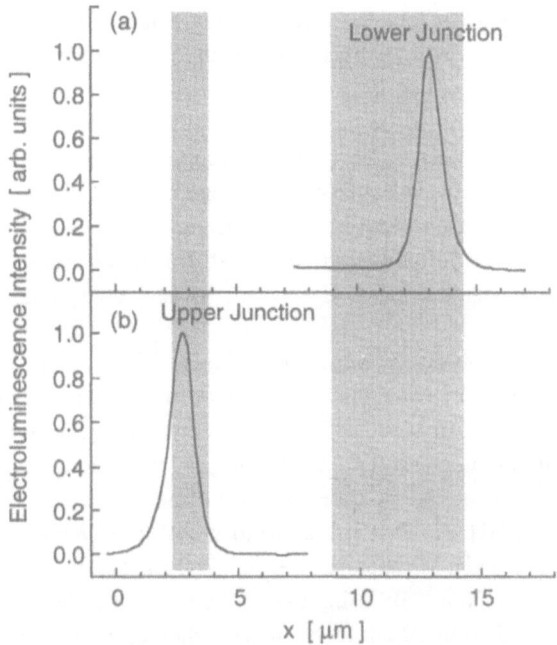

Figure 6.13. Spatial distribution of the EL intensity from the lower (a) and upper (b) ends of the slope. The shaded areas represent the transition regions of the upper and lower junctions.

hand, the width of the transition region estimated from the PL emission distribution (shaded area in these figures) represents the area in which the carrier density changes spatially.

6.2.2. Photocurrent Measurement by Multiwavelength NOM

This subsection will review near-field photocurrent imaging of the sample shown in Fig. 6.6a, using multiwavelength excitation sources (wavelength $\lambda =$ 488–830 nm) [14]. Photocurrent measurement with near-field excitation [15] is a novel technique for probing semiconductor devices including laser diodes, light-emitting diodes, and photodetectors.

The propagating modes of the optical field emitted from an I-mode NOM probe are expected to make an important contribution to the study of these internal optical and transport properties. In particular, when the probe's foot diameter d_f is larger than about 100 nm, the intensities of these propagating modes can be sufficiently high, making a larger contribution [16]. Though propagating modes into a crystal do not have any spatial resolving power, "tomographic" information of the diagnosed material can be obtained by systematically varying the optical penetration depth over a wide range, i.e., from less than the foot diameter to the order of the wavelength. Here, it should be emphasized that when the penetration depth is smaller than the foot diameter, the spatial resolution is determined only by the size of the probe. This multiwavelength imaging corresponds to the energy-dependent source size in an electron-beam-induced current (EBIC) measurement [17]. In EBIC imaging, the resolution is restricted to ~1 μm and its accuracy in tuning the excitation energy near the band edge is not so adaptable compared to optical excitation.

A schematic of the sample structure is shown in Fig. 6.14. As in the case of Fig. 6.12a, electrodes are coated in order to investigate the lower $p-n$ junction of the sample of Fig. 6.6a. An Au-coated, double-tapered fiber probe is also used which has a 200 nm foot diameter d_f as shown in Fig. 6.7b. Although the probe and its position control scheme are the same as those of Section 6.2.1, multiwavelength light sources are used here. That is, light from an Ar^+ laser ($\lambda = 488$ nm), a He–Ne laser ($\lambda = 633$ nm), and a Ti-sapphire laser ($\lambda = 780$ and 830 nm) are coupled to the probe of the I-mode NOM. By using these sources, optical penetration depths in GaAs can be varied from 80 nm to 1 μm. The photocurrent induced by the light through the probe is collected at the electrodes and amplified with a current injection preamplifier. The signal is synchronously detected with a lock-in amplifier.

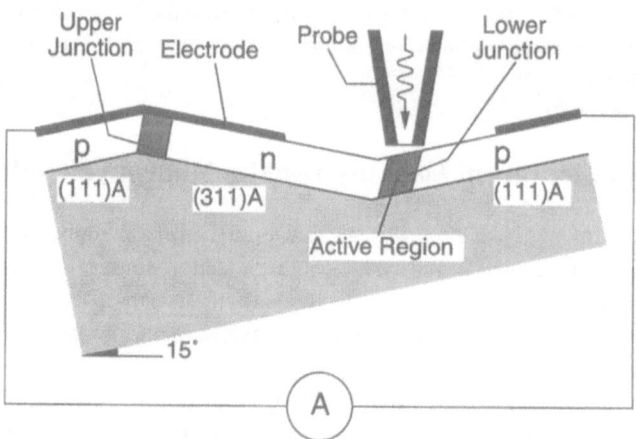

Figure 6.14. Schematic explanation of the cross-sectional structure of the lateral $p-n$ junction and diagram of the experimental geometry. The sample is tilted by 15 deg in order to avoid contact with the cladding of the fiber for the fiber probe.

The propagating modes of the excitation light cannot be neglected in the case when d_f is larger than about 100 nm. The evanescent mode distribution in the tangential wavevector $k_\parallel$ (the component parallel to the surface) space is determined by the foot radius $d_f/2$.[*] Ünlü *et al.* concluded that the cutoff wavevector of the evanescent mode distribution lies at π/d_f [16]. In free space, the evanescent modes are dominant in the region $k_\parallel > 2\pi/\lambda$ if $d_f < \lambda/2$. When the probe is close to an optically dense material, part of the evanescent mode present in the region $k_\parallel < n(2\pi/\lambda)$ (n is the refractive index of the material) is coupled into propagation modes in the material. With the current probe ($d_f = 200$ nm), all the evanescent modes are coupled into propagating modes in GaAs ($n = 3.5$) because the relation $2/d_f < n(2\pi/\lambda)$ holds over the entire wavelength range ($\lambda = 488$–830 nm). The beam spread angle ø of the light propagating in GaAs (cf. Fig. 6.17) is determined by the foot diameter. Although in the present experiment the resolution is limited not only by the probe size but also by the penetration depth, it will be demonstrated in the following paragraph that wavelength-dependent imaging with propagating light can make a large contribution to the "tomographic" analysis of the internal structure of the p–n surface.

Near-field photocurrent images at excitation wavelengths of 488 and 830 nm are shown with a shear-force topographic image in Fig. 6.15. Uniformity of photocurrent intensity is seen along the p–n active region. The full-width at half-maximum of the photocurrent signal profile is 0.6 μm at $\lambda = 488$ nm and 1.7 μm at $\lambda = 830$ nm. The increase in penetration depth with longer wavelength leads to a decrease in the spatial resolution. Figure 6.16 shows the cross-sectional profiles of photocurrent intensities. The scale is logarithmic. At the excitation wavelength of 488 nm, due to the shallow penetration depth (80 nm), resolution is determined by the probe size and the diffusion length of photoexcited carriers. The ratio of the diffusion length of the electrons to that of the holes is estimated to be about 5 (the diffusion coefficients for electrons and holes are approximately $D_e = 220$ cm^2/sec and $D_h = 10$ cm^2/sec, respectively). This large difference can be seen in the asymmetric signal profile [15] due to the slower rise of photocurrent as one approaches the junction from the p side. With increasing penetration depth (longer excitation wavelength), this asymmetry reverses. This is due to the slanted geometry of the p–n interface (shown as θ in Fig. 6.17).

The asymmetric signal behavior is analyzed using a one-dimensional model (Fig. 6.17). For this analysis, the fitting parameters are the slant angle θ of the p–n interface and the beam spread angle ø. The spatial response profile of the p–n active region can be expressed as

---

[*]The foot diameter of the present probe (see Fig. 6.7b) defined by Fig. 3.2c is nearly equivalent to the aperture diameter of the apertured probe fabricated by pulling a heated fiber. See also the footnote in Section 4.1.5.

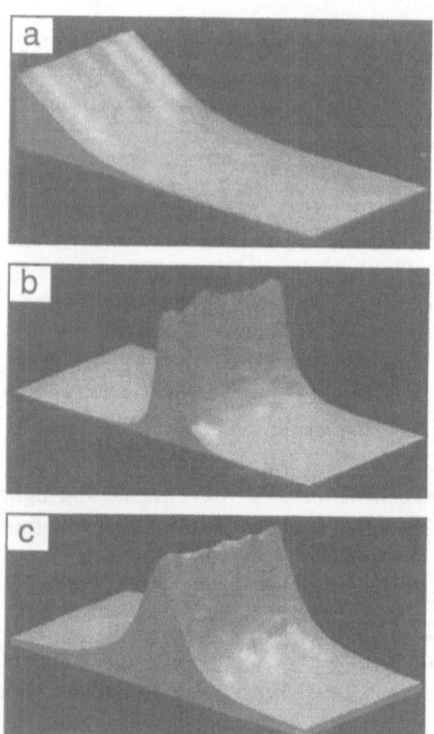

Figure 6.15. Perspective views of the shear-force topographic image in the vicinity of the lower junction (a), and the near-field photocurrent images at the excitation wavelength of 488 nm (b) and 830 nm (c). Scan area is $5 \times 10\ \mu m^2$. The height of the slope in panel (a) is 1 μm.

$$F(x, z) = \exp(x/L_n) \qquad x < -z \tan \theta$$

$$= \exp(-x/L_p) \qquad x > -z \tan\theta \qquad (6.2.1)$$

where L_n and L_p are the decay lengths. For $\lambda = 488$ nm (curve A in Fig. 6.16), the shallow penetration depth means that the measured intensity is directly proportional to the spatial response, allowing the direct determination of the characteristic decay lengths due to carrier diffusion. Thus, these were determined from curve A in Fig. 6.16 to be $L_n = 400$ nm and $L_p = 520$ nm. For $\lambda = 633, 780$, and 830 nm (curves B–D in Fig. 6.16), whose penetration depths are larger than their wavelengths in GaAs, the relation between the spatial response function and measured intensity is more complicated, as described below.

For the angular distribution of propagating light into GaAs, a Gaussian profile is assumed:

$$D(x, z) = \exp[-(x - x_p)^2/(z \tan \phi)^2]/z \qquad (6.6.2)$$

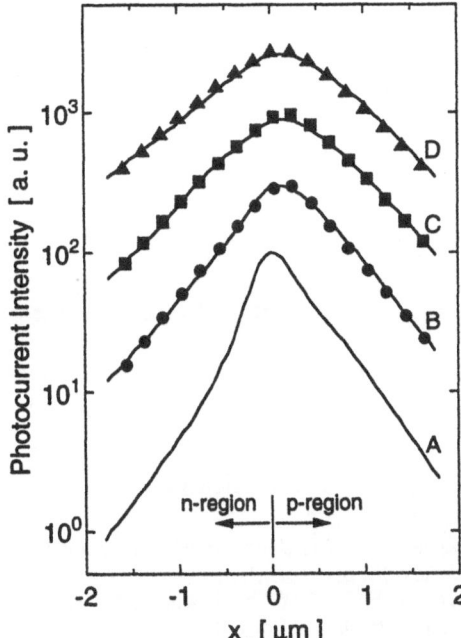

Figure 6.16. Cross-sectional profiles of near-field photocurrent signals as a function of excitation wavelength. (A) 488 nm, (B) 633 nm, (C) 780 nm, (D) 830 nm. Closed circles, squares, and triangles represent the calculated results of Eq. (6.2.4) with fitting parameters $\theta = 15$ deg and $\emptyset = 27, 37$, and 40 deg, respectively.

where x_p indicates the position of the probe and $\emptyset$ is a function of the wavelength. The effect of a wavelength-dependent penetration depth is given via

$$A(x,z) = \exp(-l/L_{pd}) \tag{6.2.3}$$

where $l^2 = x^2 + z^2$ and L_{pd} (= 0.25 μm at 633 nm, 0.65 μm at 780 nm, and 1.0 μm at 830 nm) is the penetration depth at the excitation wavelength. The intensity of the photocurrent signal is

$$I(x_p) = \int_0^d \int_{-\infty}^\infty D(x, z)A(x, z)F(x, z)\, dx\, dz \tag{6.2.4}$$

where $d = 1$ μm is the depth of the p–n interface. The experimental curves B–D in Fig. 6.16 are perfectly fitted by Eq. (6.2.4). From this calculation, one obtains the slant angle of the p–n interface $\theta = 15 \pm 8$ deg and the beam spread angle $\emptyset = 40 \pm 8$ deg for excitation at 830 nm. This angle $\emptyset$ determines the tangential wavevector of the propagating light, $n(2\pi/\lambda) \sin \emptyset = 0.017$ nm^{-1}, in good agreement with $\pi/d_f = 0.016$ nm^{-1}. This result implies that, as concluded by Ünlü $et\ al.$, the cutoff wavevector of the evanescent mode distribution lies at π/d_f.

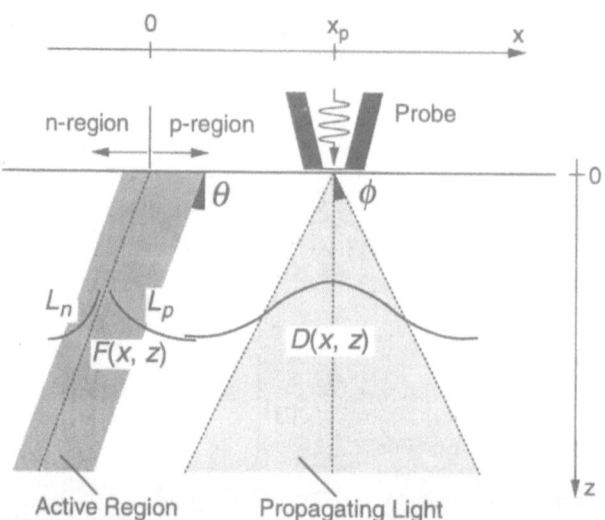

Figure 6.17. The slanted p–n interface and the propagating light into GaAs. The response profile of the p–n region is assumed to be exponential with decay lengths L_n and L_p determined by curve A in Fig. 6.16. The slant angle θ and the beam spread angle ϕ are the fitting parameters in the calculation of Eq. (6.2.4).

The total slant of the p–n interface is 30 ± 8 deg on the p side. This is the sum of the observed slant angle θ of 15 ± 8 deg and the intended tilt angle of 15 deg in the experimental setup. It can be explained by the crystal orientation dependence of the growth process [18]. The most significant factor is that the growth rate of GaAs on (311)A is faster than that on (111)A, which causes a shift of the n-type region toward the p side during growth.

Although other factors also can be examined using the experimental results from PL and EL measurements, the "tomographic" diagnostics with the near-field photocurrent measurement will provide new insight in understanding the crystal growth mechanism.

6.3. PHOTOLUMINESCENCE SPECTROSCOPY OF A SEMICONDUCTOR QUANTUM DOT

The long diffusion length of carriers results in a limited spatial resolution of near-field photoluminescence (PL) spectroscopy for conventional semiconductor materials and devices, including the surface-emitting device described in Section 6.2. This problem can be solved by confining carriers in a nanometric structure such as a quantum dot (QD). This quantum device structure enables us to investigate the attractive phenomenon of carriers in the sharp density distribution of energy states. Recently, individual spectroscopic characteristics of localized electrons in low-dimensional quantum structures have been reported using the near-field technique [19–21]. This section reviews spatially and spectrally resolved I-mode imaging and spectroscopy of GaAs QD structures [22].

For this work, the I-mode NOM is installed in a He-flow type optical cryostat for low-temperature operation (cf. Fig. 3.14). Figure 6.18 shows a cross-sectional view of the probe and GaAs QD installed in the cryostat. The probe used here is equivalent to the one shown in Fig. 6.7b. The sample is held by a tube-type PZT actuator whose three-dimensional scanning range at 10 K decreases to 10% of its value (80 μm × 80 μm × 5 μm) at room temperature. Shear force is employed to control the probe position. A multiline Ar^+ laser is used to excite carriers in the AlGaAs barriers. For shear-force monitoring, a 1.3-μm laser diode is used to avoid excitation of unwanted carriers in GaAs. In order to obtain the PL spectrum, the PL light is collected by a lens in the cryostat and focused onto a photon-counting Si avalanche photodiode in conjunction with a 20-cm-long grating monochromator.

A two-dimensional array of GaAs QDs is fabricated by selective epitaxial growth on a SiO_2-patterned GaAs (100) substrate using metal organic chemical vapor deposition [23]. The difference of the growth rate between the crystal axes gives rise to three-dimensional confinement of the carriers. A scanning electron micrograph of the sample provides information on the size of the QD (190 nm × 160 nm × 12 nm) and the separation between adjacent QDs (2 μm). The effective lateral size of the QD is estimated to be 70 nm by magneto-PL measurements. Although carriers are excited over a localized area in the present experimental configuration, the PL light comes from an extended region due to carrier diffusion

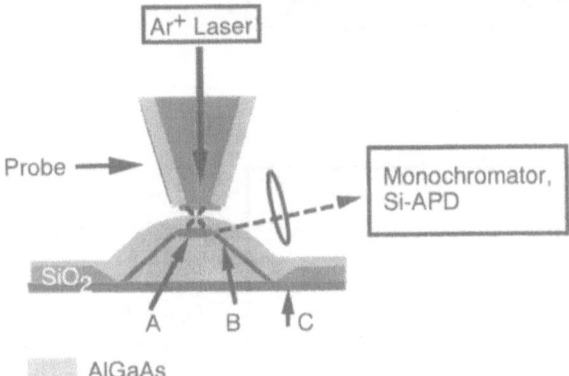

Figure 6.18. Cross-sectional structure of a quantum dot and experimental setup of PL spectroscopy using an I-mode NOM in a cryostat. (A) GaAs QD, (B) GaAs QW, (C) GaAs bulk.

into the AlGaAs barrier. In order to reduce this carrier diffusion into the barrier, the barrier thickness is decreased to 50 nm.

Figure 6.19 shows the spatially resolved PL spectra at a temperature of 18 K. Figure 6.19a shows the PL spectrum obtained by fixing the probe 200 nm above the QD. It provides the far-field PL spectrum originating from the carriers excited not only in the QD region, but also in the quantum well (QW) and substrate regions. Figures 6.19b and 6.19c are obtained by fixing the probe less than 20 nm above the QD region and at the SiO_2 mask region, respectively. They provide the near-field PL spectra. Figures 6.20a–6.20c show the spatially resolved PL images obtained

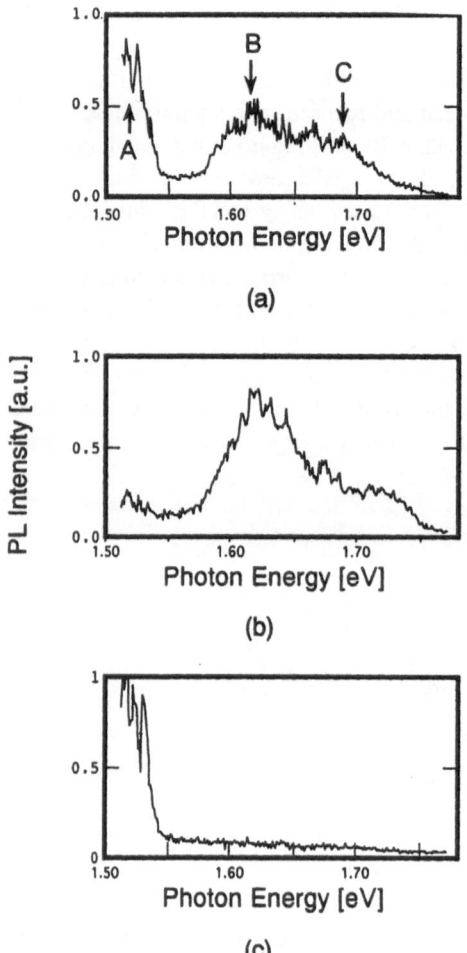

Figure 6.19. Spatially resolved PL spectra at a temperature of 18 K obtained by fixing the probe (a) 200 nm above the QD, (b) less than 10 nm above the QD, and (c) less than 10 nm above the SiO_2 mask.

at the positions marked by arrows A–C in Fig. 6.19a. In order to map the luminescent region, the topographic image is taken simultaneously by monitoring the shear force and is shown in Fig. 6.20d.

The PL signals at energies A–C in Fig. 6.19a originate from the GaAs bulk, the GaAs QD, and the GaAs QW, respectively. The photon energy at the PL peaks, the position dependence of the PL spectra, and the optical images all confirm this. Since the QW is thinner than the QD in this sample, the energy shift in the QD is smaller than that of the QW. In Fig. 6.20c, the spatial distribution of the PL intensity from the QW is asymmetrical with respect to the center of the QD structure. This is because only a part of the PL originating from the QW reaches to the collecting lens due to the pyramidal shape of the observed region of the sample.

It should be noted that Figs. 6.19b and 6.19c do not have any spectral components originating from the GaAs bulk or QD/QW, respectively. Further, Fig. 6.19a shows that the PL intensity from the QD is as large as that from the GaAs

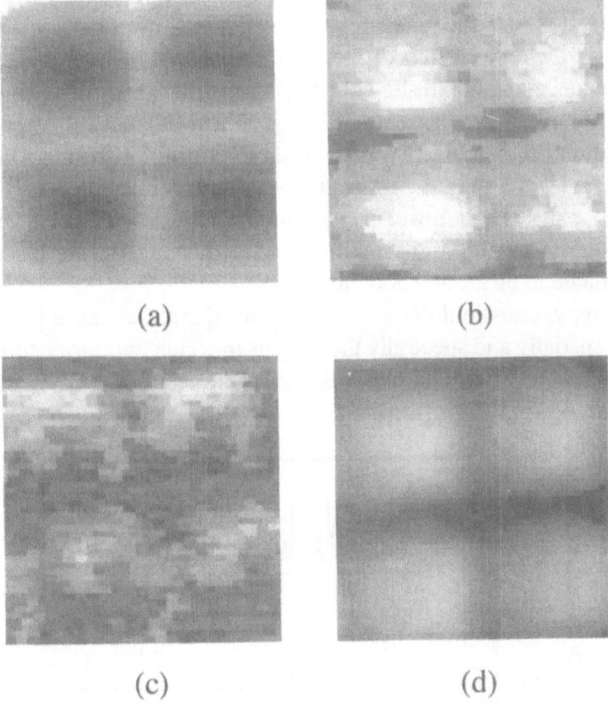

(a) (b)

(c) (d)

Figure 6.20. Spatial profiles of spectrally resolved PL intensity images at 18 K for corresponding positions marked by arrows (a) A, (b) B, and (c) C in Fig. 6.19a. The excitation power is 50 nW. Scan area is 3.5×3.5 μm^2. (d) Simultaneously observed shear-force topographic image.

bulk. These results mean that the carriers diffuse and are captured effectively in the QD and the QW regions after they are excited in the barrier region. Therefore, the sample is free from the bottleneck problem [24], and the carriers diffuse from the outer region to the QD region, or a giant oscillator strength is realized due to the three-dimensionally confined structure of the sample.

Figure 6.21 shows the PL spectrum obtained under a lower excitation power with the probe above the top of a QD. There exist a number of sharp spectral lines with linewidths as narrow as a few millielectron volts. These lines are reproducible and thus confirmed to originate from the individual QDs. Since the present sample has a disk-shaped dot structure, the carriers are confined along the direction normal to the disk plane. The ensemble of these spectral lines is therefore mainly due to either the inhomogeneity of the QD thickness or the intrinsic properties of the GaAs material.

The carrier diffusion region and the effective excited region of the QD can be monitored by varying the excited light power. Figures 6.22a and 6.22b show the PL intensity images with excitation power of 50 and 5 nW, respectively. Comparing these images, it is seen that an increase in excitation power can expand the region of the QD in which the carriers diffuse and are captured.

Figures 6.22c and 6.22d show the cross-sectional PL intensity profiles across the QD as indicated by the arrow in Figs. 6.22a and 6.22b, respectively. The full-width at half-maximum (FWHM) of these profiles is about 650 and 300 nm, respectively. A single, sharper PL spectral peak (indicated by the arrow in Fig. 6.21) is observed for the case of lower excitation power. From the FWHM of this single peak, the estimated width of the effective region in which the carriers are captured can be estimated to be around 300 nm.

In summary, near-field PL spectroscopy for QD structures at low temperature can resolve spatially and spectrally the regions in which the carriers diffuse and are captured in the QD, QW, and bulk regions. Modification of the carrier diffusion

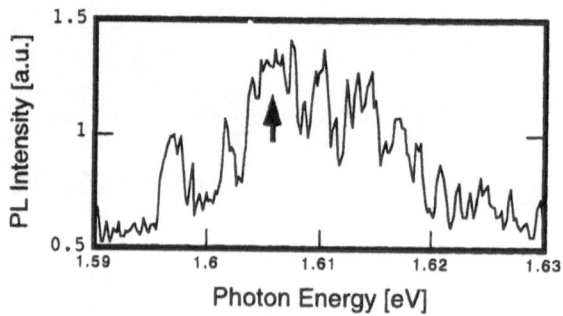

Figure 6.21. Photoluminescence spectra obtained under 5 nW excitation power by fixing the probe less than 10 nm above the QD.

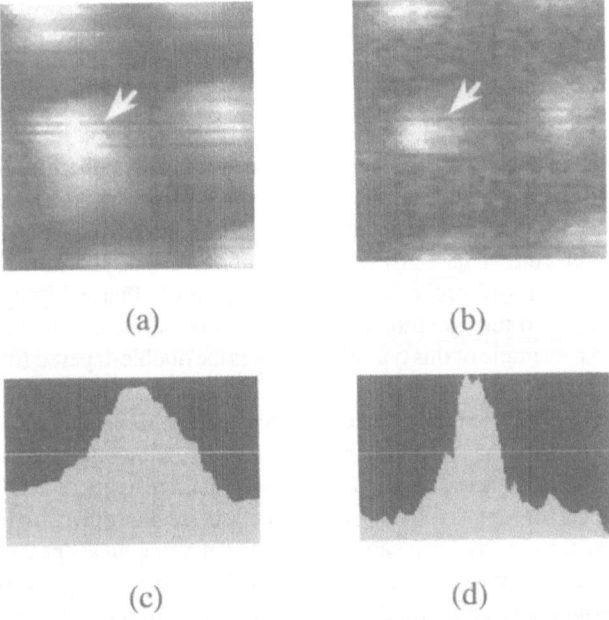

(a) (b)

(c) (d)

Figure 6.22. Photoluminescence intensity images obtained under the excitation power of (a) 50 nW and (b) 5 nW. Scan areas is $2.8 \times 2.8 \ \mu m^2$. (c, d): Cross-sectional PL intensity profiles at the QD indicated by the arrow in panel (a) and (b), respectively. Their full-widths at half-maximum are 650 and 300 nm, respectively.

region can also be observed by varying the excitation power. More information on quantum confinement phenomena and criteria for designing novel light-emitting materials/devices can be expected with further quantitative study of the near-field PL spectroscopy of smaller QDs.

6.4. IMAGING OF OTHER MATERIALS

6.4.1. Fluorescence Detection from Dye Molecules

Fluorescence from dye molecules can be detected by the NOM. The fluorescence from a single dye molecule has been detected by using an I-mode NOM, and the molecular orientation estimated. Irradiation was through a 100-nm-diameter aperture [25]. The quenching process of this fluorescence and the reversible variation of fluorescence intensity with a time interval of several tens of seconds were also measured. Furthermore, the effect of a metallic coating around the aperture on the fluorescence lifetime was also studied on a nanosecond time scale [26, 27].

It is expected that the fluorescence spectroscopy of a single dye molecule by the NOM described above will be widely applied in areas such as labeling biological samples. This is because the fluorescence spectral profiles, the lifetimes, and the quantum efficiencies of dye molecules depend strongly on their surroundings. However, the principal problem associated with fluorescence detection from a single dye molecule is its sensitivity. A solution to this problem could be the development of a novel probe for an I-mode operation that is highly efficient in guiding the light to the aperture while suppressing the heat induced by thermal dissipation. An example of this type of solution is the double-tapered fiber described in Section 4.1.5.

In the case of fluorescence detection from a single molecule, since the role of the probe is to reduce the contribution of fluorescence from other molecules, a very high spatial resolution is not required if dye molecules are dispersed on the substrate with sufficiently low density. In fact, real-time fluorescence detection from a single dye molecule has already been carried out in water using an evanescent field on a planar substrate as a pumping source to suppress fluorescence quenching [28].

As an example of applying the fluorescence detection technique, the spectral properties of sensitizing dyes deposited on nanometric silver halide tabular grains were studied. Silver halide has been used as the basic material for photographic film, on which sensitizing dyes are deposited in order to enhance spectral sensitivity. Thus, the detailed analysis of the spatial distribution and spectral properties of sensitizing dyes on silver halide grains is important for realizing higher spectral sensitivity. Conventional analytical methods have employed far-field absorption or reflection spectroscopy of emulsions of submicrometer silver halide grains. Unfortunately, their diffraction-limited spatial resolution has not provided information on the spatial distribution and spectral properties of sensitizing dyes adsorbed to individual silver halide grains. In order to obtain this information, a method using the NOM can be employed. Several preliminary experiments have been carried out [29].

An I-mode NOM is used and the sample is excited by light from an Ar^+ laser ($\lambda = 488$ nm) through the fiber probe. The fiber probe was an Au-film-coated, double-tapered fiber with a 300 nm foot diameter (Section 4.1.5). Its profile is equivalent to that of Fig. 6.7b except that the protruded tip part is longer and the apex of the tip is sharper.

Fluorescence emitted from the dye is collected at the rear side of the substrate (for high collection efficiency) and detected by a photon-counting avalanche photodiode (APD). The shear-force technique is employed to maintain constant sample–probe separation.

After the pseudoisocyanine sensitizing dyes (fluorescence at $\lambda = 590$ nm) are adsorbed onto the AgBr tabular grains, they are dispersed on a mica plate. The

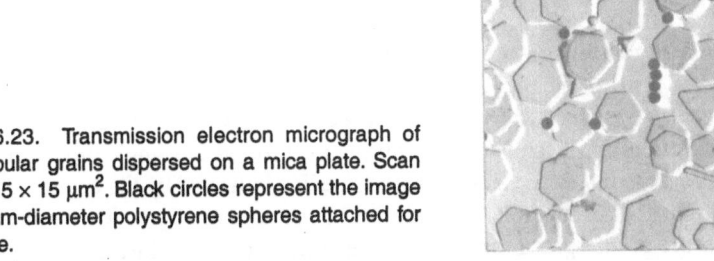

Figure 6.23. Transmission electron micrograph of AgBr tabular grains dispersed on a mica plate. Scan area is 15×15 μm^2. Black circles represent the image of 500-nm-diameter polystyrene spheres attached for reference.

density of adsorbed dye molecules covering the surface of the AgBr grain is as high as 60%. Figure 6.23 shows a scanning electron micrograph of the AgBr grains dispersed on a mica plate specially prepared for the experimental demonstration. In this figure, the hexagonal structures seen either as singly isolated or as super-posed one over the other correspond to AgBr grains, and the dark spots correspond to polystyrene spheres (diameter 500 nm) attached for reference. AgBr tabular grains are typically 100 nm in height and 500–1000 nm in cross-sectional area.

Figure 6.24 shows a simultaneously measured (a) shear-force topographic image of the AgBr grains and (b) the corresponding spatial distribution of fluores-

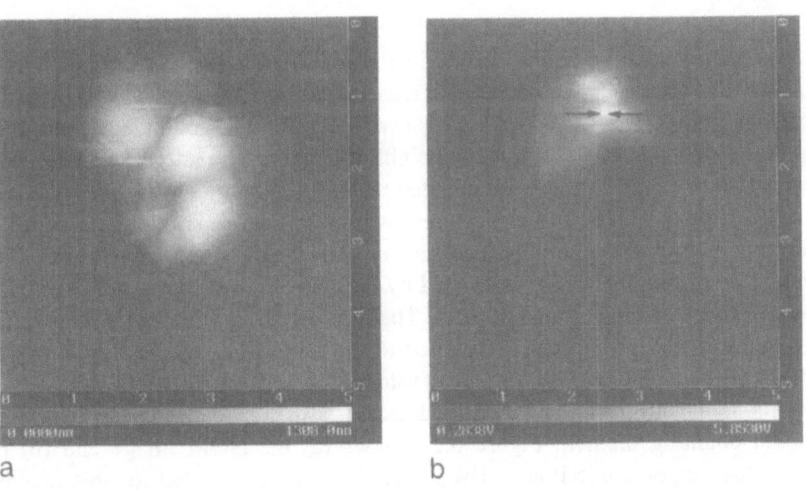

a   b

Figure 6.24. (a) Shear-force topographic image of AgBr tabular grains. Scan area is 4×4 μm^2. (b) Spatial distribution of the fluorescence intensity from the dyes adsorbed onto the AgBr tabular grains, as measured by the I-mode NOM. Scan area is 4×4 μm^2. Size of the bright spot as indicated by the arrows is 50 nm.

cence intensity. In Fig. 6.24a, more than two AgBr tabular grains are superposed over one another. The maximum counting rate of fluorescent photons was 1.2×10^5 sec^{-1} for obtaining the image of Fig. 6.24b, a sufficiently high S/N ratio. Comparing Fig. 6.24b with Fig. 6.24a, it can be seen that the fluorescence is emitted mainly from the rim of the AgBr tabular grains. This spectral inhomogeneity in the fluorescence intensity distribution is because the dyes are preferably adsorbed to the rim of the AgBr tabular grains during the process of sample preparation.

From the intensity variation across the line indicated by the arrows, the size of the bright spot in Fig. 6.24b is estimated to be around 50 nm. This value can give a measure of the spatial resolution of this measurement, which is smaller than the foot diameter (< 200 nm). This is attributed to the selective excitation of sensitizing dyes by the high-spatial-Fourier components of the evanescent field of the Ar^+ laser localized in the apex region of the protruded probe. In other words, this high spatial resolution is due to the size-dependent decay characteristics of the evanescent field intensity shown in Section 5.1.1. Although this feature was observed as a boundary effect in Section 5.2.2.2, it is worth noting that this kind of size-dependent decay is also observed in the case of the detection of the fluorescent evanescent field from excited dyes.

Since the signal-to-noise ratio of the present experiment is sufficiently high, higher spatial resolution can be expected by using a smaller probe. Further, design criteria for novel photographic films could be obtained if the detailed adsorption characteristics of sensitizing dyes were measured with commercially used AgBr tabular grains.

A second example is a C-mode NOM imaging experiment using the fluorescence-emitting probe shown in Fig. 4.32 [30]. For evaluating the performance of this probe, a compact disk is used as a reference sample. It has a grating structure period of 2 μm and a moth's-eye pattern on the grating structure. Each eye has a diameter of 0.3 μm and a height of 0.1 μm. The interval between each eye is 0.1 μm. The sample is illuminated under total internal reflection by an Ar^+ laser ($\lambda = 488$ nm). The probe is scanned across the sample surface by controlling the sample–probe separation with shear-force feedback. Rhodamine 6G attached on the top of the fiber probe is excited by the 488-nm evanescent field and emits fluorescence centered around 600 nm. The fluorescent light emitted from the dye at the top of the fiber probe is coupled to the same fiber and passed through a holographic notch filter (488-nm cutoff filter with 1×10^6 attenuation), and a 30-cm grating monochromator and finally detected with a photomultiplier using the photon counting method. Figure 6.25 shows (a) the NOM image and (b) the shear-force topographic image. Even though the time required for this imaging experiment is as long as 20 min, decay in fluorescence-emission efficiency is not observed, from which the fairly long lifetime of this probe is confirmed.

Compared with the shear-force topographic image in Fig. 6.25b which gives information on topography, the NOM image in Fig. 6.25a has low contrast for the

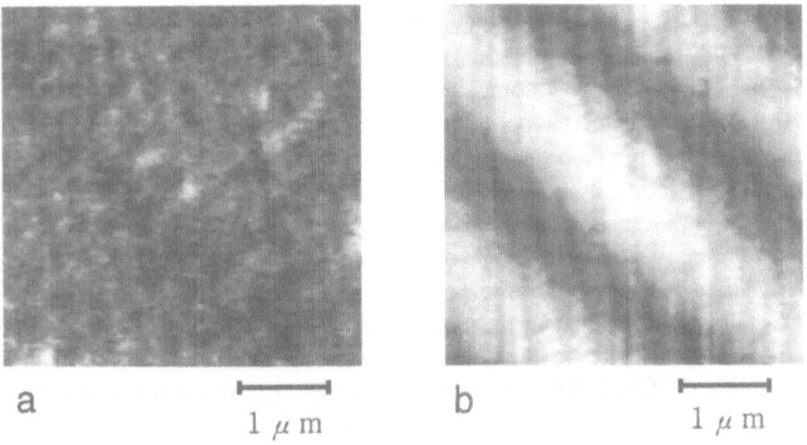

a   1 μ m   b   1 μ m

Figure 6.25. Images of a compact disk due (a) to C-mode NOM and (b) shear force with a fluorescence-emitting probe.

grating structure of 2 μm period, but high contrast for the moth's-eye pattern. Spatial power spectral analysis was applied to these images to extract the smallest structure of the interval to be 0.1 μm between each eye. The results show good agreement with the compact disk structure.

The success of this imaging experiment can be attributed mainly to the wavelength-conversion method, which realizes a very high sensitivity by rejecting the background light from the Ar^+ laser. Higher spatial resolution and sensitivity can be expected from further improvements in the performance of the fluorescence-emitting probe.

6.4.2. Spectroscopy of Solid-State Materials

The emission spectrum of ruby under mechanical stress [31] and the Raman spectrum of diamond [32] have been observed. Furthermore, detection of surface plasmon propagation on a metallic film [33] and improvement of resolution by utilizing the surface plasmon [34] have also been reported.

A surface plasmon is a quantized oscillation of an electron on a planar surface of a metallic film. Its wavelength is shorter than that of the light used for excitation. When excited by a resonant interaction with an evanescent field, the light power can be concentrated in a limited area in the proximity of the surface. Further, since the surface plasmon exhibits sharp resonance characteristics depending on the shape and topography of the materials, the three-dimensional nanometric probe used for the NOM can excite a completely different type of resonant mode of surface plasmon, considerably different from that on a planar surface of a metallic film.

By exciting a surface plasmon on a patterned planar surface of a metallic film, a short-wavelength interference pattern can be generated between the incident and reflected surface plasmons at the edge of the pattern. This interference can be used to realize passive nanometric photonic devices such as an optical switch (a Mach–Zehnder interferometer type or a resonator type) or an optical directional coupler/divider.

Preliminary studies investigating the interference of surface plasmons have been carried out [35]. Figure 6.26a is the interference pattern formed between surface plasmons incident and reflected from the edge of a silver film as measured by C-mode NOM. The Ag film is coated on a glass plate with a thickness of 50 nm, and a 6-nm Ge film is fixed between the glass and Ag film to decrease the grain size of Ag. The edge of the Ag film is formed by masking during the stage of Ag film evaporation. Figure 6.26b shows the cross-sectional profile of the interference pattern. These results show that the surface plasmon propagates on the Ag surface for over 17 μm, as long as 28 times that of the plasmon's wavelength. Furthermore, the reflectivity for the surface plasmon at the edge of the Ag film is estimated from these results to be as high as 25%. Such a long propagation length and high reflectivity have not been realized without fixing a Ge film; the effectiveness of the Ge film is confirmed. From the results of these experiments, it is expected that a carefully evaporated Ag film can be used as a low-loss waveguide for a surface plasmon, to be applied to nanometric photonic devices in the future.

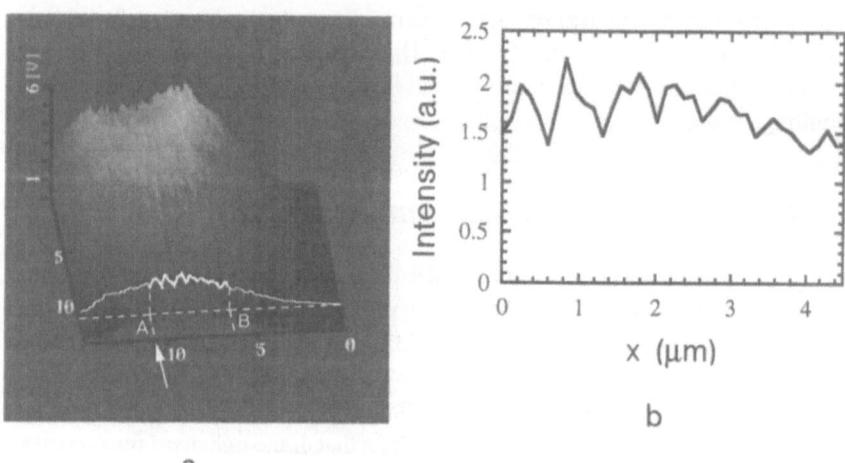

Figure 6.26. (a) Image of a surface plasmon taken by a C-mode NOM at the edge of a planar surface of an Ag film. Scan area is 15 × 15 μm². Arrow represents the position of the edge of the Ag film. (b) Cross-sectional profile of the surface plasmon intensity plotted along the line A–B of panel (a).

6.5. REFERENCES

1. A. Miki, Y. Okamura, and S. Yamamoto, Optical waveguide directional coupler measurements using a microcomputer-assisted TV camera system, *J. Lightwave Technol.* 7: 1912–1918 (1989).

2. Y. Toda and M. Ohtsu, High spatial resolution diagnostics of optical waveguides using a photon-scanning tunneling microscope, *IEEE Photon. Technol. Lett.* 7: 84–86 (1995).

3. K. Tada, T. Nakabayashi, T. Iwashima, and T. Ishikawa, Fabrication of $LiTaO_3$ optical waveguide by H^+ exchange method, *Jpn. J. Appl. Phys.* 26: 503–504 (1987).

4. I. P. Kaminov and L. W. Stulz, Loss in cleaved Ti-diffused $LiNbO_3$ waveguides, *Appl. Phys. Lett.* 33: 62–64 (1978).

5. W. P. Huang and C. L. Xu, Simulation of three-dimensional optical waveguides by a full-vector beam propagation method, *IEEE J. Quantum Electron.* 29: 2639–2649 (1993).

6. T. Saiki, S. Mononobe, M. Ohtsu, N. Saito, and J. Kusano, Spatially-resolved photoluminescence spectroscopy of lateral $p–n$ junctions prepared by Si-doped GaAs using a photon scanning tunneling microscope, *Appl. Phys. Lett.* 67: 2191–2193 (1995).

7. D. Bimberg, J. Christen, A. Steckenbom, G. Weimann, and W. Schlapp, Injection, intersubband relaxation and recombination in GaAs multiple quantum wells, *J. Luminescence* 30: 562–579 (1985).

8. J. M. Ballingall and C. E. C. Wood, Crystal orientation dependence of silicon autocompensation in molecular beam epitaxial gallium arsenide, *Appl. Phys. Lett.* 41: 947–949 (1982).

9. W. I. Wang, E. E. Mendez, T. S. Kuan, and L. Esaki, Crystal orientation dependence of silicon doping in molecular beam epitaxial AlGaAs/GaAs heterostructures, *Appl. Phys. Lett.* 47: 826–828 (1985).

10. N. Saito, M. Yamada, F. Sato, I. Fujimoto, M. Inai, T. Yamamoto, and T. Watanabe, Light emission from lateral $p–n$ junctions on patterned GaAs(111)A substrates, in *Proceedings 1993 International Symposium on GaAs and Related Compounds, Freiburg,* Institute of Physics Conference Series, Vol. 136, 1994, pp. 601–606.

11. M. Fujii, T. Yamamoto, M. Shigeta, T. Takebe, K. Kobayashi, S. Hiyamizu, and I. Fujimoto, Lateral $p–n$ junctions on GaAs(111)A substrates patterned with equilateral triangles, *Surface Sci.* 267: 26–28 (1992).

12. Y. Okano, M. Shigeta, H. Seto, H. Katahama, S. Nishine, and I. Fujimoto, Incorporation behavior of Si atoms in the molecular beam epitaxial growth of GaAs on misoriented (111)A substrates, *Jpn. J. Appl. Phys.* 29: L1357–L1359 (1990).

13. M. Inai, T. Yamamoto, M. Fujii, T. Takebe, and K. Kobayashi, Electrical characterization of lateral $p–n$ junctions grown on (111)A GaAs nonplanar substrates by molecular beam epitaxy, *Jpn. J. Appl. Phys.* 32: 523–527 (1993).

14. T. Saiki, N. Saito, J. Kusano, and M. Ohtsu, Determination of slant angle of $p–n$ interface by multiwavelength near-field photocurrent measurement, *Appl. Phys. Lett.* 69: 644–646 (1996).

15. S. K. Buratto, J. W. P. Hsu, E. Betzig, J. K. Trautman, R. B. Bylsma, C. C. Bahr, and M. J. Cardillo, Near-field photoconductivity: Application to carrier transport in InGaAsP quantum well lasers, *Appl. Phys. Lett.* 65: 2654–2656 (1994).

16. M. S. Ünlü, B. B. Goldberg, W. D. Herzog, D. Sun, and E. Towe, Near-field optical beam induced current measurements on heterostructures, *Appl. Phys. Lett.* 67: 1862–1864 (1995).

17. H. J. Leamy, Charge collection scanning electron micrography, *J. Appl. Phys.* 53: R51–R80 (1982).

18. T. Nishinaga, K. Mochizuki, H. Yoshinaga, C. Sasaoka, and M. Washiyama, Growth induced compositional non-uniformity in (Ga,Al) As and thermodynamics analysis, *J. Crystal Growth* 98: 98–107 (1989).

19. R. D. Grober, T. D. Harris, J. K. Trautman, E. Betzig, W. Wegscheider, L. Pfeiffer, and K. West, Optical spectroscopy of a GaAs/AlGaAs quantum wire structure using near-field scanning optical microscopy, *Appl. Phys. Lett.* **64**: 1421–1423 (1994).

20. H. F. Hess, E. Betzig, T. D. Harris, L. Pfeiffer, and K. West, Near-field spectroscopy of the quantum constituents of a luminescent system, *Science* **264**: 1740–1745 (1994).

21. U. Mohideen, M. J. Yoo, H. Hess, W. S. Hobson, F. Ren, R. Kopf, and R. E. Slusher, GaAs/AlGaAs quantum-dot near-field scanning opticalmicroscopy, *Tech. Digest Quantum Electron. Laser Sci.* **16**: 85 (1995).

22. Y. Toda, M. Kourogi, M. Ohtsu, Y. Nagamune, and Y. Arakawa, Spatially and spectrally resolved imaging of GaAs quantum-dot structures using near-field optical technique, *Appl. Phys. Lett.* **69**: 827–829 (1996).

23. Y. Nagamune, S. Tsukamoto, M. Nishioka, and Y. Arakawa, Growth process and mechanism of nanometer-scale GaAs dot-structure using MOCVD selective growth, *J. Crystal Growth* **126**: 707–717 (1993).

24. P. D. Wang, C. M. S. Torres, H. Benisty, C. Weisbush, and S. P. Beaumont, Radiative recombination in GaAs–Al$_x$Ga$_{1-x}$As quantum dots, *Appl. Phys. Lett.* **61**: 946–948 (1992).

25. J. K. Trautman, J. J. Macklin, L. E. Brus, and E. Betzig, Near-field spectroscopy of single molecules at room temperature, *Nature* **369**: 40–42 (1994).

26. W. P. Ambrone, P. M. Goodwin, J. C. Martin, and R. A. Keller, Single molecule detection and photochemistry on a surface using near-field optical excitation, *Phys. Rev. Lett.* **72**: 160–163 (1994).

27. X. S. Xie and R. C. Dunn, Probing single molecule dynamics, *Science* **265**: 361–364 (1994).

28. T. Funatsu, Y. Harada, M. Tokunaga, K. Saito, and T. Yanagida, Imaging of single fluorescent molecules and individual ATP turnovers by single myosin molecules in aqueous solution, *Nature* **374**: 555–559 (1995).

29. J. K. Rogers, R. Toledo-Crow, M. Vaez-Iravani, G. Di Francesco, T. Zhao, and R. Hallstone, Correlative near-field optical direct/fluorescence imaging and spectroscopy of a sensitizing dye on single microcrystals of silver halide, *J. Imaging Sci. Technol.* **39**: 205–209 (1995).

30. K. Kurihara, K. Watanabe, and M. Ohtsu, Photon scanning tunneling microscopy with light-emitting probes, in *Conference Proceedings of the Eleventh International Conference on Optical Fiber Sensors, Sapporo, May 1996*, pp. 694–697.

31. P. J. Moyer, C. L. Jahncke, M. A. Paesler, R. C. Reddick, and R. J. Warmack, Spectroscopy in the evanescent field with an analytical photon scanning tunneling microscope, *Phys. Lett. A* **145**: 343–347 (1990).

32. D. P. Tsai, A. Othonos, M. Woskwits, and D. Uttamachandani, Raman spectroscopy using a fiber optic probe with subwavelength aperture, *Appl. Phys. Lett.* **64**: 1768–1770 (1994).

33. P. Dawson, F. de Fornel, and J.-P. Goudonnet, Imaging of surface plasmon propagation and edge interaction using a photon scanning tunneling microscope, *Phys. Rev. Lett.* **72**: 2927–2930 (1994).

34. M. Specht, J. D. Pedaring, W. M. Heckl, and T. W. Hänsch, Scanning plasmon near-field microscope, *Phys. Rev. Lett.* **68**: 476–479 (1992).

35. M. Ashino, Kanagawa Academy of Science and Technology, unpublished work.

SEVEN

FABRICATION AND MANIPULATION

7.1. FABRICATION OF PHOTONIC DEVICES

Nonlinear optical phenomena can be induced by utilizing the high optical energy density of the evanescent field, which can lead to the processing of nanometric areas of material surfaces, the fabrication of nanometric optical functional devices, and nanometric photolithography. As an example of realizing photonic devices with nanometric dimensions, experiments on high-density optical storage using the I-mode NOM have been carried out.

Due to the rapid improvement of storage technology, the storage densities of optical and magnetic memories have reached about 1 Gb/inch^2 and 500 Mb/inch^2, respectively. Although it should be possible to increase optical storage density to several tens of Gb/inch^2 within the next two decades, it may not be possible to realize a density higher than this value using conventional methods because of the diffraction limit of the propagating light. In order to realize higher storage densities, several methods using an atomic force microscope [1] or solid immersion lens [2] have been proposed. Use of an NOM probe offers an alternative and promising possibility to surpass through this limit.

Thermal-mode optical storage has been carried out for a Pt/Co magnetooptical storage medium by heating its surface to a temperature higher than the Curie point (400°C) with the evanescent field from the apertured probe of the I-mode NOM [3]. A pit diameter of 60 nm was obtained, corresponding to a storage density of 45 Gb/inch^2 (about 20 times that of the conventional optical memory). In order to reduce the size of the experimental system, the Ar$^+$ laser has since been replaced by a semiconductor laser [4].

In these experiments, the aperture diameter was larger than the grain size of the storage medium. Recalling that the highest energy transfer efficiency is achieved when these sizes are equal (refer to Fig. 2.7 demonstrating the resonance phenomenon with respect to the sizes of the probe and sample), reducing the aperture diameter to the grain size will result in a significant improvement in storage speed and readout sensitivity.

A phase-change medium is better for thermal-mode optical storage because of the lower temperature (200°C) required for the phase transition between the crystalline and amorphous states. A GeSbTe film has been used as the storage medium, with the local change of its reflectivity as the readout signal [5]. A stored pit diameter as small as that of the magnetooptical storage has been realized.

Photon-mode optical storage has also been carried out utilizing the local photochemical reaction of an amphiphilic azobenzene derivative with the evanescent field [6]. Figure 7.1a shows that the *trans* isomer of this photochromic molecule is transformed into the *cis* isomer under ultraviolet irradiation and this photochemical transformation corresponds to storing a memory. This memory is erasable because the *cis* isomer can be transformed back to the *trans* isomer by heating or irradiating with visible light. Since the optical absorption of the *cis* isomer is about half that of the *trans* isomer at 360 nm wavelength, measurement of the difference in absorption by the NOM corresponds to reading out the memory.

A 180-nm-thick Langmuir–Blodgett (LB) film of this photochromic material was fixed on a glass substrate, and an Ar^+ laser (360 nm wavelength) was used as the light source of the I-mode NOM for storage and readout. The evanescent field power used for the storage was 30 nW. The difference in the evanescent field powers

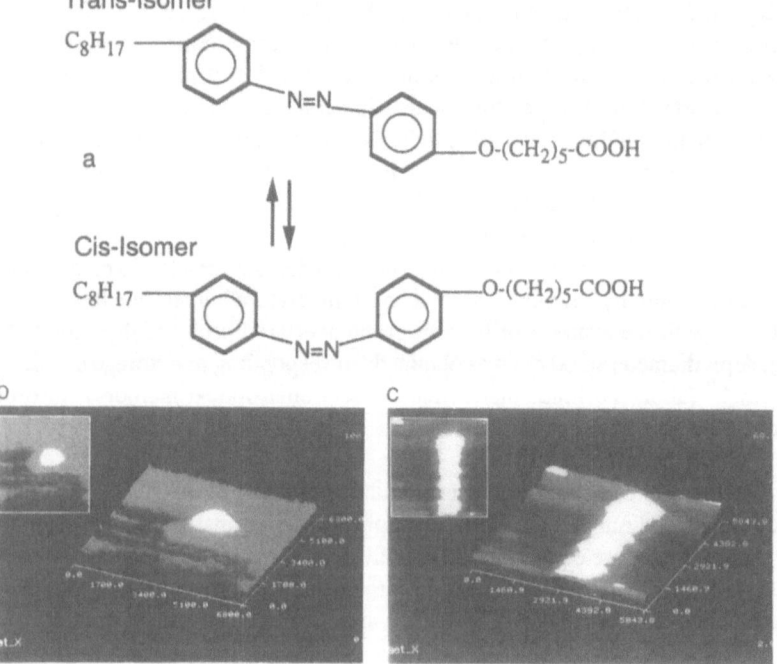

Figure 7.1. (a) Molecular structures of amphiphilic azobenzene derivative. (b) Near-field optical image of a stored circular pit. The circle at the center represents part of the *cis* isomer with a diameter of 50–100 nm. (c) Near-field optical image of stored stripe with a width 50–100 nm.

absorbed by the *trans* and *cis* isomers was 0.3 nW (measured while reading out). Figure 7.1b is the image of a stored circular pit read out by the I-mode NOM. Its diameter falls between 50 and 100 nm, depending on experimental conditions, the same size as stored by the thermal-mode system described above. Figure 7.1c shows the image of a stripe which is stored by scanning the probe along one direction. The stripe width is almost equal to the circular diameter of Fig. 7.1b. This result implies that the present method of storage can also be used for fabricating a nanometric fine grating. The further significance of the present photon-mode storage experiment is that it demonstrates the possibility of inducing a nanometric photochemical reaction by an evanescent field.

To increase the efficiency of the storage process, it is effective to utilize the resonance phenomenon with respect to the sizes of the probe tip and the sample, as shown in Fig. 2.7. However, it is rather difficult to expect such a resonance for the homogeneous surface of the LB film. In order to realize this resonance, a preliminary layer of 42-nm-diameter polystyrene spheres is densely packed on a glass plate to form a two-dimensional array as shown by the SEM micrograph in Fig. 7.2 [7]. The deposition is done using an evaporation method, controlling the temperature and the humidity. The evanescent field from the probe will be coupled more efficiently to the LB film formed on the array. It promises optical storage with increased sensitivity and higher density. This kind of two-dimensional array of nanometric particles should be explored further in order to find ways to fix the LB film on it in a stable manner.

Although the technology of high-density optical storage by the NOM is still at the initial stage, a very high density (as high as about 1 Tb/inch^2) can be expected in the future, depending on the sizes of the probe and of the grains of the storage medium. However, several problems have to be solved in order to realize a practical storage and readout system using near-field optical technology. Figure 7.3 illus-

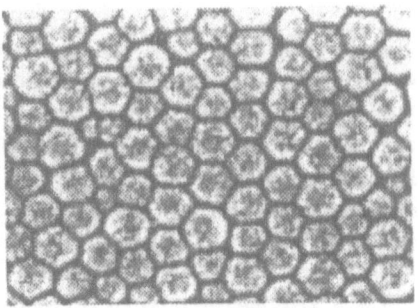

300 nm

Figure 7.2. SEM micrograph of a two-dimensional array of 42-nm-diameter polystyrene spheres fixed on a glass plate.

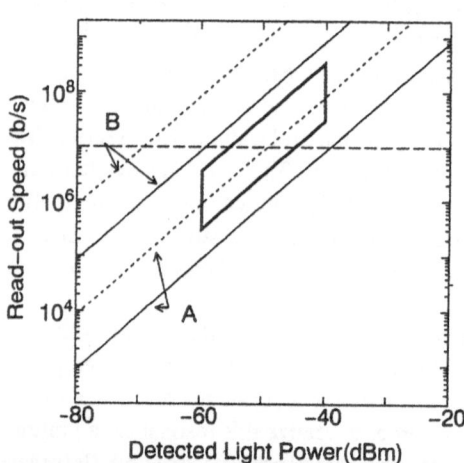

Figure 7.3. Calculated relation between the detected light power for reading out and the readout speed for maintaining the bit error rate as low as 1×10^{-9}. For this calculation, the photon energy is assumed to be 4×10^{-19} J, which corresponds to a propagating light wavelength of 500 nm. Solid and broken lines represent the results for a duty ratio of 0.1 and 1.0, respectively. Lines A and B are for a quantum efficiency of the photodetector of 0.1 and 1.0, respectively. The horizontal broken line represents the 10 Mb/sec readout speed. The area surrounded by bold lines represents the values which can be realized by solving the problems discussed in Sections 7.1.1.

trates the key problem. The readout speed for maintaining the bit error rate at 1×10^{-9} is plotted against the detected light power. Other issues to be considered are the quantum efficiency of the photodetector and the duty ratio of the stored medium (i.e., the ratio of the reflected light power from the pit to that from the unstored area of the medium surface). In the following we discuss requirements to be met for maintaining a 10 Mb/sec readout speed (indicated by the horizontal broken line in Fig. 7.3).

7.1.1. Development of a High-Efficiency Probe

Nanometric fiber-probe fabrication processes are required to realize higher efficiency in delivering the incident light to the apex region of the probe. This is because the detected light power depends on the generation efficiency of the evanescent field. One possibility is to realize a double-tapered fiber as described in Section 4.1.5. Since the delivery efficiency of this probe is at least ten times higher than those of the probes described in Sections 4.1.1 and 4.1.2 having a similar cone angle, a detected light power of 1–100 nW (–60 to –40 dBm) is expected. Furthermore, it is necessary to develop a probe which is less sensitive to optical damage by the high-power laser light.

7.1.2. Development of a Highly Sensitive Storage Medium

The experimental results reviewed above have been obtained by using conventional optical storage media, whose duty ratios are only 0.1. Higher sensitivity to heating or photochemical reactions are required during the storage process. Smaller grain size is also needed. Improvements in the flatness of the medium surface and development of a thin protective layer (i.e., 10 nm) are required because the sample–probe separation has to be less than 50 nm.

7.1.3. Fast Scanning of the Probe

The area surrounded by bold lines in Fig. 7.3 represents the values which can be expected to be realized by solving the above two problems. A readout speed of 10 Mb/sec can be realized. However, in order to realize this, the probe has to be scanned with a sufficiently high readout speed. For this high-speed scanning, one has to expand drastically the bandwidth of the servocontrol loop to regulate the sample–probe separation by developing a novel technique such as scanning by a flying head slider on a hard disk drive. Techniques of micromachining will be also required to integrate a probe tip with a microactuator on the flying head [8]. Further, the development of a precise tracking method is also indispensable for stable scanning.

In addition to the requirements listed above, appropriate software for reading out the ultrahigh density optical memory will be required. For example, the readout system must be able to skip nanometric dislocations on the medium surface.

7.2. MANIPULATING ATOMS

One of the ultimate goals of the nanometric fabrication technique is the handling of atoms. By using the dipole force of the nanometric evanescent field, it is possible to control the thermal motions of atoms flying freely in a vacuum. Conventional methods of laser cooling and trapping have controlled the position of atoms three-dimensionally using propagating light [9]. However, the accuracy of their spatial control is limited by diffraction. Future development of laser cooling and trapping will have to (1) decrease the residual thermal energy of atoms and increase the density of trapped atoms for application to, for example, the study of Bose–Einstein condensation, (2) decrease the density of trapped atoms in order to manipulate a single atom [10], and (3) decrease the spatial dimension of the manipulation.

It is expected that a three-dimensional nanometric evanescent field can be used for realizing requirements 2 and 3. It should be noted that the component of the wavevector of the evanescent field parallel to the material surface is much larger than that of the propagating light because the normal component takes an imaginary value. This means that the photon momentum of the parallel component of the evanescent field is much larger than that of the propagating photon. Thus, the evanescent field can impose a very large mechanical force on a small particle such as an atom. Further, one-dimensional and also zero-dimensional atom manipulations below the diffraction limit are possible because of the nanometric volume of the evanescent field; requirement 3 is thus realized.

It is also possible to trap and manipulate nanometric particles using an evanescent field. For example, in order to trap biological particles in water, the conventional approach has used a focused laser beam whose wavevector is always in parallel with the direction of the energy propagation. However, with an evanescent field, a variety of modes for manipulating nanometric particles can be expected (not only translation, but also rotation and twisting) because the direction of the wavevector of the evanescent field is not parallel to the direction of the field strength gradient.

7.2.1. Zero-Dimensional Manipulation

Use of the evanescent field on the probe tip of the I-mode NOM has been proposed to realize zero-dimensional atom trapping [11], utilizing the resonant interaction of an atom with the evanescent field (Fig. 7.4). An atom coming into the evanescent field scatters an evanescent photon into a propagating photon via resonant absorption followed by spontaneous emission, resulting in an extremely large recoil momentum parallel to the probe surface. As a consequence, the atom feels a strong dipole force normal to the probe surface due to the short decay length of the evanescent field. These features result in a mechanical interaction between the atom and the probe which is analogous to the interatomic van der Waals force. A novel technique of trapping atoms can thus be realized by using the localized evanescent field on the probe tip of the I-mode NOM.

As a simple consideration of the possibility of trapping atoms, let us assume that the evanescent field at the probe tip is red-detuned from the atomic resonance frequency. In this case, an atom coming into the evanescent field resonantly absorbs the counterpropagating evanescent photon due to a positive Doppler shift against the red detuning and spontaneously emits a propagating photon. The atom strongly

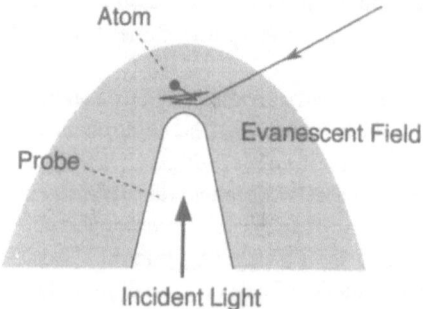

Figure 7.4. Trapping atoms by using the evanescent field at the probe tip of the I-mode NOM.

recoils in the direction parallel to the probe surface in accordance with Heisenberg's uncertainty principle, since the absorbed momentum of the evanescent photon is much larger than the emitted momentum of the propagating photon. The momentum difference acts as a cooling force until the momentum of the atom becomes comparable to the recoil momentum. Because of the spatial variation of the evanescent field momentum along the curvature of the probe, the cooling force has a centrifugal component. This component is compensated for by the strong dipole force [12] in the near-field region (where the decay length of the evanescent field intensity is as short as the apex diameter of the probe). The balance between the centrifugal component of the cooling force and the dipole force produces a potential valley along the probe surface. The atom will move back and forth in this potential valley like a pinball in the optical standing wave. The trapping condition is achieved when the cooled atomic motion is reversed by each recoil. Such scattering occurs several times per passage at the proximity of the probe tip. This condition determines the optimum apex diameter of the probe and the equivalent temperature of the trapped atoms.

As an example, in the case of the Rb atom (D_2 line resonance at $\lambda = 780$ nm), the trapping condition is satisfied by a probe tip with an apex diameter of about 30 nm. The equivalent temperature of the depth of the potential valley for trapping will be about 80 μK when the red detuning is fixed to be the optimized value (10 MHz). Since an evanescent field intensity comparable to the saturation parameter of the Rb atom is sufficient for trapping, a light source power of about several milliwatts is sufficient. Atoms prepared by a conventional laser cooling technique are sufficiently cold that they can be trapped using this method. As a preliminary experiment to confirm the large momentum transfer from evanescent photons to atoms, Doppler-free pump-probe laser spectroscopy of Rb has been carried out [13].

This trapping corresponds to realizing an "atomic quantum dot" because of the zero degrees of freedom of the trapped atomic motion. As an alternative method of realizing an atomic quantum dot, trapping has been proposed using the evanescent field generated at a nanometric dip on a dielectric surface [14]. Atom trapping using an evanescent field at a subwavelength-size circular hole in a thin plate has also been proposed [15].

If such a controlled atom is transferred to the surface of a cold crystal, it can be precisely placed on a certain position of the crystal surface, allowing atomic-level crystal growth and fabrication of atomic-level devices. For crystal growth and device fabrication, it is necessary to increase the number of atomic species that can be trapped by the present method. This is possible by using coherent light resonant to the relevant atom. Table 7.1 lists the resonant wavelengths for various atoms. Although the resonance wavelengths of several atoms are as short as 200 or 300 nm, coherent light at these wavelengths can be generated by nonlinear optical frequency conversions of compact semiconductor lasers because the required ultraviolet power is as low as several milliwatts [16].

Table 7.1. Resonant Wavelengths for Several Atoms

Atom	Wavelength (nm)	Atom	Wavelength (nm)
Ag	328	K	767
Al	394	Li	671
Au	268	Mg	285
B	250	Na	589
Be	235	O	778
Ca	423	Pb	368
Cs	852	Rb	780
Cu	327	Si	252
Ga	403	Sr	461
Hg	254		

7.2.2. One-Dimensional Manipulation

The thermal motion of atoms can be controlled so that the residual atomic motion becomes one-dimensional. In order to realize such an "atomic quantum wire," a method using a cylindrical evanescent field generated on the inner wall of a hollow fiber has been proposed [17, 18], as is schematically explained in Fig. 7.5. (Another configuration, using a bundle of four eccentric-core fibers, has been proposed [19].) This scheme of atom manipulation has been called "atom guidance." It will enable novel studies of quantum effects [20], cavity quantum electrodynamics [21], and topological phenomena [22]. It should be noted that the blue-detuned evanescent field on the inner wall of the hollow fiber is used here in order to impose an intense repulsive force on the atoms. Hence atoms are guided along the hollow fiber while being repelled by this force. In contrast to the traditional thermal or chemical methods involving a process treating a large number

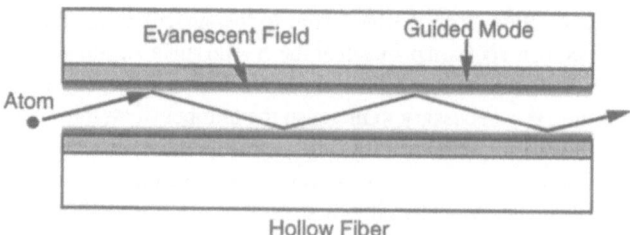

Figure 7.5. Guiding atoms through a hollow fiber by using the dipole force of the evanescent field on the inner wall.

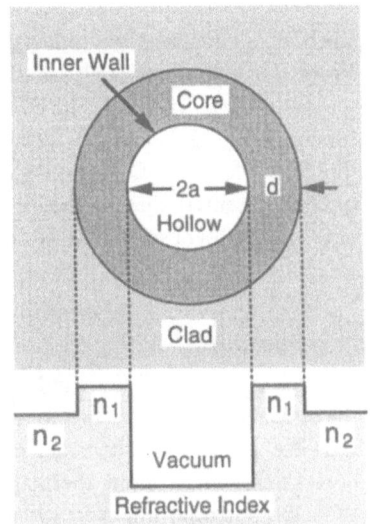

Figure 7.6. Cross-sectional profile of a cylindrical-core hollow fiber with its refractive index distribution.

of atoms, the mechanical approach using atom guidance enables one to handle even a single atom.

Figure 7.6 shows the cross-sectional profile of a cylindrical-core hollow fiber used for atom guidance. For calculations, a stepped-index fiber is assumed with refractive indices of core and cladding being n_1 and n_2, respectively. The radius of the hollow vacuum region and the thickness of the cylindrical core are a and d, respectively. The guided modes of this fiber can be analyzed based on Maxwell's equations expressed in terms of the cylindrical coordinate system (r, θ, z).

7.2.2.1. Theoretical Considerations

Although the exact solution of the guided modes of the fiber and the evaluation of the optical potential for guiding Rb atoms have been obtained [18], this subsection presents another analysis based on the weakly guiding approximation (WGA) to study atom guidance [23]. This approximation simplifies the estimation of the potential barrier and offers a picture that is easy to understand.

The WGA assumes that the relative refractive index difference Δn between the core and cladding is much smaller than unity. For the practical stepped-index cylindrical-core hollow fiber shown in Fig. 7.6, $a = 3.5$ μm, $d = 3.8$ μm, and $\Delta n = (n_1^2 - n_2^2)^{1/2}/2n_1^2 = 0.18\%$. It has been confirmed by comparison with the exact solution and the experimental observation of the mode pattern (see Fig. 7.12 of the next subsection) that the WGA is still valid in spite of a considerable refractive index difference between the core and the hollow vacuum region. In the WGA

Maxwell's equations are reduced to scalar ones, which leads to LP modes [24]. From the numerical analysis of the dispersion relation (relation between the wavenumber and propagation constant of the guided mode), one finds that three LP modes (LP_{01}, LP_{11}, and LP_{21}) can be excited for the Rb D_2 line ($\lambda = 780$ nm).

The evanescent field leaking into the hollow center of the fiber produces a deep optical potential due to the position-dependent Stark effect. By applying a two-energy-level scheme to atoms, the optical potential can be expressed as [25]

$$U_0(r, \theta) = \frac{1}{2}\hbar\Delta \ln\left\{1 + \frac{I(r, \theta)/I_0}{1 + 4\Delta^2/\Gamma^2}\right\} \tag{7.2.1}$$

where Γ is the natural linewidth, I_0 is the saturation intensity, Δ ($= \omega - \omega_0 - \beta v_z$) is the optical detuning, ω is the optical frequency, ω_0 is the atomic resonant frequency, β is the propagation constant of the guided mode, and v_z is the longitudinal component of the atomic thermal velocity. In the case of blue-detuning ($\omega > \omega_0 - \beta v_z$), the potential imposes a repulsive force on the atom. Under the WGA, the intensity $I(r,\theta)$ of the evanescent field can be derived from the z component of the time-averaged Poynting vector and is expressed as [23]

$$I(r, \theta) = \frac{\beta}{2\omega\mu_0} C^2 I_m^2(vr) \cos^2(m\theta) \tag{7.2.2}$$

where $v = \sqrt{\beta^2 - k^2}$, μ_0 is the magnetic permeability, and $I_m(vr)$ is a modified Bessel function of the first kind of mth order. The constant C is determined from the boundary conditions and the guiding power. Figure 7.7 shows the calculated contours of the field intensity for the lowest three LP modes given by Eq. (7.2.2). These agree with the experimental results (see Fig. 7.12).

Figure 7.8 shows the profile of the optical potential barriers corresponding to these modes calculated using Eqs. (7.2.1) and (7.2.2). The excitation of the LP_{01} mode is necessary because a nodeless evanescent field distribution is required for the stable guidance of atoms. For Rb atoms, the equivalent temperature representing the height of the potential barrier is evaluated to be 1 mK for a guided mode power of a few milliwatts. This height corresponds to a transverse component of the thermal velocity of 30 cm/sec. This can be achieved using either a well-collimated atomic beam or laser-cooled atoms. (It has been demonstrated that laser cooling of atoms can cool alkali metal atoms to a few microkelvins.)

Several effects which lower the potential barrier and thus decrease the atomic guidance efficiency have to be examined when designing an actual experimental system. Three principal factors decrease the atomic guidance efficiency: (1) cavity QED effects, (2) tunneling of atoms through the potential barrier, and (3) spontaneous emission effects [17]. These effects have been investigated under the WGA for the present cylindrical-core hollow fiber:

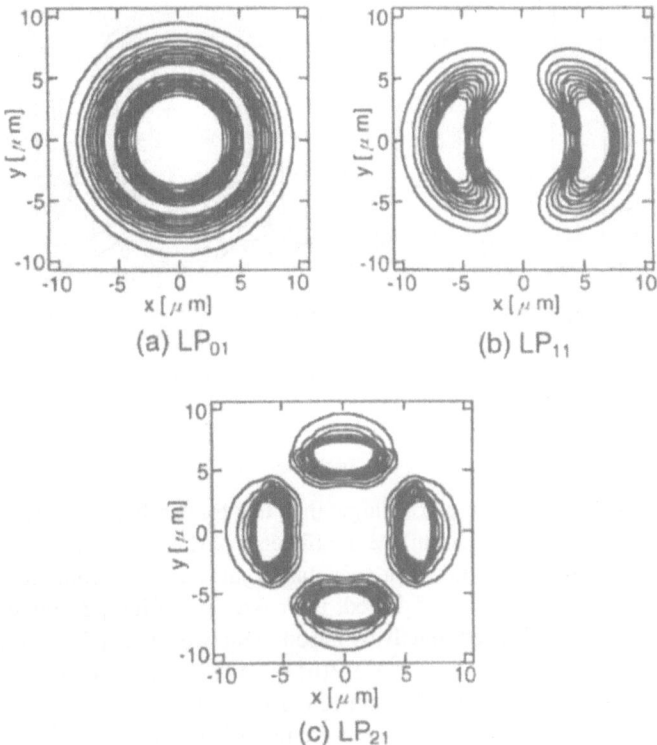

(a) LP$_{01}$ (b) LP$_{11}$

(c) LP$_{21}$

Figure 7.7. Calculated contours of the field intensity for the (a) LP$_{01}$, (b) LP$_{11}$, and (c) LP$_{21}$ modes.

1. The potential barrier given by Eq. (7.2.1) is modified by the cavity QED effect. The attractive force arising from this effect reduces the height of the potential barrier in the vicinity of the inner surface. In the case of two parallel mirrors, the cavity potential for the ground-state atom can be expressed as [20, 26]

$$U_c(r) = -\frac{\pi}{48\varepsilon_0 a^3} \sum_i p_{ig}^2 \int_0^\infty \frac{\cosh(\pi\rho r/a)}{\sinh(\pi p)} \tan^{-1}\left(\frac{\rho\lambda_{ig}}{4a}\right)\rho^2 d\rho \qquad (7.2.3)$$

where p_{ig} and λ_{ig} are the matrix element of the dipole moment and the resonance wavelength, respectively, for the transition between the ith excited state and the ground state. Unfortunately, the exact cavity potential near a dielectric cylindrical surface has not yet been calculated. (The case of a dielectric sphere has been recently calculated [27].) However, given similar conditions, the cavity potential for a

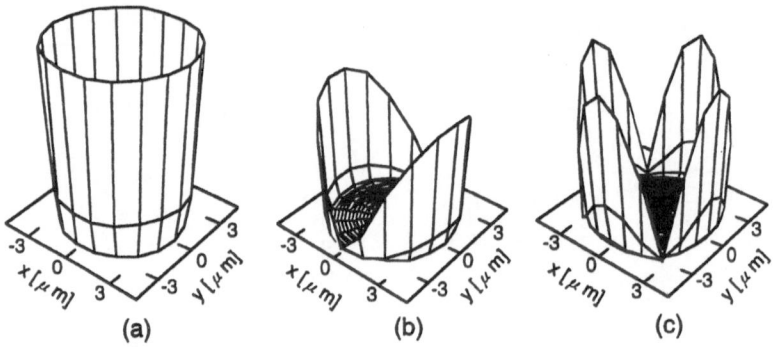

Figure 7.8.   Calculated profiles of the optical potential barriers corresponding to the (a) LP$_{01}$, (b) LP$_{11}$, and (c) LP$_{21}$ modes.

cylindrical surface is expected to be larger than that for parallel surfaces but smaller than that for a sphere. In addition, the atomic energy shift near a dielectric surface is generally smaller than that near a metallic surface. Therefore, if resonance behavior due to the cavity is neglected, the cavity potential near the cylindrical surface can be treated approximately by introducing two adjustable factors into Eq. (7.2.3); the dielectric factor $\xi = (n_1^2 - 1)/(n_1^2 + 1))$ [17] and the geometrical factor η for the cylinder ($1 < \eta < 10$; $\eta = 1$ and 10 correspond to two parallel surfaces and a sphere, respectively). Figure 7.9 shows the modified potential barrier in terms of the equivalent temperature representing the transverse kinetic energy of ^{85}Rb under excitation by the 20-mW LP$_{01}$ mode ($\Delta = 1000\Gamma$ and $\eta = 5$). Within 100 nm from the inner surface of the hollow fiber, the height of the potential barrier is decreased by the cavity QED effect [26].

2.   In order to estimate the tunneling rate γ_T of the atoms for the potential $U(r,\theta) = U_0(r,\theta) + U_c(r)$, the WKB formula can be applied [28]:

$$\gamma_T = \frac{\exp\left[-(2/\hbar)\int_{r_{CT}}^{r_0} \sqrt{2M(U(r,\theta) - E_t)}\, dr\right]}{\left[d/dE_t \int_{-r_{CT}}^{r_{CT}} \sqrt{2M(E_t - U(r,\theta))}\, dr\right]_{E_t = E_{CT}}} \qquad (7.2.4)$$

where r_0 and r_{CT} are the position of the maximum of the potential and the classical turning point, respectively, M is the mass, and E_t is the transverse component of the kinetic energy. The numerator on the right-hand side of this equation is the tunneling probability per bounce and the denominator is the number of seconds per bounce [17]. Taking the classical turning point to be 40 nm from the surface of the inner wall for the potential barrier shown in Fig. 7.9, we plot the tunneling rate as a function of the transverse velocity component v_t of the thermal motion of ^{85}Rb in Fig. 7.10a. For $v_z = 300$ m/sec and a fiber of length $L = 3$cm, the loss of atom guidance ($L\gamma_T/v_z$) can be kept as low as 1% for $v_t = 50$ cm/sec.

3. The optimum detuning is that which maximizes the height of the potential barrier (i.e., the coupled power of the laser beam to the hollow fiber). In general, the optimum detuning increases with excitation power. For a low power of about several milliwatts, it is about 10Γ. In this case, however, spontaneous emission cannot be ignored. The spontaneous emission rate γ_{sp} can be estimated using the ratio of the transverse component of the kinetic energy E_t to the detuning energy $\hbar\Delta$.

$$\gamma_{sp} \cong \frac{\Gamma}{va}\frac{E_t}{\hbar\Delta} \tag{7.2.5}$$

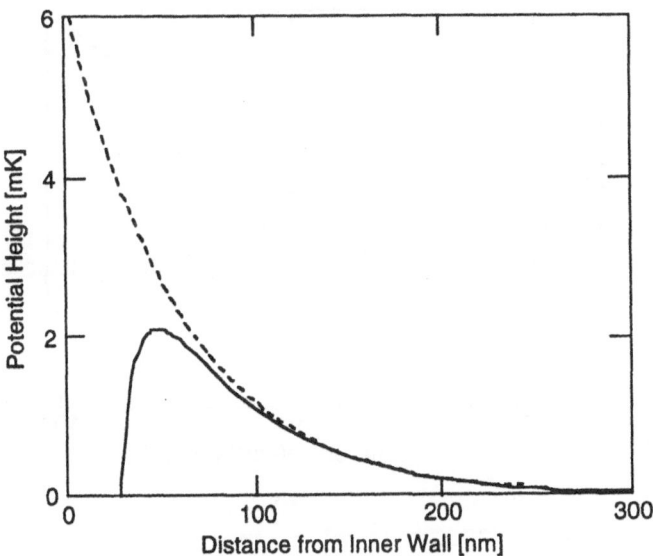

Figure 7.9. The cross-sectional potential barrier under 20-mW excitation of the LP$_{01}$ mode with $\Delta = 1000\Gamma$, as a function of the distance from the surface of the inner wall, in terms of the temperature corresponding to the transverse kinetic energy of the ^{85}Rb atom. The solid curve represents the total potential made up of the optical potential and the cavity potential in the case of $n_1 = 1.45$ and $\eta = 5$. The broken curve represents the optical potential.

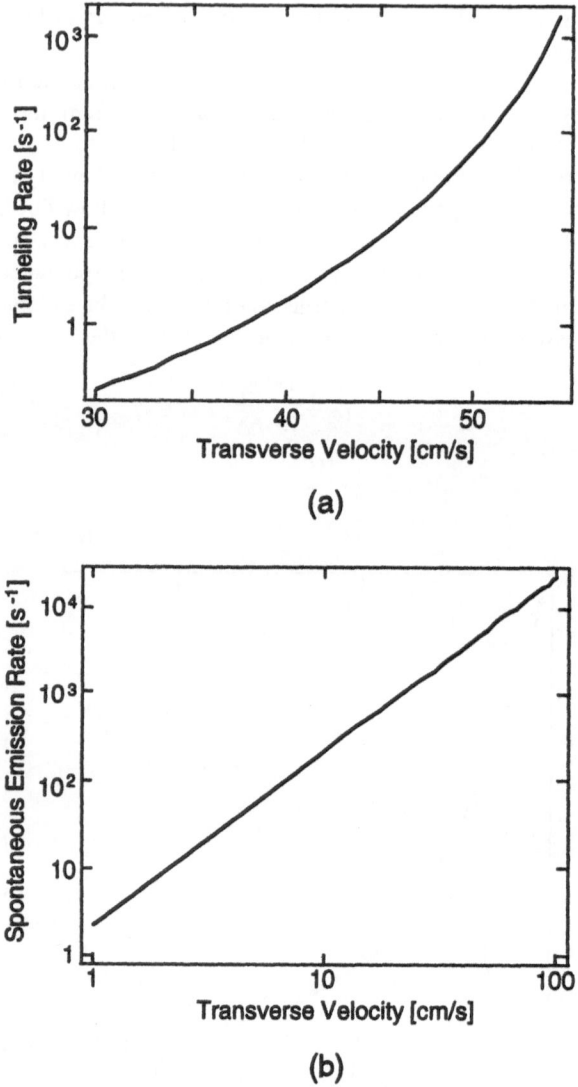

(a)

(b)

Figure 7.10. (a) Tunneling rate of the potential barrier under 20-mW LP_{01} mode excitation. (b) Spontaneous emission rate. These rates are plotted as a function of the transverse velocity of ^{85}Rb in the case of $\Delta = 1000\Gamma$.

where the factor va appears due to the spatially localized nature of the evanescent field. Figure 7.10b shows the relation between the transverse velocity component of ^{85}Rb and γ_{sp} for $\Delta = 1000\Gamma$. It is confirmed from this figure that about 10% of atoms with $v_z = 300$ m/sec and $v_t = 50$ cm/sec emit spontaneously while in the 3-cm-long fiber. The spontaneous emission rate does not necessarily decrease the atomic guidance efficiency even if some heated atoms may go over the potential barrier or be pumped to noninteractive hyperfine energy levels. This spontaneous emission is, however, serious for some applications such as atomic interferometers [29] since the atoms experiencing spontaneous emission reduce the cross-correlation between atomic wave functions. In order to suppress spontaneous emission, a low transverse velocity component and large detuning are required. That is, additional transverse cooling and an increase of the excitation power (so as not to lower the potential barrier) are required. It should be noted that a minimum for the longitudinal velocity component exists because the number of atoms emitting spontaneously increases with decreasing longitudinal velocity.

7.2.2.2. Experimental

Two types of atom guidance using optical fibers have been proposed, one using a blue-detuned evanescent field (described in section 7.2.2.1) and the other using a red-detuned propagating light field utilizing a Gaussian laser beam [25]. An experimental demonstration of the latter has been reported recently [30], using a hollow-core capillary fiber ($a = 20$ μm). The intense Gaussian laser beam propagating inside the hollow core can guide atoms by attracting them to the center of the hollow region, where the laser intensity is maximum. However, it causes strong heating of the guided atoms due to spontaneous emission.

In the former case, the spontaneous emission effect is relatively negligible. Since the evanescent field penetrates into the hollow fiber center only a few hundred nanometers in the visible wavelength region, the guided atoms interact with the evanescent field only when they approach the inner wall—otherwise they are free. This scheme of atom guidance enables a hollow fiber with a hollow diameter as small as 1 μm to be used.

The fiber used for the first atom guidance experiment [31], using the evanescent field in a cylindrical-core hollow fiber, is shown in Fig. 7.11 [23]. Its hollow radius a was 3.5 μm, core thickness d was 3.8 μm, and length L was 3 cm. Figure 7.12 shows the observed guided mode patterns. The experimental setup is shown in Fig. 7.13. A thermal atomic beam from a heated Rb oven (collimated to within 1 mrad) is introduced into the hollow fiber. Atoms which do not enter the hollow region are blocked by the large fiber holder so that they are not detected at the fiber exit. The coupling efficiency of the atomic beam into the hollow fiber is about 4×10^{-4}. A mirror with a hole that allows the atoms to pass through is used to couple light from a $Ti:Al_2O_3$ laser into the fiber core with 40% efficiency. At the 780 nm wavelength

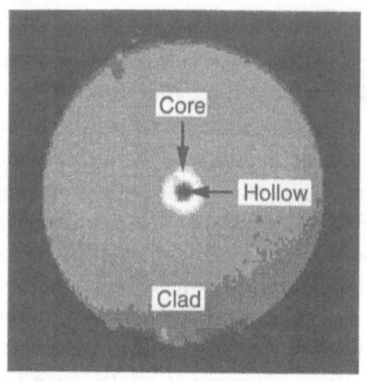

Figure 7.11. Optical microscope image of the cross section of a cylindrical-core hollow fiber. Hollow radius a and the thickness d of the cylindrical core are 3.5 and 3.8 μm, respectively. The relative refractive index difference Δn is 0.18%.

for guiding Rb atoms in the 3-cm-long fiber, a doughnut-shaped mode can be excited even though the lowest three guided modes (LP_{01}, LP_{11}, and LP_{21}) are not clearly distinguished when monitored by a CCD camera.

Of the several possible techniques for detecting atoms transmitted through the hollow fiber (i.e., a hot wire), the two-step photoionization (PI) technique is employed, in which the beams of two lasers are overlapped near the fiber exit. The ionized atoms are then detected by a high-voltage channel electron multiplier (CEM). The two-step PI technique is illustrated schematically in Fig. 7.14. A grating-feedback diode laser tuned to 780 nm saturates the $5S_{1/2} \rightarrow 5P_{3/2}$ transition. The overlapping Ar^+ laser beam ionizes atoms in the $5P_{3/2}$ state. The PI efficiency of this scheme can be estimated by using the PI cross section σ_i from the $5P_{3/2}$ state to the ionized state [= $(1-2) \times 10^{-17}$ cm^2] [32] and the condition of efficient ionization [33]

$$P_i \cong P_0(\sigma_0/\sigma_i) \tag{7.2.6}$$

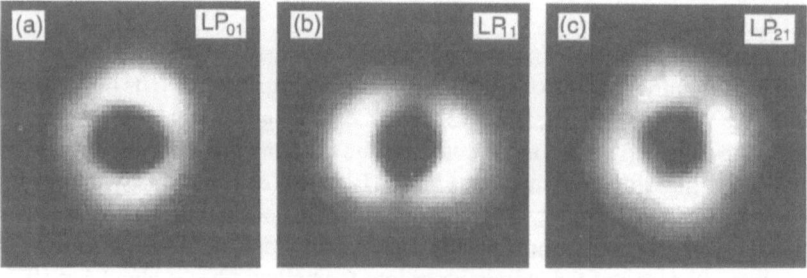

Figure 7.12. Guided mode patterns for the hollow fiber of Fig. 7.11. These modes are excited by a 780-nm-wavelength diode laser.

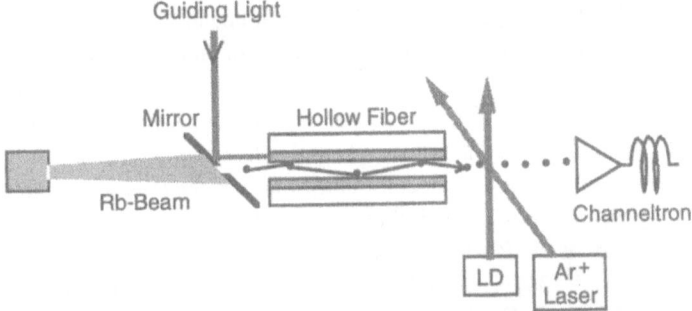

Figure 7.13. Experimental setup for guiding Rb atoms.

where P_i is the optimum laser intensity for ionization, and P_0 and σ_0 are the saturation power and the resonant excitation cross section for the $5S_{1/2} \rightarrow 5P_{3/2}$ transition, respectively. For a 4-W Ar^+ laser beam ($\lambda = 476.5$ nm) focused to a 100-μm spot diameter, the PI efficiency is estimated to be 32%. Assuming the quantum efficiency of the CEM to be 0.9, the total detection efficiency can be as high as 29%. (Although PI efficiency is lower at the wavelengths of 457.9 and 488 nm, these popular Ar^+ laser wavelengths can also be used for the present scheme of detection by increasing the bias voltage of the CEM.)

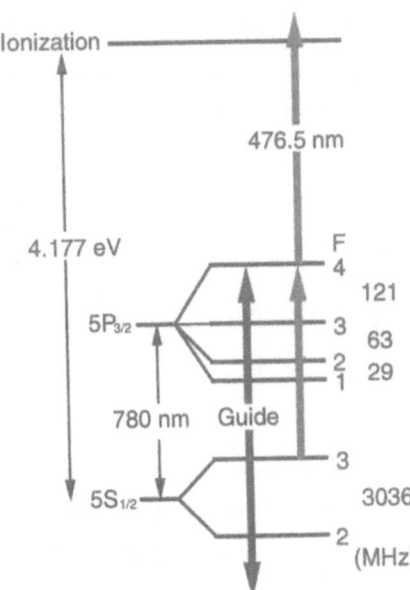

Figure 7.14. Energy level diagram of the Rb atom relevant to guidance and photoionization.

Figure 7.15 shows a PI signal for the ^{85}Rb atoms in the $F = 3$ hyperfine level of the ground state as a function of the frequency detuning Δ of the guiding light with a coupled power of 130 mW. Note that the detuning is measured with respect to the transition frequency of the $F = 3$ level of the ground state and does not include the Doppler shift because of the present Doppler-free excitation scheme. The oven temperature of 160°C and present collimation angle of the atomic beam provide an average longitudinal velocity of 328 m/sec and transverse velocity of 0.23 m/sec. The incident flux of the Rb atoms impinging on the hollow entrance is estimated to be 1.3×10^6 sec^{-1} [34] (73% of the incident atoms are the ^{85}Rb isotope). A diode laser tuned to the $(5S_{1/2}, F = 3) \rightarrow (5P_{3/2}, F = 4)$ transition and on Ar$^+$ laser are used to ionize the guided atoms.

Curve A of Fig. 7.15 illustrates the effect of detuning on atomic guidance. The peak is at a blue-detuning of 2.7 GHz. The ion-counting signal extended over 20 GHz detuning. The maximum atom flux at the transmission peak is 3×10^4 sec^{-1}. The flux without the guiding light is 1.5×10^3 sec^{-1}, as shown by curve B. Therefore, the enhancement factor is about 20. This large value of the enhancement factor is due to the use of an atomic beam and a small-diameter hollow fiber, and the high sensitivity of detection.

In the red-detuning region, the atomic flux is decreased below the background level due to adsorption on the inner wall of the hollow fiber by the attractive optical force. Moreover, in the vicinity of the $(5S_{1/2}, F = 3) \rightarrow 5P_{3/2}, F = 4)$ resonant transition, it seems that ^{85}Rb atoms in the $F = 3$ ground-state hyperfine level are pumped into the lower $F = 2$ level due to the interaction with the resonant evanescent

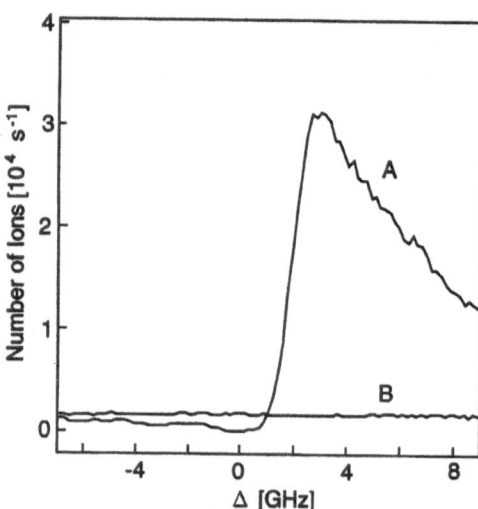

Figure 7.15. Counting rate of ions for the transmitted atoms, plotted as the frequency detuning of the guiding light (130 mW). Curves A and B represent results with and without the guiding light, respectively. The detuning is measured with respect to the resonance frequency of the $(5S_{1/2}, F = 3) \rightarrow (5P_{3/2}, F = 4)$ transition of ^{85}Rb.

field or the guiding light which leaks into the hollow region at the entrance. This effect is confirmed by detecting the guided atoms in the lower hyperfine level by tuning the diode laser to the $(5S_{1/2}, F = 2) \rightarrow (5P_{3/2}, F = 1)$ transition. A small increase (about 30% of the background signal) of ^{85}Rb atoms in the $F = 2$ level is observed near the resonant guidance frequency of the $(5S_{1/2}, F = 3) \rightarrow (5P_{3/2}, F = 4)$ transition.

Figure 7.16 shows the transmitted ^{85}Rb atom flux in the $F = 3$ level as a function of a guiding light power from 1 to 280 mW at the optimum detuning of 2.7 GHz and an oven temperature of 174°C. The flux (ion counts per second) is plotted after subtracting the background signal level (1.6×10^4 sec^{-1}; recorded in the absence of the guiding light). Therefore, the values in this figure represent the net transmission flux due only to optical guidance. As shown in this figure, a saturation behavior in the transmission can be observed as the coupled power is increased to about 150 mW. A maximum net flux of 9.5×10^4 sec^{-1} was obtained at 280 mW. Assuming a total detection efficiency of 29%, an incidence flux of 9×10^5 sec^{-1} for the ^{85}Rb isotope, and adding the background signal, the total efficiency of atom guidance is estimated as about 43% for the ^{85}Rb atom in the $F = 3$ level, whereas the net optical guidance efficiency is 37%.

By taking into consideration the fact that the ^{87}Rb isotope (see curve A of Fig. 7.17) as well as the ^{85}Rb atoms in the lower $F = 2$ level are guided, the total efficiency is estimated to be higher than 50%. This very high efficiency is realized because a collimated atomic beam is used. In the case of the excitation of the LP_{01} mode at 300 mW power and a 3-GHz blue-detuning, one obtains from Eqs. (7.2.1) and (7.2.2) an optical potential as high as 120 mK in terms of the equivalent temperature

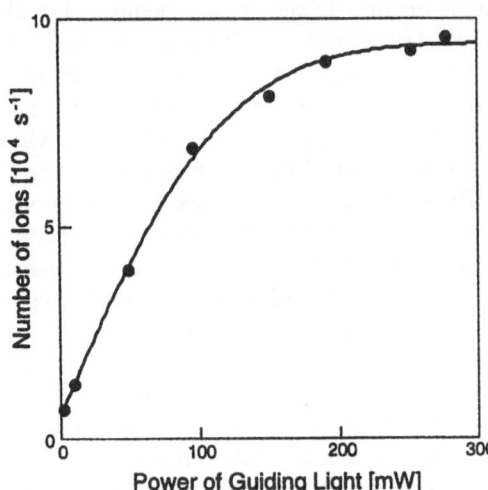

Figure 7.16. Counting rate of ions for the transmitted atoms, plotted as a function of guiding light power.

of the atomic kinetic energy. This corresponds to a maximum transverse velocity of 3.4 m/sec. Therefore, most of the atoms entering the hollow fiber are expected to be guided since the average transverse velocity due to the present beam collimation is 0.23 m/sec.

Using the results for two parallel dielectrics [21, 26], one can roughly estimate the effects of the cavity potential $U_c(r)$, which reduce the height of the optical potential barrier [23]. For the excitation of the LP_{01} mode at 280 mW, a 30% reduction in the potential barrier is estimated using Eq. (7.2.3). With 1 mW incident power, the potential barrier almost disappears due to the attractive cavity potential. Rb atoms can be guided above the coupled guiding power of 1 mW as shown in Fig. 7.16. Below 1 mW, a rapid decrease of the guided flux is observed. Note that the maximum transverse velocity due to the optical potential $U_0(r, \theta)$ for 1 mW power is about 0.23 m/sec, which is equivalent to that determined by the beam collimation. Therefore, the observed threshold behavior near 1 mW power may be an indirect indication of the cavity QED effects on the cylindrical dielectric surface. For a quantitative study of the cavity QED effects, similar experiments using several hollow fibers with various hollow radii of less than 2 μm is required. It should be pointed out that atom guidance experiments using a fiber of 1 μm inner hollow radius have already succeeded [35].

There are several possibilities for using this kind of atomic guidance, i.e., an "atomic quantum wire," in materials science, such as single-atom-level crystal growth. As a first step in this direction, an isotope-separation experiment was carried out using the two stable Rb isotopes in the upper hyperfine levels of the ground state. A state-selective guidance of atoms in the hollow fiber is employed for this purpose. Based on the dispersive nature of the guidance shown in Fig. 7.15, a specific isotope is selected by adjusting the detuning frequency of the guiding light. Figure 7.17 shows the experimental result. The upper hyperfine levels of the ground states of the two isotopes ^{85}Rb and ^{87}Rb are separated by about 1 GHz. For curve A of Fig. 7.17, the frequency of the guiding light is fixed at a large blue-detuning, similar to that in Figs. 7.15 and 7.16, for both the $5S_{1/2}$, $F = 2$ level of ^{87}Rb and the $5S_{1/2}$, $F = 3$ level of ^{85}Rb. On the other hand, for curve B of Fig. 7.17, the guiding light is blue-detuned for the ^{87}Rb atoms, but red-detuned for the ^{85}Rb atoms. As is clear in this figure, the transmission flux of the ^{85}Rb isotope is greatly suppressed to 10% of the background level. This result confirms that the hollow fiber works as an in-line isotope separator or an atomic-state filter.

One possible future direction for this type of experiment could be the deposition of transmitted atoms onto a cold crystal substrate, realizing atomic-level direct-write photolithography. Use of atom guidance for this process can improve the spatial accuracy of the deposition beyond the diffraction limit of the conventional far-field "optical crystal" technique [36].

In order to use atomic guidance technology in future applications, it is important to improve the efficiency of coupling colder atoms to the hollow fiber. For this

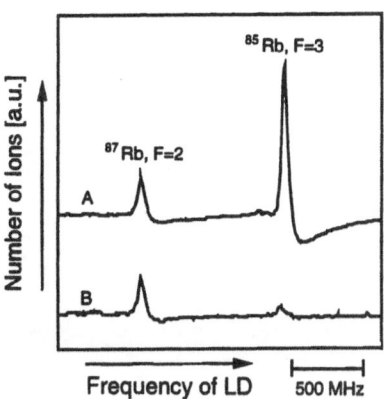

Figure 7.17. Experimental results of in-line separation of the two stable Rb isotopes. Counting rates of ions for the transmitted atoms are plotted as a function of the frequency of the diode laser used for photoionization. (A) The frequency of the guiding light is fixed at a large blue-detuning for both the $5S_{1/2}$, $F = 2$ level of ^{87}Rb and the $5S_{1/2}$, $F = 3$ level of ^{85}Rb. (B) The guiding light is blue-detuned for the ^{87}Rb atoms, but nearly red-detuned for the ^{85}Rb atoms.

purpose, the use of an "atomic funnel" has been proposed [37]. Figure 7.18 shows the schematic explanation of such an atomic funnel. It consists of a conical hollow prism illuminated by an intense, doughnut-shaped laser beam from the mouth of the funnel. Total internal reflection of the doughnut beam generates an evanescent field on the inner wall of the prism. With blue-detuning, the evanescent field exerts a repulsive force on atoms approaching the inner wall, resulting in their reflection.

Conversion of a Gaussian beam into a doughnut beam has been performed using holographic techniques with conversion efficiencies up to 50% [38, 39]. However, a simpler method using a double-cone prism (Fig. 7.19a) exhibits nearly 100% efficiency. A Gaussian beam is divided into two by the first refraction into

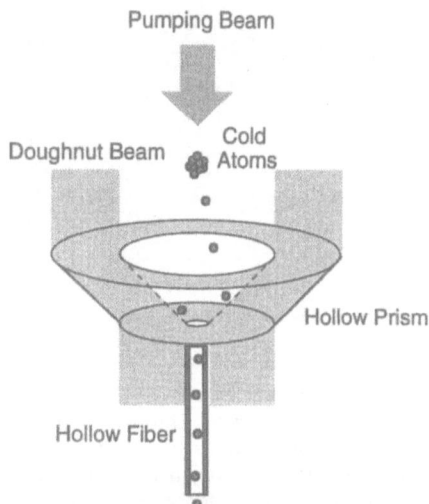

Figure 7.18. An atomic funnel made up of a conical hollow prism through which cold atoms are dropped from a magnetooptical trap into a hollow fiber for atom guidance.

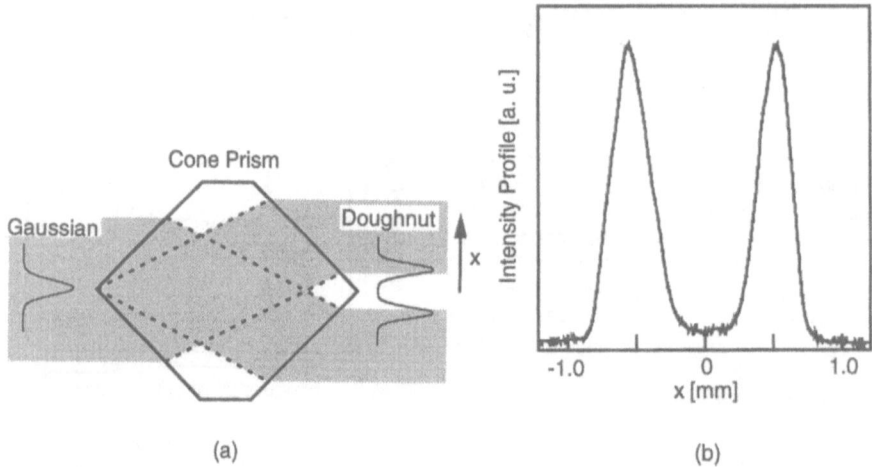

(a)   (b)

Figure 7.19. (a) Conversion of a Gaussian beam to a doughnut-shaped beam through a double-cone prism. (b) Intensity profile of an experimentally generated doughnut beam. A Gaussian beam with a width of 0.85 mm is converted by a double-cone prism with a length of 4.25 mm and a diameter of 2.5 mm.

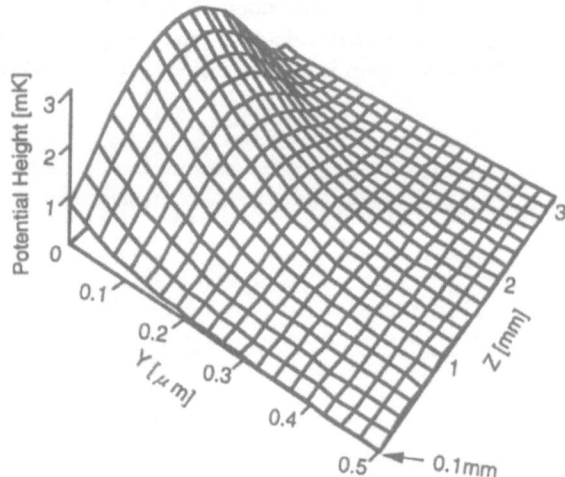

Figure 7.20. Potential barrier for Rb atoms, expressed in term of temperature, produced by a 500-mW doughnut beam. The blue-detuning Δ is taken to be 160 times the natural linewidth Γ. The slope angle and exit hole of the conical hollow prism are assumed to be 60 deg and 0.2 mm, respectively. The y axis represents the distance from the surface of the inner wall, while the z axis is the distance along the slope from the center of the exit hole.

the prism and a nondivergent doughnut beam generated by the second refraction out of the prism. Figure 7.19b shows the intensity profile of the doughnut beam formed by a 4.25-mm-long, 2.5-mm-diameter double-cone prism. The hollow diameter and ring thickness of the doughnut beam are determined by the prism length and the width of the input Gaussian beam, respectively.

The funnel can be used to reflect cold atoms falling from a magnetooptical trap [9]. Figure 7.20 shows the potential barrier for Rb atoms expressed in terms of temperature; assuming a funnel prism with a 0.2-mm exit hole and a 60 deg slope, a doughnut beam converted from a 500-mW Gaussian beam with a width of 1.6 mm produces a potential barrier higher than 1 mK through a distance of 3 mm along the slope of the inner wall of the funnel prism. Unless the funnel includes some cooling mechanism to compensate for acceleration due to gravity, atoms with an initial velocity of more than 10 cm/sec cannot escape from the exit hole. Since addition of a weak pumping beam to cool atoms down to the recoil limit has been recently reported [40], use of the additional beam propagating downward can make up for the lack of cooling mechanisms. The controllability of atomic thermal motion and coupling efficiency can be improved by using this funnel.

7.3. REFERENCES

1. H. J. Mamin, L. S. Fan, S. Hoen, and D. Rugar, Tip-based storage using micromechanical cantilevers, *Sensors Actuators A* **48**: 215–219 (1996).
2. B. D. Terris, H. J. Mamin, and D. Rugar, Near-field optical data storage, *Appl. Phys. Lett.* **68**: 141–143 (1996).
3. E. Betzig, J. K. Trautmann, R. Wolfe, E. M. Gyorgy, P. L. Finn, M. H. Kryder, and C.-H. Chang, Near-field magneto-optics and high density data storage, *Appl. Phys. Lett.* **61**: 142–144, (1992).
4. S. Hosaka, T. Shintani, Y. Maruyama, K. Nakamura, A. Kikukawa, and R. Imura, Nanometer recording using a scanning near-field optical microscope with a laser diode, in *Proceedings Symposium on Optical Memory 1994—Technical digest*, Waseda, Tokyo, 1994, pp. 21–22.
5. S. Hosaka, T. Shintani, M. Miyamoto, A. Hirotsune, M. Terao, M. Yoshida, K. Fujita, and S. Krammer, Nanometer-sized phase-change recording using a scanning near-field optical microscope with a laser diode, *Jpn. J. Appl. Phys.* **35**: 443–447 (1996).
6. S. Jiang, J. Ichihashi, H. Monobe, M. Fujihira, and M. Ohtsu, Highly localized photochemical processes in LB films of photochromic material by using a photon scanning tunneling microscope, *Opt. Commun.* **106**: 173–177 (1994).
7. R. Micheletto, H. Fukuda, and M. Ohtsu, A simple method for the two-dimensional, ordered array of small latex particles, *Langmuir* **11**: 3333–3336 (1995).
8. W. Tang, V. Temesvary, R. Miller, A. Desai, Y.-C. Tai, and D. K. Miu, Silicon micromachined electromagnetic microactuators for rigid disk drives, *IEEE Trans. Magnetics* **31**: 2964–2966 (1995).
9. E. L. Raab, M. G. Prentiss, A. E. Cable, S. Chu, and D. E. Pritchard, Trapping of neutral sodium atoms with radiation pressure, *Phys. Rev. Lett.* **59**: 2631–2634 (1987).
10. M. H. Anderson, J. R. Ensher, M. R. Matthews, C. E. Wieman, and E. A. Cornell, Observation of Bose–Einstein condensation in a dilute atomic vapor, *Science* **269**: 198–201 (1995).

11. H. Hori, S. Jiang, M. Ohtsu, and H. Ohsawa, A nanometric-resolution photon scanning tunneling microscope and proposal of single atom manipulation, *Technical Digest of the 18th International Quantum Electronics Conference, Vienna, June 1992*, pp. 48–49.

12. J. P. Gordon and A. Ashkin, Motion of atoms in a radiation trap, *Phys. Rev. A* **21**: 1606–1617 (1980).

13. M. Kozuma, S. Jiang, T. Pangaribuan, M. Ohtsu, and H. Hori, Analysis and experimental evaluation of a localized evanescent field by using Yukawa potential, *Tech. Digest Quantum Electron. Laser Sci.* **16**: 227–228 (1995).

14. J. P. Dowling and J. Gea-Banacloche, Atomic quantum dots, in *Technical Digest of the 19th International Quantum Electronics Conference, Washington D.C., May 1994*, pp. 185–186.

15. V. V. Klimov and V. S. Letokhov, New atom trap configurations in the near field of laser radiation, *Opt. Commun.* **121**: 130–136 (1995).

16. M. Ohtsu, K. Nakagawa, M. Kourogi, and W. Wang, Frequency control of semiconductor lasers, *J. Appl. Phys.* **73**: R1–R17 (1993).

17. S. Marksteiner, C. M. Savage, P. Zoller, and S. L. Rolston, Coherent atomic waveguides from hollow optical fibers: Quantized atomic motion, *Phys. Rev. A* **50**: 2680–2690 (1994).

18. H. Ito, K. Sakaki, T. Nakata, W. Jhe, and M. Ohtsu, Optical potential for atom guidance in a cylindrical-core hollow fiber, *Opt. Commun.* **115**: 57–64 (1995).

19. W. Jhe, M. Ohtsu, H. Hori, and S. R. Friberg, Atomic waveguide using evanescent waves near optical fibers, *Jpn. J. Appl. Phys.* **33**: L1680–L1682 (1994).

20. D. J. Harris and C. M. Savage, Atomic gravitational cavities from optical fibers, *Phys. Rev. A* **51**: 3967–3971 (1995).

21. W. Jhe, QED level shifts of atoms between two mirrors, *Phys. Rev. A* **43**: 5795–5803 (1991).

22. K. Sangster, E. A. Hinds, S. M. Barnett, and E. Riis, Measurement of the Aharonov–Casher phase in an atomic system, *Phys. Rev. Lett.* **71**: 3641–3644 (1993).

23. H. Ito, K. Sakaki, T. Nakata, W. Jhe, and M. Ohtsu, Optical guidance of neutral atoms using evanescent waves in a cylindrical-core hollow fiber: Theoretical approach, *Ultramicroscopy* **61**: 91–97 (1995).

24. D. Marcuse, *Theory of Dielectric Optical Waveguides*, 2nd ed., Academic Press, New York, 1991.

25. M. A. Ol'Shanii, Yu. B. Ovchinnikov, and V. S. Lethokov, Laser guiding of atoms in a hollow optical fiber, *Opt. Commun.* **98**: 77–79 (1993).

26. C. I. Sukenik, M. G. Boshier, D. Cho, V. Sandoghdar, and E. A. Hinds, Measurement of the Casimir–Polder force, *Phys. Rev. Lett.* **70**: 560–563 (1993).

27. W. Jhe and J. W. Kim, Atomic energy-level shifts near a dielectric microsphere, *Phys. Rev. A* **51**: 1150–1153 (1995).

28. E. Merzbacher, *Quantum Mechanics*, 2nd. ed., Wiley, New York, 1970.

29. Special issue, *Appl. Phys. B* **54** (5) (1992).

30. M. J. Renn, D. Montgomery, O. Vdovin, D. Z. Anderson, C. E. Wieman, and E. A. Cornell, Laser-guided atoms in hollow-core optical fibers, *Phys. Rev. Lett.* **75**: 3253–3256 (1995).

31. H. Ito, T. Nakata, K. Sakaki, M. Ohtsu, K. I. Lee, and W. Jhe, Laser spectroscopy of atoms guided by evanescent waves in micron-sized hollow optical fibers, *Phys. Rev. Lett.* **76**: 4500–4503 (1996).

32. T. P. Dinneen, C. D. Wallace, K.-Y. N. Tan, and P. L. Gould, Use of trapped atoms to measure absolute photoionization cross sections, *Opt. Lett.* **17**: 1706–1708 (1992).

33. V. S. Letokhov, *Laser Photoionization Spectroscopy*, Academic Press, New York, 1987.

34. N. F. Ramsey, *Molecular Beams*, Oxford University Press, Oxford, 1956.

35. H. Ito, T. Nakata, K. Sakaki, W. Jhe, and M. Ohtsu, Spectroscopy of atoms guided by evanescent waves in cylindrical-core hollow fibers in *Technical Digest, Workshop on Atom Optics and Atom Interferometry, Cairns, Australia, July 1996*, paper PW1.

36. R. W. McGowan, D. M. Giltner, and S. A. Lee, Light force cooling, focusing and nanometer-scale deposition of aluminum atoms, *Opt. Lett.* **20:** 2535–2537 (1995).
37. H. Ito, K. Sakaki, T. Nakata, W. Jhe, and M. Ohtsu, Atomic funnel with evanescent waves, in *Technical Digest, Quantum Electronics and Laser Science Conference*, Anaheim, 1996, pp. 91–92.
38. H. S. Lee, B. W. Stewart, K. Choi, and H. Fenichel, Holographic nondiverging hollow beam, *Phys. Rev. A* **49:** 4922–4927 (1994).
39. S. Friebel, R. Deutschmann, M. Schffer, G. Wokurka, and W. Ertmer, Transverse confinement of atoms in a gravitational cavity by a doughnut beam, in *Technical Digest, Quantum Electronics and Laser Science Conference*, Baltimore, 1995, p. 196.
40. J. Soding, R. Grimm, and Yu. B. Ovchinnikov, Gravitational laser trap for atoms with evanescent-wave cooling, *Opt. Commun.* **119:** 652–662 (1995).

EIGHT

OPTICAL NEAR-FIELD THEORY

8.1. INTRODUCTION

In this chapter, we study the theoretical background of optical near-field problems involving near-field optical microscopy (NOM). As discussed in previous chapters, NOM involves complicated optical processes consisting of subsystems of different characteristic scales of the electromagnetic interaction. In other words, in NOM both microscopic and macroscopic electromagnetic interactions play roles on the mesoscopic scale between atomic sizes and optical wavelengths. At present, no theoretical treatment has been established that evaluates the entire NOM system and reproduces exactly the NOM images obtained from experiments. Experimental effort is being made to establish standard samples and probes which can provide NOM images with which to test theoretical descriptions. However, an interesting aspect of mesoscopic problems is not their generality, but the wide variety of experimental possibilities and interpretations they involve. Thus, optical near-field problems involve a variety of theoretical ways of description and understanding. We have to keep our view as wide as possible in order to obtain a clear understanding as well as useful insights for applications.

Parts of this and the following chapter are concerned with basic electromagnetic theory. They involve many complicated expressions, including definitions of functions, usage of subscripts, and so on. Although we try to introduce this material in a self-contained way, the reader should also refer to standard textbooks, such as Jackson [1], Born and Wolf [2], Sommerfeld [3], and Morse and Feshbach [4, 5].

For general mesoscopic physics, an empirical or intuitive model description serves as an indispensable part of a more comprehensive theory. Indeed, a powerful way to solve complicated electromagnetic problems is with a computer-aided numerical calculation, but the underlying physics is not clearly revealed. In this book, we concentrate on developing analytical treatments, and introduce approximations and model descriptions for general optical near-field phenomena. At present, neither a rigorous theory nor a well-established model has been developed for the rapidly developing discipline of near-field optics. However, this does not imply ambiguity, but rather shows the wide variety of interesting physics and potential applications.

The theoretical treatment in this chapter is as follows. The basis of the electromagnetic theory is reviewed in Section 8.2. We discuss first the response of matter to an electromagnetic field. This provides some of the most important ideas in near-field optics. The correlation and screening of electromagnetic interactions in a macroscopic system are discussed in relation to the spatial scale. On this basis, we consider near-field electromagnetic interactions. We introduce the idea of system response or system susceptibility for a material system of subwavelength size. We then present relations between the boundary conditions and the material response. In particular we examine electromagnetic boundary conditions in terms of the vector potential in order to study their relation to those in quantum mechanics. This approach provides insight into the vectorial nature of electromagnetic boundary conditions. Treatment of Fresnel's formula using evanescent waves shows a tunneling effect in the one-dimensional case of the optical problem.

In Section 8.3 the optical near-field theory is studied on the basis of scattering theory. We first present a brief review of the conventional self-consistent approach to electromagnetic scattering problems. An internal many-body interaction is involved in a macroscopic material system. A unique approach to these problems is introduced for the case of a small nonmagnetic system in terms of the polarization potential description and assumed magnetic current. We show an example in which an assumed source field replaces the electromagnetic boundary conditions. A very useful formulation is presented for optical near-field problems by introducing surface and bulk terms in the scattering potential. This provides a way of setting near-field criteria by which we can make use of near-field approximations and treat problems in a simple way using the idea of a local subsystem. Several numerical results are presented which provide a very useful and informative basis for evaluating and understanding the near-field properties, such as intensity distribution, polarization dependence, the effect of a probe tip, and an associated optical waveguide. These results provide an intuitive understanding of NOM images.

In Section 8.4 we briefly review another important approach to the optical near-field problem in terms of diffraction theory. We first introduce conventional diffraction theory based on the Kirchhoff integral and related far-field treatment in order to demonstrate the basic ideas of diffraction theory and the difficulty of applying it in the near-field regime. Next we review the famous problem of diffraction from a subwavelength-size aperture developed extensively by Bethe, Bowkamp, and Leviatan. A remarkable feature related to the optical near-field is demonstrated, namely the size-dependent localization of the optical near-field in light scattering from a small three-dimensional object.

Finally, in Section 8.5 near-field problems are studied in terms of the scaling properties of the electromagnetic correlation produced in electric polarizations of a material system. We discuss extensively the nature of the screening effects, which are dependent on the scale of the scatterer. We study the dispersion relation of photons and its relation to optical near-field problems for a system involving multiple scattering of optical fields. An intuitive model description of screened quasistatic interactions is introduced with the use of a Yukawa-type screened potential. As usual, it is difficult to provide support for the model from rigorous theories; however, the model description provides a number of insights for near-field problems. The validity and importance of the model is demonstrated by a comparison with the so-called "exact" solutions for the near-field of a small aperture discussed in Section 8.4. The meaning of the model description is studied in terms of a virtual photon with effective mass. This approach shows another way of describing the character of spatial localization or screening of the electromagnetic interaction taking place in a subwavelength-size material system. The physical and mathematical meaning of the model is also considered in relation to the quantum theory of the evanescent wave developed by Carniglia and Mandel through a comparison with elementary excitations such as the polariton description developed first by Hopfield. We use the model description to discuss the meaning of the signal transfer function and the related NOM resolution.

8.2. ELECTROMAGNETIC THEORY AS THE BASIS OF TREATING NEAR-FIELD PROBLEMS

8.2.1. Microscopic Electromagnetic Interaction and Averaged Field

Optical near-field problems involve mesoscopic electromagnetic phenomena. It is instructive to start with a microscopic consideration of the material response to a classical electromagnetic field.

Let us consider the microscopic Maxwell equations written in SI units,

$$\nabla \cdot \mathbf{E} = \frac{\rho}{\varepsilon_0}, \qquad \nabla \cdot \mathbf{B} = 0 \qquad (8.2.1)$$

$$\nabla \times \mathbf{E} = -\frac{\partial \mathbf{B}}{\partial t}, \qquad c^2 \nabla \times \mathbf{B} = \frac{\mathbf{j}}{\varepsilon_0} + \frac{\partial \mathbf{E}}{\partial t} \qquad (8.2.2)$$

Here $\mathbf{E}$ and $\mathbf{B}$ stand, respectively, for the electric and magnetic fields defined in relation to the Lorentz force exerted on a charged particle with charge q and velocity $\mathbf{v}$ as

$$\mathbf{F} = q (\mathbf{E} + \mathbf{v} \times \mathbf{B}) \qquad (8.2.3)$$

where $\varepsilon_0 c^2 = 10^7$ and c is the velocity of light. The microscopic behavior of the sources, charge density ρ, and current density $\mathbf{j}$ are determined from the quantum mechanical motion of the charges in response to the electromagnetic field. Maxwell's equations impose the continuity relation for the sources,

$$\frac{\partial \rho}{\partial t} + \nabla \cdot \mathbf{j} = 0 \tag{8.2.4}$$

and the source distributions are determined consistently with the electromagnetic fields.

However, when we consider the electromagnetic response of a material system, we are not always interested in the microscopic behavior of individual charges. Instead, we consider the material response averaged over a certain volume in space-time which characterizes our interest and the equipment used for the measurement. In other words, we trace out irrelevant variables by taking an average.

Let us consider as a simple example tracing out the microscopic behavior of an electron gas in a metallic crystal. This leads to a cumulative motion of the electron gas described in terms of plasma oscillations. The residual microscopic electron interactions are screened when we consider a macroscopic spatial distance.

Consider the density operator of the electronic gas $\hat{\rho}_{\mathbf{q}}$ described by spatial Fourier analysis with wavevector $\mathbf{q}$. The quantum behavior of the density operator obeys the Heisenberg equation [6],

$$i\hbar \frac{d\hat{\rho}_{\mathbf{q}}}{dt} = \left[\hat{\rho}_{\mathbf{q}}, \hat{\mathcal{H}}\right], \qquad \hat{\rho}(\mathbf{r}) = \sum_{\mathbf{q}} \hat{\rho}_{\mathbf{q}} e^{i\mathbf{q}\cdot\mathbf{r}} \tag{8.2.5}$$

where the Hamiltonian $\hat{\mathcal{H}}$ describes the Coulomb interaction of the electron gas in a continuum of positive charge [7],

$$\hat{\mathcal{H}} = \sum_i \frac{\hat{\mathbf{p}}_i^2}{2m} + \frac{1}{2}\sum_{\mathbf{q}} \frac{e^2}{\varepsilon_0 |\mathbf{q}|^2}\left[V_L \hat{\rho}_{\mathbf{q}}^\dagger \hat{\rho}_{\mathbf{q}} - n\right] \tag{8.2.6}$$

Here V_L is the volume of a unit lattice cell, and the second term in the Hamiltonian describes the Coulomb interaction minus the self-energy.

Substituting the temporal derivative of $\hat{\rho}$, $\hat{\rho}$, again into the Heisenberg equation and separating the term with $\mathbf{q} = \mathbf{q}'$, we get the equation of motion for $\rho_{\mathbf{q}}$ as

$$\ddot{\hat{\rho}}_{\mathbf{q}} = -\omega_p^2 \hat{\rho}_{\mathbf{q}} - \sum_i \left[\frac{\mathbf{q}\cdot\hat{\mathbf{p}}_i}{m} + \frac{\hbar^2 \mathbf{q}^2}{2m}\right]^2 e^{-i\mathbf{q}\cdot\mathbf{r}}$$

$$+ \sum_{\mathbf{q}'\neq\mathbf{q}} \frac{e^2}{\varepsilon_0 m |\mathbf{q}'|^2} \mathbf{q}\cdot\mathbf{q}' \hat{\rho}_{\mathbf{q}} \sum_i e^{i(\mathbf{q}-\mathbf{q}')\cdot\mathbf{r}} \tag{8.2.7}$$

The first term corresponds to the plasma oscillation with frequency $\omega_p = \sqrt{ne^2/\varepsilon_0 m}$. The residual terms show the screened interaction, which vanishes as the average over a scale larger than the characteristic spatial range λ due to the randomness assumed in the microscopic Coulomb scattering. For a metallic crystal, the value of the screening distance is similar to the single-electron screening length, which is defined for electrons lying on the Fermi surface as $\lambda \simeq \sqrt{a_{\text{Bohr}}/k_{\text{Fermi}}}$ [7].

The result tells us that the electron gas behaves as an oscillator when we observe the electron interaction relevant to smaller q ($|q| < 1/\lambda$) on a spatial scale larger than the screening distance λ. This type of cumulative electronic behavior is described in terms of plasmons in the second quantization framework. In contrast, the electronic interaction observed within the screening distance results in local fluctuations relevant to higher spatial frequencies, $|q| < 1/\lambda$.

Similar to the plasma oscillation in a metallic medium, we can consider a long-range correlation in an induced electric polarization p in a dielectric medium by tracing out microscopic interactions. Then the collective motion induced by an incident electromagnetic wave can be written by Fourier analysis as

$$\ddot{p}_\omega + \omega^2 p_\omega = \beta[E_\omega, B_\omega] \tag{8.2.8}$$

If we study this equation together with Maxwell's equations we find an electromagnetic wave propagating in the dielectric medium as a coupled mode of an electromagnetic field with a polarization wave. The quantum mechanical description of coupled modes involves the idea of the polariton as introduced by Hopfield [8]. For a material medium showing resonance behavior, the collective motion is of great interest from the viewpoint of solid-state physics. On the other hand, a shorter range interaction is of interest in near-field physics. This will be discussed later in this chapter.

As shown above, we will see different behaviors of electronic systems according to the characteristic spatial scale, or spatial frequency, relevant to the measurement apparatus:

$$\langle \rho \rangle_V = \langle \rho \rangle_{\text{averaged over the volume } V \text{ considered}} \tag{8.2.9}$$

In some cases, it is convenient to describe charge distributions in a material system in terms of the multipole expansion given by

$$\langle \rho \rangle_V = \rho_{\text{net}} - \nabla \cdot p_{\text{Multipole}} \tag{8.2.10}$$

$$= \rho_{\text{net}} - \nabla \cdot p + \nabla \cdot (\nabla \cdot Q) - \cdots \tag{8.2.11}$$

where the terms p and Q stand for the electric dipole and quadrupole moments, respectively. The physical quantities p, Q, and so on then provide the representation of the electromagnetic response of the material system viewed on a certain scale of space-time.

We usually discuss the electromagnetic response of matter on a scale larger than the electronic correlation length in the system:

$$V^{1/3} >> \lambda_{\text{electronic correlation length}} \tag{8.2.12}$$

In this case we can resort to a *local response* approximation for a description of the material response to an incident electromagnetic field. In the weak-field regime, we assume a local and linear response, so that we can write down induced multipoles and an averaged electromagnetic field $\mathbf{D}$ with the use of a polarizability tensor a and dielectric tensor ε, respectively, as

$$\mathbf{p} = a \cdot \mathbf{E} \tag{8.2.13}$$

$$\mathbf{D} = \varepsilon \cdot \mathbf{E} = \varepsilon_0 \mathbf{E} + (\mathbf{p} - \nabla \cdot \mathbf{Q} + \cdots) \tag{8.2.14}$$

On the other hand, when we study the electronic behavior on the scale of the microscopic electron correlation distance,

$$V^{1/3} << \lambda_{\text{electronic correlation length}} \tag{8.2.15}$$

we need to assume a *nonlocal response.* Then we describe the averaged source and field in terms of integral equations given by

$$\mathbf{p}_\alpha(\mathbf{r}, t) = \sum_\beta \int_{V'} dV' \int_{t'} dt' \, a_{\alpha\beta}(\mathbf{r}', t') \cdot \mathbf{E}_\beta(\mathbf{r} - \mathbf{r}', t - t') \tag{8.2.16}$$

$$\mathbf{D}_\alpha(\mathbf{r}, t) = \sum_\beta \int_{V'} dV' \int_{t'} dt' \, \varepsilon_{\alpha\beta}(\mathbf{r}', t') \cdot \mathbf{E}_\beta(\mathbf{r} - \mathbf{r}', t - t') \tag{8.2.17}$$

It is useful to employ Fourier analysis with respect to spatial and temporal frequencies $\mathbf{K}$ and ω, respectively, given by

$$f(\mathbf{K}, \omega) = \int dV \int dt \, f(\mathbf{r}, t) \, e^{-i\mathbf{K}\cdot\mathbf{r} + i\omega t} \tag{8.2.18}$$

Then the expressions become much simpler:

$$\mathbf{p}_\alpha(\mathbf{K}, \omega) = \sum_\beta a_{\alpha\beta}(\mathbf{K}, \omega) \cdot \mathbf{E}_\beta(\mathbf{K}, \omega) \tag{8.2.19}$$

$$\mathbf{D}_\alpha(\mathbf{K}, \omega) = \sum_\beta \varepsilon_{\alpha\beta}(\mathbf{K}, \omega) \cdot \mathbf{E}_\beta(\mathbf{K}, \omega) \tag{8.2.20}$$

The nonlocal behavior or microscopic electronic correlation becomes important when we consider electromagnetic interactions of a small material system, such as a small metallic object with a size comparable to the screening distance of the Coulomb interaction. Such theoretical treatments are available [9–12].

From the electromagnetic point of view the material response appears as the source of a macroscopic electromagnetic field, the so-called *displacement vector* **D** and *magnetic field* **H**. As we have seen in the above, both the sources and the macroscopic field are considered to be functionals of the microscopic electromagnetic field, $E(r')$ and $B(r')$,

$$\rho(r) = \mathcal{F}_\rho\,[E(r'), B(r')] \tag{8.2.21}$$

$$j(r) = \mathcal{F}_j\,[E(r'), B(r')] \tag{8.2.22}$$

$$D_i(r) = \mathcal{F}_D\,[E(r'), B(r')] \tag{8.2.23}$$

$$H_i(r) = \mathcal{F}_H\,[E(r'), B(r')] \tag{8.2.24}$$

These relations are referred to as the constitutive equations. Maxwell's equations together with the constitutive equations determine the electromagnetic field and the sources self-consistently. This results in difficulties for an explicit treatment of electromagnetic interactions in the system except for the case where a certain symmetry in the given problem significantly simplifies the solution. Therefore in order to solve an electromagnetic problem in the presence of a material system, we need to resort to special techniques which replace the explicit treatment of the electromagnetic sources.

8.2.2. Optical Response of Macroscopic Matter

In the study of the optical response of a macroscopic material system one needs to consider the fact that the electromagnetic response of the material produces a depolarization field which screens the applied field [13]. To see this, let us consider the electromagnetic response of a dielectric sphere in the linear and local response regime. The polarization density **p** induced in the sphere due to the applied field **E** is described in terms of the susceptibility χ by

$$p = \alpha E = \varepsilon_0 \chi E \tag{8.2.25}$$

where the scalar polarizability is assumed to be uniform. Then the electric displacement vector or macroscopic field is defined by

$$D = \varepsilon_0 E + p = \varepsilon_0 \varepsilon E = (1 + \chi)\varepsilon_0 E \tag{8.2.26}$$

Due to the discontinuity of the induced polarization on the sphere surface there appears a surface charge density σ given by

$$\sigma = -\nabla \cdot \mathbf{p} = \hat{\mathbf{n}} \cdot \mathbf{p} \tag{8.2.27}$$

which in turn produces a depolarization field in the sphere. For a dielectric sphere the depolarization field $\mathbf{E}_1$ is given by

$$\mathbf{E}_1 = -\frac{1}{3\varepsilon_0}\mathbf{p} \tag{8.2.28}$$

The effective field is the external field plus depolarization field,

$$\mathbf{E} = \mathbf{E}_0 + \mathbf{E}_1 = \mathbf{E}_0 - \frac{\mathbf{p}}{3\varepsilon_0} \tag{8.2.29}$$

Due to the consistency with the induced polarization $\mathbf{p}$ we have

$$\mathbf{p} = \varepsilon_0 \chi \mathbf{E} = \frac{3\chi\varepsilon_0}{3 + \chi}\mathbf{E}_0 = \frac{3\varepsilon_0(\varepsilon - 1)}{\varepsilon + 2}\mathbf{E}_0 \tag{8.2.30}$$

and we obtain the effective polarizability due to the external field including the screening factor $(\varepsilon - 1)/(\varepsilon + 2)$. Integrating the polarization density over a radius of sphere a, we obtain the polarizability of the dielectric sphere as

$$\alpha_{\text{sphere}} = 4\pi\varepsilon_0 a^3 \frac{\varepsilon - 1}{\varepsilon + 2} \tag{8.2.31}$$

These results are pictured in Fig. 8.1.

In a similar way, it can be shown that the local effective field in a homogeneous dielectric medium is given by the sum of the macroscopic field $\mathbf{E}$ in the medium, which consists of an external field plus a depolarization field $\mathbf{E} = \mathbf{E}_0 + \mathbf{E}_1$, and a

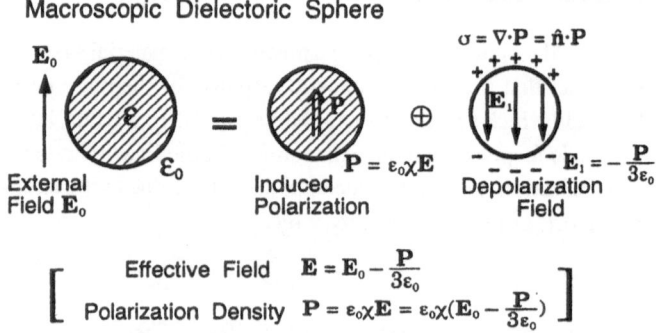

Figure 8.1. Dielectric response of a sphere; induced polarization produces a depolarization field due to discontinuities of the dielectric medium at the surface. The induced macroscopic polarization and the effective field are determined in a self-consistent way.

field E_2 in a spherical hollow bored around the local polarization p (Fig. 8.2). Considering the reversal of the depolarization in the spherical dielectric, the field in the hollow center is given by $p/3\varepsilon_0$. Then the polarization density in the medium is described in terms of the polarizability and the local field by

$$p = \alpha E_{local} = \alpha \left(E + \frac{p}{3\varepsilon_0} \right) \tag{8.2.32}$$

which serves as the self-consistent equation for the polarization density. Let the dielectric constant be ε for the homogeneous medium; then the polarization density is written as

$$p = \frac{\alpha}{1 - \alpha/(3\varepsilon_0)} E \tag{8.2.33}$$

$$= \varepsilon_0 \chi = \varepsilon_0(\varepsilon - 1) \tag{8.2.34}$$

We obtain the expression for polarizability in terms of the dielectric constant as

$$\alpha = 3\varepsilon_0 \frac{\varepsilon - 1}{\varepsilon + 2} \tag{8.2.35}$$

which is the well-known Lorentz–Lorenz formula. Here we again find the screening factor $(\varepsilon - 1)/(\varepsilon + 2)$.

It should be noted that if we consider the local field on the atomic scale, the above discussion holds only for a crystal site with good symmetry. Otherwise the

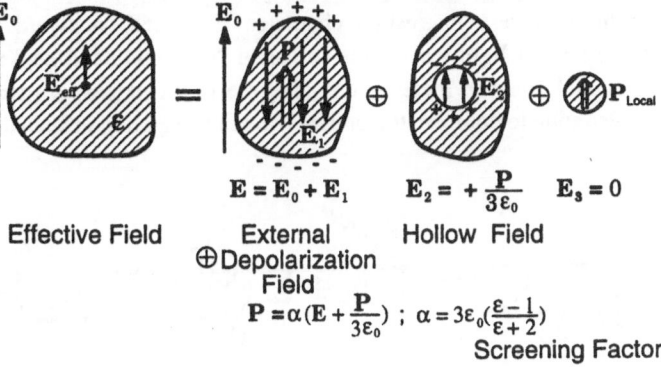

Figure 8.2. Effective field in a homogeneous dielectric medium; a small dielectric sphere in a dielectric medium feels external minus depolarization plus hollow fields. Then the effective field in the medium is screened by $(\varepsilon - 1)/(\varepsilon + 2)$.

internal local field is strongly dependent on the crystalline configuration and the site considered [13].

8.2.3. Optical Response of Small Objects and the Idea of System Susceptibility

The idea of a macroscopic material response is applicable also for an oscillating electromagnetic field provided that an average is taken over a spatial volume on the scale of one wavelength. Here an interesting situation arises when the optical response of the system consists of subwavelength-size elements. Compared to the homogeneous case, the averaged field should be modified to include the multiple scattering taking place in such a system. That is, we need to account for both the optical response of the individual elements and the interaction between the elements, which depends on the geometrical arrangement, as shown in Fig. 8.3. We can then introduce the important idea of system susceptibility [14–16].

The optical response of individual small elements is described in terms of a polarizability which is proportional to the macroscopic dielectric constant and the volume of each element to the first-order approximation. However, the polarizability of the system is not simply given by the sum of the individual elements. The multiple scattering process enhances or reduces the optical response of the system. For a system that is small compared to the incident optical wavelength, the interactions create a near-field effect where evanescent or inhomogeneous components play important roles. We give a theoretical treatment of these effects in Section 8.3.

It is worth noting that even for a macroscopic system which consists of regular material elements of near-wavelength size subject to small fluctuations, the optical response exhibits a very interesting character. Enhanced backward scattering, strong interference of light waves (the so-called "photonic band gap") [17], and three-dimensional "photon localization" [18] are under extensive research. It is especially intriguing to compare the optical waves propagating in such a "photonic

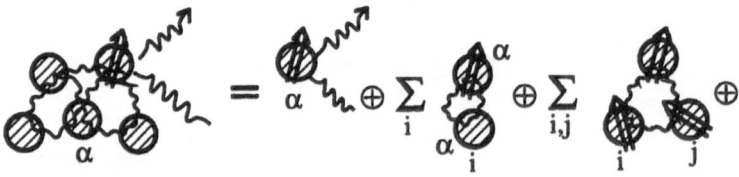

Figure 8.3. Optical response of a dielectric system and multiple scattering processes. The optical response of the system is determined by a direct scattering and associated internal interactions described as higher order many-body processes.

crystal" with quantum mechanical waves, electron waves, propagating in an atomic crystal [19].

With respect to the near-field optics, the analogy between electron tunneling and "photon tunneling" is useful in basic physical research as well as in cavity QED. The analogy is also instructive when we consider quantum electronic devices such as the quantum well, quantum wire, and quantum dot.

8.2.4. Electromagnetic Boundary Value Problem

8.2.4.1. Macroscopic Maxwell Equations and Electromagnetic Boundary Conditions

In order to avoid an explicit treatment of the sources we usually start with the macroscopic Maxwell equations in which the electromagnetic response of the medium is included in the macroscopic field variables. The consistency of the field variables and material response is described by electromagnetic boundary conditions. Let us briefly review the macroscopic Maxwell equations for the nonmagnetic case.

A microscopic consideration of a nonmagnetic medium shows that the discontinuity and temporal variation of the polarization density act as a microscopic charge density and current, respectively.

With the use of a macroscopic average of the polarization density over the very small volume under consideration, these sources are described by

$$\rho_{pol} = -\nabla \cdot \mathbf{p}, \qquad \mathbf{j}_{pol} = \frac{\partial \mathbf{p}}{\partial t} \tag{8.2.36}$$

The averaged electric charge density ρ_E and current density $\mathbf{j}_E$ are described by

$$\rho_E = \rho_{ext} + \rho_{pol} \tag{8.2.37}$$

$$\mathbf{j}_E = \mathbf{j}_{cond} + \mathbf{j}_{pol} \tag{8.2.38}$$

where ρ_{ext} and $\mathbf{j}_{cond}$ are the isolated extra charge density and conduction current density, respectively.

We can include the source fields in the macroscopic field variables as

$$\mathbf{D} = \varepsilon_0 \mathbf{E} + \mathbf{p}, \qquad \mathbf{H} = \varepsilon_0 c^2 \mathbf{B} \tag{8.2.39}$$

where $\mathbf{D}$ and $\mathbf{H}$ are the electric displacement and magnetic field, respectively. Then the macroscopic Maxwell equations are written as

$$\nabla \cdot \mathbf{D} = \rho_{ext}, \qquad \nabla \cdot \mathbf{H} = 0 \tag{8.2.40}$$

$$\nabla \times \mathbf{E} = -\frac{\partial \mathbf{B}}{\partial t}, \qquad \nabla \times \mathbf{H} = \mathbf{j}_{cond} + \frac{\partial \mathbf{D}}{\partial t} \qquad (8.2.41)$$

For a magnetic medium we need to include the so-called "molecular current" term $\mathbf{j}_{mol}$ which is described in terms of the magnetization density $\mathbf{M}$ as $\mathbf{j}_{mol} = \nabla \times \mathbf{M}$. We can then include this in the macroscopic magnetic field as $\mathbf{H} = \varepsilon_0 c^2 \mathbf{B} - \mathbf{M}$ [20].

Next the consistency between the macroscopic field and included sources is described by the boundary conditions. In order to evaluate the macroscopic behavior of the electromagnetic field, we start with the integral form of the Maxwell equations written for macroscopic field variables,

$$\oint_{\text{Surface}} \mathbf{D} \cdot \mathbf{n} \, dS = \int_V \rho \, dV \qquad (8.2.42)$$

$$\oint_{\text{Surface}} \mathbf{B} \cdot \mathbf{n} \, dS = 0 \qquad (8.2.43)$$

and

$$\oint_{\text{Loop}} \mathbf{E} \cdot d\mathbf{s} = -\int_{\text{Surface}} \frac{\partial \mathbf{B}}{\partial t} \cdot \mathbf{n} \, dS \qquad (8.2.44)$$

$$\oint_{\text{Loop}} \mathbf{H} \cdot d\mathbf{s} = \int_{\text{Surface}} \left(\mathbf{j} + \frac{\partial \mathbf{D}}{\partial t} \right) \cdot \mathbf{n} \, dS \qquad (8.2.45)$$

Applying the integration to the infinitesimal volume and loop across a boundary surface of media 1 and 2, we obtain the continuity and discontinuity relations, or the electromagnetic boundary conditions, as

$$(\mathbf{D}_2 - \mathbf{D}_1) \cdot \mathbf{n} = \sigma_{\text{surface}} \qquad (8.2.46)$$

$$(\mathbf{B}_2 - \mathbf{B}_1) \cdot \mathbf{n} = 0 \qquad (8.2.47)$$

where σ_{surface} stands for the surface charge density, and

$$\mathbf{n} \times (\mathbf{E}_2 - \mathbf{E}_1) = 0 \qquad (8.2.48)$$

$$\mathbf{n} \times (\mathbf{H}_2 - \mathbf{H}_1) = \mathbf{j}_{\text{surface}} \qquad (8.2.49)$$

where $\mathbf{j}_{\text{surface}}$ stands for the surface current density.

8.2.4.2. Electromagnetic Boundary Value Problems and Green's Theorem

An effective way to solve electromagnetic problems is to find an optimal mode function appropriate for describing a given material system. When one finds a set of basis mode functions, one can construct a field distribution under arbitrary initial conditions by means of a linear combination of these mode functions.

As discussed above, electromagnetic problems, i.e., the interaction of an electromagnetic field with matter, are described by the Maxwell equations and the constitutive equations,

$$\mathbf{D} = \mathcal{F}_{\mathbf{D}} \, [\mathbf{E}, \mathbf{B}] \tag{8.2.50}$$

$$\mathbf{H} = \mathcal{F}_{\mathbf{H}} \, [\mathbf{E}, \mathbf{B}] \tag{8.2.51}$$

which describe the material response consistent with the electromagnetic field. It would seem to be very difficult to solve such a set of functional equations. However, we often encounter electromagnetic problems in which the macroscopic material behavior is well known and can be replaced by a set of continuity relations of the field variable and its derivative. These boundary value problems provide the basic approach for finding a set of optimal mode functions.

A more complicated problem is to find the optimal mode functions which take into account the microscopic or nonlocal response of matter to an electromagnetic field. This requires a Hartree–Fock type method and will be discussed later.

There are several different ways of setting boundary value problems [1, 3–5]. Green's theorem provides one of the most intuitive ways. Applying Gauss' law for the vector field $\phi\nabla\psi$ as

$$\int_V \nabla \cdot (\phi\nabla\psi) \, dV = \oint_S \phi\nabla\psi \cdot \mathbf{n} \, dS \tag{8.2.52}$$

we get the first Green's identity

$$\int_V (\phi\nabla^2\psi + \nabla\phi \cdot \nabla\psi) \, dV = \oint_S \phi \, \frac{\partial\psi}{\partial n} \, dS \tag{8.2.53}$$

Taking the antisymmetric sum of this identity for ψ and ϕ, we get a second identity, which is usually referred to as "Green's theorem,"

$$\int_V (\phi\nabla^2\psi - \psi\nabla^2\phi) \, dV = \oint_S \left(\phi \, \frac{\partial\psi}{\partial n} - \psi \, \frac{\partial\phi}{\partial n} \right) dS \tag{8.2.54}$$

This relation tells us that for a field which satisfies the Poisson equation

$$\nabla^2\phi(\mathbf{r}) = -s(\mathbf{r}) \tag{8.2.55}$$

by taking the Green's function $G(\mathbf{r}, \mathbf{r}')$ for $\psi(\mathbf{r})$

$$\nabla^2 \psi(\mathbf{r}) = \nabla^2 G(\mathbf{r}, \mathbf{r}') = -4\pi\delta\,(\mathbf{r} - \mathbf{r}') \tag{8.2.56}$$

we find that the entire field $\phi(\mathbf{r})$ in the volume is described by the boundary values together with the sources inside as follows:

$$\phi(\mathbf{r}) = \frac{1}{4\pi}\int_V s(\mathbf{r}')G(\mathbf{r},\,\mathbf{r}')\,dV + \frac{1}{4\pi}\oint_S \left[G(\mathbf{r},\,\mathbf{r}')\frac{\partial\phi}{\partial n} - \phi\,\frac{\partial G(\mathbf{r},\,\mathbf{r}')}{\partial n}\right] dS \tag{8.2.57}$$

The Green's function for the Poisson equation is usually written as

$$G(\mathbf{r},\,\mathbf{r}') = \frac{1}{|\mathbf{r} - \mathbf{r}'|} + \mathbf{f}(\mathbf{r},\,\mathbf{r}') \qquad \text{with} \qquad \nabla^2 f(\mathbf{r},\,\mathbf{r}') = 0 \tag{8.2.58}$$

In the case of a vector field, the idea of the Green's function is extended to the Green's dyadic $\mathbf{G}(\mathbf{r},\,\mathbf{r}')$, which determines the relation of the vector field to a vector point source.

We can solve two types of boundary value problems with this result. One is of the Dirichlet type, in which (a) the value of the field variable ϕ is specified on the surface S, and (b) the restriction $G_D(\mathbf{r},\,\mathbf{r}') = 0$ is imposed for $\mathbf{r}'$ lying on the surface S. The other is the Neumann-type problem, in which (a) the value of the normal derivative of the field variable $\partial\phi/\partial n$ is specified on the surface S, and (b) the restriction $\partial G_D\,(\mathbf{r},\,\mathbf{r}')/\partial n' = 0$ is imposed for $\mathbf{r}'$ lying on the surface S.

For both the Dirichlet and Neumann problems, the first Green's identity ensures the uniqueness of the solution. It is known that the specification of both the value and the normal derivative of the field variable is overconstraining for any problem on a closed surface. This type of boundary value problem is referred to as the Cauchy problem.

One can also set a mixed boundary value problem in which the value of the field variable is specified for part of the surface S and the normal derivative for the rest of S. Such a problem often arises in near-field optics. An important example is the diffraction of light by a small aperture in a conducting plane. For example, consider a circular hole with radius smaller than the incident wavelength. In this problem the values of the potential are given on the conducting surface, and the continuity relations of the field, i.e., the derivatives of the potential, are set in the aperture. However, in such a case, it is not easy to find an appropriate Green's function [1]. One approach is to set a well-defined boundary value problem for which either the Dirichlet or the Neumann Green's function applies and replace the continuity relation which does not apply the assumed boundary problem by imaginary sources on the boundary [21]. For the small-aperture problem one first considers a perfect-conducting plane and includes the effect of the aperture by replacing the continuity relation in the aperture region by a set of assumed magnetic currents running on the surface around the aperture. Then one can determine both the field and the source in a self-consistent manner [22, 23]. We study this problem in Section 8.4, since the derived solution highlights an important feature of near-field optics.

It is instructive to give an example of a popular boundary-value problem in the electrostatic regime. Consider a point electric charge placed near the planar boundary of a dielectric medium with different dielectric constants ε_1 and ε_2 as shown in Fig. 8.4a, where the field exhibits a singularity in medium 1 due to the charge q. The electric field in medium 1 which fulfills the given boundary conditions is

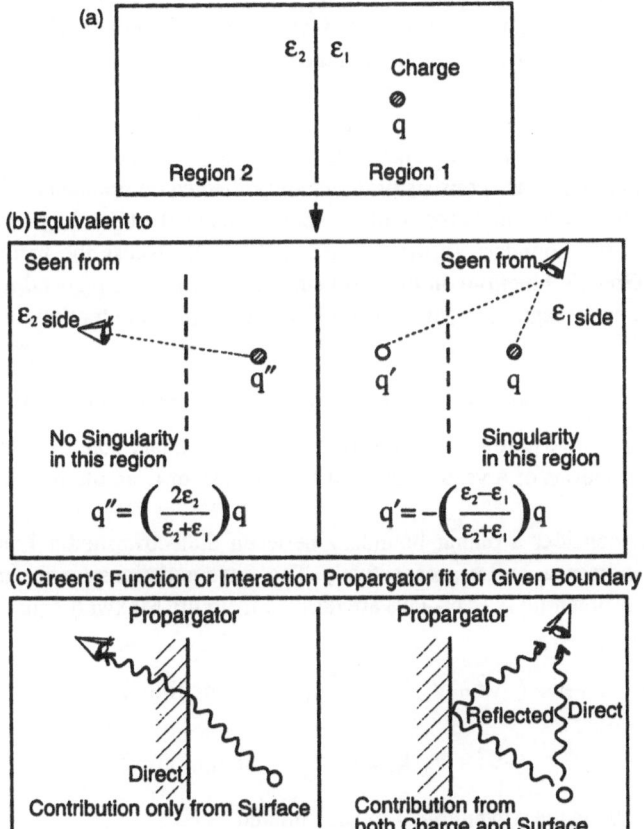

Figure 8.4. Interaction of a point charge with a dielectric boundary. (a) A charged particle is placed in region 1 near the dielectric boundary. (b) Depending on the boundary conditions, there arise two different problems. An additional image charge appears when it is viewed from region 1, whereas a screened charge is seen from region 2. (c) Corresponding Green's functions, or interaction propagators. There arise two different paths in the effective Green's function when the problem is viewed from region 1.

equivalent to that due to the real charge q plus an image charge q' in a uniform medium with dielectric constant ε_1 (Fig. 8.4b). The image charge q' is opposite in sign and reduced by a screening factor $(\varepsilon_2 - \varepsilon_1)/(\varepsilon_2 + \varepsilon_1)$ and placed at the symmetry point with respect to the boundary plane. On the other hand, the field in medium 2, where no singularity arises, is equivalent to that due to a screened charge $q'' = 2q\varepsilon_2/(\varepsilon_2 + \varepsilon_1)$ at the original position in a uniform medium with dielectric constant ε_2 (Fig. 8.4b).

This result shows several important features of boundary value problems. First, the boundary condition can be replaced by an appropriate source distribution. Second, the field in the medium containing a source or singularity has contributions both from the real charge and from its image, which is actually the contribution from the boundary surface. The field in the medium with no singularity only has a contribution from the boundary surface. This is shown in Fig. 8.4c. This demonstrates how the Green's function or interaction propagator appropriate for a given material system is constructed. That is, an appropriate Green's function for a system with a boundary surface consists of direct and reflected components. The Green's function therefore accounts for all of the sources inside the boundary as well as the response of the boundary to these sources. We will discuss the formal way to construct Green's functions in the following section. Before preceeding, we first study the electromagnetic boundary conditions in terms of potentials.

8.2.4.3. Electromagnetic Boundary Conditions in Terms of Potentials

It is instructive to rewrite the electromagnetic boundary conditions in terms of continuity relations of a vector potential for the case of a simple planar dielectric boundary.

Let us consider a planar boundary between dielectric media 1 and 2 with refractive indices n_1 and n_2, respectively. The wave equations for the monochromatic vector potential $\mathbf{A}\exp(-i\omega t)$ are derived from the Maxwell equations in the Coulomb gauge as

$$(-\nabla^2 + V_1)\mathbf{A}_1 = K^2\mathbf{A}_1 \qquad \text{in medium 1} \qquad (8.2.59)$$

$$(-\nabla^2 + V_2)\mathbf{A}_2 = K^2\mathbf{A}_2 \qquad \text{in medium 2} \qquad (8.2.60)$$

where $K = \omega/c$ is the wavenumber in vacuum and

$$V_i = K^2(1 - n_i^2) \qquad (8.2.61)$$

is the uniform potential in medium $i = 1, 2$. The electromagnetic boundary conditions are represented by the continuity relations for the field value and its derivative on the boundary plane at $z = 0$. Let us consider a nonmagnetic homogeneous medium [24].

For s-polarized waves, the continuity relations are given by

$$\begin{cases} \mathbf{A}_{1s}\Big|_{z=0} = \mathbf{A}_{2s}\Big|_{z=0} \\ \dfrac{\partial}{\partial z}\mathbf{A}_{1s}\Big|_{z=0} = \dfrac{\partial}{\partial z}\mathbf{A}_{2s}\Big|_{z=0} \end{cases} \tag{8.2.62}$$

This indicates that both the field value and its derivative are continuous at the boundary for the case of s-polarized waves. The problem is therefore equivalent to the scattering of quantum mechanical waves at a potential step $V_2 - V_1$, so that one can consider the wave equations as a vector Schrödinger equation.

For p-polarized waves, the continuity relations are given by

$$\begin{cases} n_1\mathbf{A}_{1p}\Big|_{z=0} = n_2\mathbf{A}_{2p}\Big|_{z=0} \\ \dfrac{1}{n_1}\dfrac{\partial}{\partial z}\mathbf{A}_{1p}\Big|_{z=0} = \dfrac{1}{n_2}\dfrac{\partial}{\partial z}\mathbf{A}_{2p}\Big|_{z=0} \end{cases} \tag{8.2.63}$$

This shows that there exists a discontinuity at the boundary for p-polarized waves, so that the boundary condition itself depends on the ratio n_1/n_2. This shows the distinguishing feature of electromagnetic boundary value problems. However, we will see in Section 8.3.2 that such a discontinuity can be removed if we resort to a different expression for the macroscopic field by introducing scattering potentials.

It is instructive to study the problems of reflection and refraction of a plane incident wave at a planar dielectric boundary using the continuity relations discussed above. Let us assume an incident plane wave at an angle θ_1 to the normal of the boundary traveling from the $z < 0$ side of dielectric medium 1. Medium 1 has a refractive index n_1 larger than the index n_2 of medium 2, which occupies $z > 0$ (Fig. 8.5). Due to the spatial translational symmetry along the boundary surface, the problem is basically one-dimensional, so that the wavevector parallel to the surface $K_\parallel = K\, n_1 \sin \theta_i$ is continuous. Because we consider the boundary conditions separately for s- and p-polarizations, it is enough to consider the one-dimensional scalar wave equation

$$\left(-\frac{d^2}{dz^2} + V_i\right)\varphi_i(z) = K_{zi}^2\varphi_i(z), \qquad i = 1, 2 \tag{8.2.64}$$

where the z components of the wavenumbers are determined from the dispersion relation $K^2 = K_\parallel^2 + K_{zi}^2$ as

$$K_{z1} = Kn_1 \cos \theta_i \tag{8.2.65}$$

$$K_{z2} = K \sqrt{n_2^2 - n_1^2 \sin^2 \theta_i} \tag{8.2.66}$$

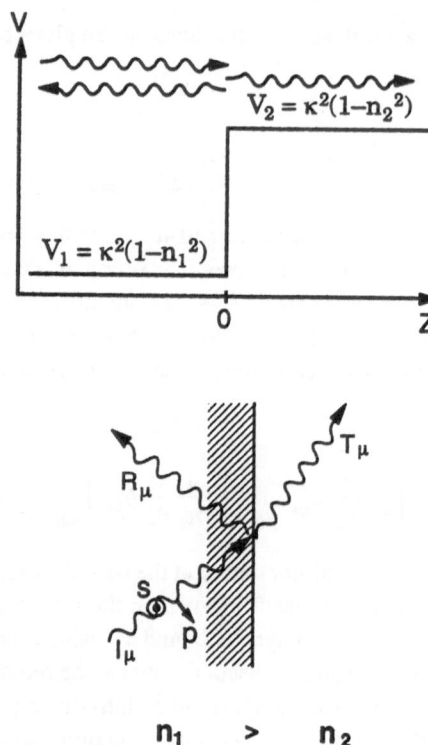

Figure 8.5. Reflection and refraction of light waves and their one-dimensional reduction viewed as a scattering process at a step potential.

The general solution is then given in the form

$$\varphi_i(z) = A_i e^{iK_{zi}z} + B_i e^{-iK_{zi}z} \tag{8.2.67}$$

According to the condition of incidence, $A_1 = I_\mu$ corresponds to the μ-polarized incident wave, $B_1 = R_\mu$ to the reflected wave, $A_2 = T_\mu$ to the refracted or transmitted wave, and $B_2 = 0$.

Using the continuity relations on the boundary, we get the following relations for the field amplitudes:

- For the s-polarized waves

$$I_s + R_s = T_s \tag{8.2.68}$$

$$K_{z1}(I_s - R_s) = K_{z2}T_s \tag{8.2.69}$$

- For the p-polarized waves

$$n_1(I_p + R_p) = n_2 T_p \tag{8.2.70}$$

$$\frac{K_{z1}}{n_1}(I_p - R_p) = \frac{K_{z2}}{n_2} T_p \tag{8.2.71}$$

These equations yield, respectively,

$$T_s = \frac{2K_{z1}}{K_{z1} + K_{z2}} I_s \tag{8.2.72}$$

$$T_p = \frac{2K_{z1}}{(n_2/n_1)\,K_{z1} + (n_1/n_2)\,K_{z2}} I_p \tag{8.2.73}$$

which correspond to the well-known Fresnel formulas. The reflected amplitudes are then

$$R_s = \frac{K_{z1} - K_{z2}}{K_{z1} + K_{z2}} I_s \tag{8.2.74}$$

$$R_p = \frac{n_2^2 K_{z1} - n_1^2 K_{z2}}{n_2^2 K_{z1} + n_1^2 K_{z2}} I_p \tag{8.2.75}$$

For the incident angle corresponding to $\sin \theta_i > n_2/n_1$ (assumed <1), the wavenumber of the transmitted wave K_{z2} is real and therefore a normal (homogeneous) reflection–transmission event takes place. The conservation relation for the energy flow in the homogeneous transmitted and reflected waves is then easily obtained by multiplying the continuity equations,

$$K_{z1}(I_\mu^2 - R_\mu^2) = K_{z2} T_\mu^2 \qquad (\mu = s, p) \tag{8.2.76}$$

For $\sin \theta_i < n_2/n_1$, K_{z2} becomes pure imaginary, so that the transmitted wave exhibits exponential decay in medium 2, which corresponds to the evanescent wave of Fresnel. The reflection and refraction exhibit total internal reflection. For example, assuming $n_1 = 1.5$ (glass) and $n_2 = 1$ (air), the critical angle for total internal reflection is $\theta_c = \sin^{-1}(n_2/n_1) = 42$ deg. The corresponding scattering process is shown in Fig. 8.6.

It should be noted that there exists another critical angle at which the ratio of transmission coefficients for s- and p-polarized evanescent waves is equal to unity. For incident amplitudes $I_s = I_p$, the ratio of transmission coefficients for s and p is given by

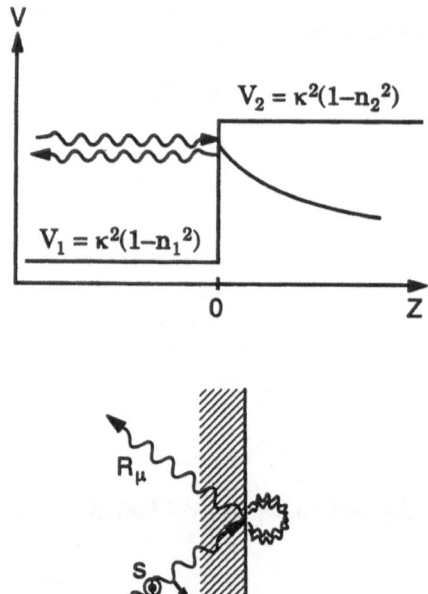

Figure 8.6. Total internal reflection of light waves and their one-dimensional reduction viewed as a scattering process at a step potential. The evanescent wave in the lower refractive index side corresponds to a penetrating wave function into the potential barrier. Due to this we can compare the one-dimensional reduction of total internal reflection with tunneling effects.

$$\left| \frac{T_s}{T_p} \right|_{I_s = I_p} = \left| \frac{(n_2/n_1)\, K_{z1} + (n_1/n_2)\, K_{z2}}{K_{z1} + K_{z2}} \right|$$

$$= \left| \frac{n_2 \cos \theta_i + i\, (n_1/n_2)\, \sqrt{n_1^2 \sin^2 \theta_i - n_2^2}}{n_1 \cos \theta_i + i\sqrt{n_1^2 \sin^2 \theta_i - n_2^2}} \right| \tag{8.2.77}$$

This yields

$$\left| \frac{T_s}{T_p} \right| \geq 1 \quad \text{for} \quad \frac{\pi}{2} > \theta_1 \geq \sin^{-1} \sqrt{\frac{2n_2^2}{n_1^2 + n_2^2}} \tag{8.2.78}$$

For $n_1 = 1.5$ and $n_2 = 1$ this gives the second critical angle $\theta_{c'} = 51.7$ deg. For incident angles above this critical value the amplitude of the s-polarized evanescent wave exceeds that of the p-polarized wave for equal incident amplitudes.

8.3. OPTICAL NEAR-FIELD THEORY AS AN ELECTROMAGNETIC SCATTERING PROBLEM

In this section we discuss the nonmagnetic scattering process for a vector field in the near-field regime. Let us consider a monochromatic vector field $E(r, \omega)$ with temporal evolution $\exp[-i\omega t]$ described by the wave equation

$$-\nabla \times \nabla \times E(r, \omega) + k^2\varepsilon(r, \omega)E(r, \omega) = 0 \qquad (8.3.1)$$

where $k = \omega/c$ is the wavenumber in vacuum. The wave equation is reduced to the Helmholtz equation

$$\left[\nabla^2 + k^2\varepsilon(r, \omega)\right] E(r, \omega) = 0 \qquad (8.3.2)$$

when the gauge condition $\nabla \cdot E(r, \omega) = 0$ is imposed.

A scattering system is described in terms of a position-dependent dielectric tensor $\varepsilon_s(r, \omega)$ buried in a homogeneous medium with $\varepsilon_h(r, \omega)$. Let the total dielectric tensor be $\varepsilon(r, \omega) = \varepsilon_h(r, \omega) + \varepsilon_s(r, \omega)$. When we rewrite the wave equation as

$$-\nabla \times \nabla \times E(r, \omega) + k^2\varepsilon_h(r, \omega)E(r, \omega) = -k^2\varepsilon_s(r, \omega)E(r, \omega) \qquad (8.3.3)$$

the polarizability tensor of the scatterer $V_s(r, \omega) = -k^2\varepsilon_s(r, \omega)$ is considered a scattering potential.

In the following we first study the general treatment of a multiple scattering problem using a self-consistent approach. Next, we consider scattering theory in the near-field regime, which provides the basis for understanding the nature of near-field optics as well as the NOM, including the interpretation of NOM images.

8.3.1. Self-Consistent Approach for Multiple Scattering Problems

8.3.1.1. Green's Dyadic and Self-Consistent Equation

Instead of finding the appropriate mode functions let us consider the tensor called the "true Green's dyadic" $G(r, r', \omega)$, which satisfies the tensor equation

$$[-\nabla \times \nabla \times + k^2\varepsilon_h(r, \omega) + k^2\varepsilon_s(r, \omega)]G(r, r', \omega) = -\delta(r - r')1 \qquad (8.3.4)$$

and all of the boundary conditions imposed on the system (1 stands for the identity tensor). When the true Green's dyadic is obtained one can immediately write down the solution of the vector field in the form of an integral equation

$$E(r, \omega) = E^0(r, \omega) + \int G(r, r', \omega) \cdot k^2\varepsilon_s(r', \omega) \cdot E^0(r', \omega) \, d^3r' \qquad (8.3.5)$$

where the incident field $E^0(r, \omega)$ satisfies the vector wave equation in a homogeneous space

$$-\nabla \times \nabla \times E^0(r, \omega) + k^2\varepsilon_h(r, \omega)E^0(r, \omega) = 0 \qquad (8.3.6)$$

Here we assume a local linear response for the material system. The second term in this integral equation represents the scattered field from the whole dielectric system including any multiple interaction processes taking place in the system. That is, the microscopic internal interactions involved in the system itself are included, or renormalized, in the true Green's dyadic $G(r, r', \omega)$.

However, in practice, it is hard to obtain the true Green's dyadic which represents the entire response of the material system including the complete boundary conditions. An exception is the case of a scattering system with a high degree of spatial symmetry, such as a spherical, cylindrical, or planar system. A practical approach for general scattering problems is to find an appropriate form of the Green's dyadic by means of an iterated integral, or the so-called self-consistent approach.

Although an important part of the self-consistent approach lies in the numerical computation of the field variables, this is beyond the scope of this book. Instead, we briefly review the theoretical framework of the self-consistent approach to scattering problems.

In general, we can find a homogeneous Green's dyadic $G^0(r, r', \omega)$ which satisfies the equation

$$[-\nabla \times \nabla \times + k^2\varepsilon_h(r, \omega)] \, G^0(r, r', \omega) = -1\delta(r - r') \qquad (8.3.7)$$

as well as the imposed boundary conditions for a reference system with a simple geometrical configuration. Then we can account for the effect of the scatterer in terms of the following self-consistent equation, which is generally referred to as the Lippmann–Schwinger equation:

$$E(r, \omega) = E^0(r, \omega) + \int G^0(r, r', \omega) \cdot k^2\varepsilon_s(r', \omega) \cdot E(r', \omega) \, d^3r' \qquad (8.3.8)$$

This implies that the true field $E(r, \omega)$ consists of the incident field $E^0(r, \omega)$ and contributions from the induced source field $\varepsilon_s(r', \omega) \cdot E(r', \omega)$ distributed in the reference system. The homogeneous Green's dyadic $G^0(r, r', \omega)$ serves as an interaction propagator from r' to r in the reference system.

A case which one often encounters in near-field problems is the Green's dyadic for the reference system of a half-space bounded by a planar dielectric surface. In this case, Fresnel evanescent waves with the half-space coordinate $r = (r_\parallel, z) = (x, y, z)$ $(z > 0)$ given by

$$f(r_\parallel, z) = \hat{\varepsilon}e^{(-k_z z + ik_\parallel \cdot r_\parallel)} \qquad (8.3.9)$$

with polarization $\hat{\varepsilon}$ and wavenumber

$$|\mathbf{k}|_\parallel^2 - k_z^2 = k^2 \qquad (8.3.10)$$

are the appropriate basis for describing $\mathbf{G}^0(\mathbf{r}, \mathbf{r}', \omega)$ in the near-field regime:

$$\mathbf{G}^0(\mathbf{r}, \mathbf{r}', \omega) = \int \frac{d^2 k_\parallel}{k} \mathbf{f}(\mathbf{r}_\parallel, z) \mathbf{f}^*(\mathbf{r}'_\parallel, z'). \qquad (8.3.11)$$

8.3.1.2. Formal Solutions for Self-Consistent Equations

Instead of using the field variables, we can describe the self-consistent equation in terms of Green's dyadic. The corresponding tensor equation is usually referred to as the Dyson equation:

$$\mathbf{G}(\mathbf{r}, \mathbf{r}', \omega) = \mathbf{G}^0(\mathbf{r}, \mathbf{r}', \omega) + \int \mathbf{G}^0(\mathbf{r}, \mathbf{r}'', \omega) \cdot k^2 \varepsilon_s(\mathbf{r}'', \omega) \cdot \mathbf{G}(\mathbf{r}'', \mathbf{r}', \omega)\, d^3 r'' \qquad (8.3.12)$$

This implies that the true Green's dyadic, or equivalently the interaction propagator, consists of the direct process $\mathbf{G}^0(\mathbf{r}, \mathbf{r}', \omega)$ from $\mathbf{r}'$ to $\mathbf{r}$ and scattered processes $\mathbf{G}^0(\mathbf{r}, \mathbf{r}'', \omega) \cdot k^2 \varepsilon_s(\mathbf{r}'', \omega) \cdot \mathbf{G}(\mathbf{r}'', \mathbf{r}', \omega)$ via a point $\mathbf{r}''$ in the reference system.

If we use the operator formalism of the Green's dyadic

$$[-\nabla \times \nabla \times + k^2 \varepsilon_h(\mathbf{r}, \omega) + k^2 \varepsilon_s(\mathbf{r}, \omega)] \hat{\mathbf{G}}_\omega = -1 \qquad (8.3.13)$$

and

$$[-\nabla \times \nabla \times + k^2 \varepsilon_h(\mathbf{r}, \omega)] \hat{\mathbf{G}}_\omega^0 = -1 \qquad (8.3.14)$$

we can rewrite the Dyson equation as

$$\hat{\mathbf{G}}_\omega = \hat{\mathbf{G}}_\omega^0 + \hat{\mathbf{G}}_\omega^0 \cdot k^2 \varepsilon_s(\mathbf{r}, \omega) \cdot \hat{\mathbf{G}}_\omega \qquad (8.3.15)$$

Here we have utilized the operator relations

$$\left[\hat{\mathcal{D}} + k^2 \varepsilon_h(\mathbf{r}, \omega)\right] \hat{\mathbf{G}}_\omega = [\hat{\mathcal{D}} + k^2 \varepsilon_h(\mathbf{r}, \omega)] \hat{\mathbf{G}}_\omega^0 (\hat{\mathbf{G}}_\omega^0)^{-1} \hat{\mathbf{G}}_\omega \qquad (8.3.16)$$

$$= -(\hat{\mathbf{G}}_\omega^0)^{-1} \hat{\mathbf{G}}_\omega$$

$$= -1 - k^2 \varepsilon_s(\mathbf{r}, \omega) \cdot \hat{\mathbf{G}}_\omega \qquad (8.3.17)$$

where $\hat{\mathcal{D}}$ stands for the differential operator $-\nabla \times \nabla \times$.

A formal solution for this operator equation is simply given by

$$\hat{\mathbf{G}}_\omega = \left[1 - \hat{\mathbf{G}}_\omega^0 k^2 \varepsilon_s(\mathbf{r}, \omega)\right]^{-1} \hat{\mathbf{G}}_\omega^0 \qquad (8.3.18)$$

In order to solve this iterated integral equation one can resort to a numerical computation using a discretized N-object system. In such an approach, the main concern is to find a way to reduce the number of calculation steps and get a good algorithm for calculating $3N \times 3N$ matrix equations. There has been extensive work in this direction [25–28]. The theoretical formulations for the self-consistent approach are rather concrete and straightforward; however, the associated numerical calculations involve many problems which need to be carefully examined, such as the convergence and numerical stability of the iterative calculations. As a result it is hard to obtain a physical understanding of the problem with only a self-consistent approach.

8.3.1.3. Interaction of Dielectric Systems and the Idea of System Susceptibility

To serve as an NOM theory, the purpose of a self-consistent calculation is to evaluate the induced polarization on the probe tip as a result of the sample–probe interaction (Fig. 8.7a) in the reference system (Fig. 8.7b). The probe-tip polarization then serves as a source producing light waves propagating into the photodetector via a certain waveguide or signal collection scheme.

To this end let us write down an iterated integral equation representing the self-consistent polarization on the probe tip,

$$\mathbf{p}(\mathbf{r}_p, \omega) = \mathbf{p}^0(\mathbf{r}_p, \omega) + \alpha(\omega) \int_{\text{probe}} \mathbf{S}(\mathbf{r}_p, \mathbf{r}', \omega) \cdot \mathbf{p}(\mathbf{r}', \omega) \, d^3 r' \qquad (8.3.19)$$

Here $\mathbf{p}^0(\mathbf{r}_p, \omega)$ represents the polarization induced at the point $\mathbf{r}_p$ of the probe tip by the self-consistent field of the sample system and $\alpha(\omega)$ is the linear polarizability. $\mathbf{S}(\mathbf{r}, \mathbf{r}', \omega)$ represents the round-trip interaction propagator by which the induced polarization at position $\mathbf{r}_p$ in the probe tip modifies the self-consistent field via multiple interactions with the sample system as well as the probe itself (Fig. 8.7c).

The round-trip interaction propagator $\mathbf{S}(\mathbf{r}, \mathbf{r}', \omega)$ involves contributions from the probe tip itself, i.e., a self-interaction $\mathbf{S}_{\text{probe}}$, and from the sample system, $\mathbf{S}_{\text{sample}}$. The overall propagator $\mathbf{S}(\mathbf{r}, \mathbf{r}', \omega) = \mathbf{S}_{\text{probe}} + \mathbf{S}_{\text{sample}}$ then determines an effective susceptibility or "system susceptibility" of the probe tip which is coupled to the entire sample–probe system via a near-field interaction (Figs. 8.7d and 8.7e). The system susceptibility is therefore modified by the sample–probe geometry, i.e., the relative position of the probe tip with respect to the sample system. Similar formulations accounting for multipolar behavior has also been developed using the multipolar expansions of the field mode and Green's dyadic [25].

The idea of system susceptibility shows one of the most important physical features of near-field optics and photonics. When the probe tip is replaced by an atomic particle, the idea of system susceptibility implies that the light scattering properties of the atom are modified by the presence of a material system nearby.

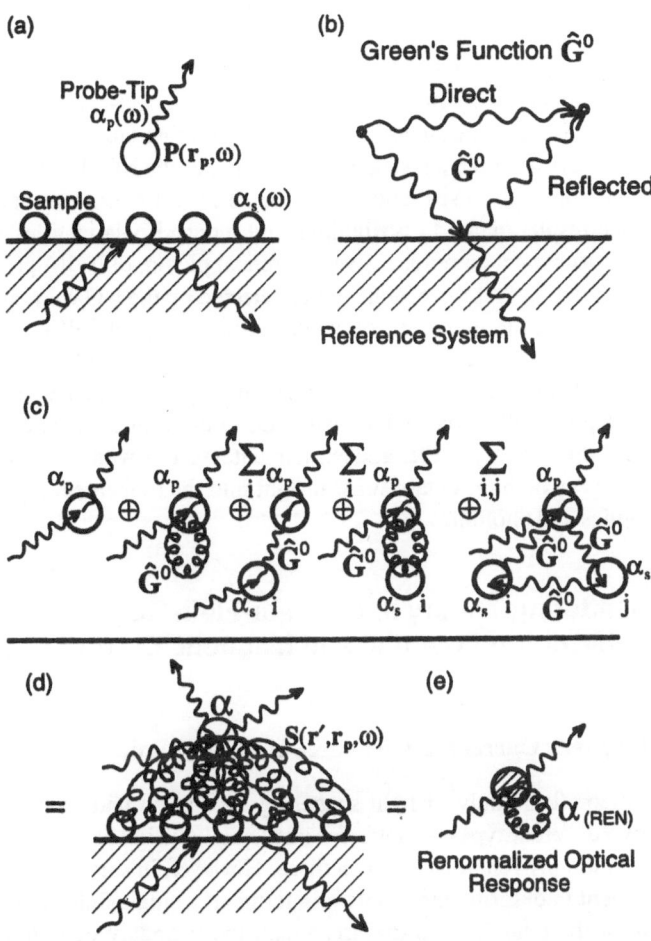

Figure 8.7. Self-consistent calculation for a measurement of the optical near-field scattered by a corrugated surface by using a dielectric sphere as a probe tip. (a) Optical near-field measurement of a corrugated surface irradiated by an evanescent wave. (b) A reference system for which the zeroth-order Green's function is determined. Here the reference system of a planar dielectric boundary is assumed. (c) Multiple scattering taking place in the object–probe interactions. The amplitude of each process is evaluated by using the dielectric response function $\alpha(\omega)$ and the Green's function for the reference system. (d) The sum of multiple scattering processes determines the system response of the sample-plus-probe configuration. (e) The dielectric response function of the probe tip is described by a renormalized value $\alpha_{(REN)}$ which is determined by the placement of the probe relative to the sample system.

Such phenomena are known as cavity QED effects because the atomic interaction with the oscillating electromagnetic field is described in terms of the annihilation and creation of photons, which represent the quantum mechanical excitations of the electromagnetic mode in a cavity. In general the cavity defines the boundaries in which we are considering the electromagnetic problem.

In contrast to a clear mathematical formulation, the self-consistent treatment in practice requires elaborate numerical computations of iterated integral equations. It is hard to achieve a simple physical understanding of the NOM process with this self-consistent approach to scattering problems. The self-consistent theories provide reference results of field distributions and induced polarizations which are useful for constructing a comprehensive theoretical picture of NOM problems.

It is worth noting that the overall signal detected by the NOM system usually depends strongly on the nature of the waveguiding or signal collection schemes employed. Therefore one must be careful not to compare NOM images directly with the near-field intensity distributions and induced polarizations without paying attention to signal collection and transfer. Of course, through such a careful comparison of theoretical pictures and experimental results with elaborate numerical studies one can perhaps obtain some insight into the fundamental processes in near-field optics and photonics.

8.3.2.   Scattering Theory in the Near-Field Regime Based on Polarization Potential and Magnetic Current

8.3.2.1.   Magnetic Current as the Source of Electric Induction

In the theoretical study of light scattering by a steep boundary, we usually introduce two different types of fields for inside and outside of the boundary and discuss the continuity relations between these. However, in some cases it is much more convenient to describe the problem in terms of a single field for both inside and outside of the boundary. In this treatment, the boundary conditions are described in terms of the scattering potential. We explore this theoretical approach in this section.

The important point is that the scattering potential is classified into bulk and surface terms, so that one can clearly see the size effect and obtain a criterion for the near-field approximation [29]. The result gives a pictorial understanding of the near-field intensity, polarization properties, and coupling efficiency between two scatterers involved in general NOM processes. This provides us with an intuitive interpretation of NOM images.

Let us consider macroscopic Maxwell equations for a nonmagnetic medium with no external source field:

$$\nabla \cdot \mathbf{D} = 0, \qquad \nabla \cdot \mathbf{H} = 0 \qquad (8.3.20)$$

$$\nabla \times \mathbf{E} = -\frac{\partial \mathbf{B}}{\partial t}, \qquad \nabla \times \mathbf{H} = \frac{\partial \mathbf{D}}{\partial t} \qquad (8.3.21)$$

where the macroscopic field variables are defined by

$$\mathbf{D} = \varepsilon_0 \mathbf{E} + \mathbf{p}, \qquad \mathbf{H} = \varepsilon_0 c^2 \mathbf{B} \qquad (8.3.22)$$

and the intrinsic sources are described in terms of the polarization by

$$\rho_{pol} = -\nabla \cdot \mathbf{p}, \qquad \mathbf{j}_{pol} = \frac{\partial \mathbf{p}}{\partial t} \qquad (8.3.23)$$

If we add both sides of the third equation to the term $\nabla \times \mathbf{p}$ and multiply by ε_0 we get

$$\nabla \times \mathbf{D} = \nabla \times (\varepsilon_0 \mathbf{E} + \mathbf{p}) = \nabla \times \mathbf{p} - \frac{1}{c^2}\frac{\partial \mathbf{H}}{\partial t} \qquad (8.3.24)$$

If we define the term $\nabla \times \mathbf{p}$ as an assumed magnetic current density by

$$\mathbf{j}_M = \nabla \times \mathbf{p} \qquad (8.3.25)$$

we can transform the macroscopic Maxwell equations into a form with duality for field variables $\mathbf{D}$ and $\mathbf{H}$ as

$$\nabla \times \mathbf{D} = \mathbf{j}_M - \frac{1}{c^2}\frac{\partial \mathbf{H}}{\partial t} \qquad (8.3.26)$$

Duality signifies the replacement of field variables and sources as

$$\mathbf{D} \leftrightarrow \mathbf{B}, \qquad -\mathbf{H} \leftrightarrow \mathbf{E}, \qquad \varepsilon_0 c^2 \mathbf{j}_M \leftrightarrow \mathbf{j}_E \qquad (8.3.27)$$

The magnetic current is rewritten in terms of the position-dependent dielectric constant $\varepsilon(\mathbf{r})$ as

$$\mathbf{j}_M = \nabla \times \mathbf{p} \qquad (8.3.28)$$

$$= \nabla \times [(\varepsilon(\mathbf{r}) - \varepsilon_0)\, \mathbf{E}] \qquad (8.3.29)$$

$$= \nabla \times \left[\left(1 - \frac{\varepsilon_0}{\varepsilon(\mathbf{r})}\right) \mathbf{D} \right] \qquad (8.3.30)$$

Using the macroscopic Maxwell equations, we can rewrite the $\mathbf{j}_M$ term in an instructive form as

$$\mathbf{j}_M = \nabla \times [(\varepsilon(\mathbf{r}) - \varepsilon_0)\mathbf{E}] = \left(1 - \frac{\varepsilon(\mathbf{r})}{\varepsilon_0}\right) \frac{1}{c^2} \frac{\partial \mathbf{H}}{\partial t} + \nabla [\ln \varepsilon(\mathbf{r})] \times \mathbf{D} \quad (8.3.31)$$

The first term on the right-hand side is attributed to the contribution from the bulk dielectric due to a retardation effect. This term is negligible when we consider a scattering object of subwavelength size in near-field problems of a quasistatic nature (an exception is the case of perfect spherical symmetry). The second term, $\nabla [\ln \varepsilon(\mathbf{r})] \times \mathbf{D}$, in the scattering potential is proportional to the gradient of the dielectric constant, and can be attributed to scattering processes at the dielectric surface (Fig. 8.8). It is especially important in the near-field regime [30].

Hereafter we consider the case of a dielectric with $\rho_{ext} = 0$ and $\mathbf{j}_E = \mathbf{0}$ and study the macroscopic Maxwell equations

$$\nabla \cdot \mathbf{D} = 0, \qquad\qquad \nabla \cdot \mathbf{H} = 0 \qquad\qquad (8.3.32)$$

$$\nabla \times \mathbf{D} = \mathbf{j}_M - \frac{1}{c^2} \frac{\partial \mathbf{H}}{\partial t}, \qquad \nabla \times \mathbf{H} = \frac{\partial \mathbf{D}}{\partial t} \qquad (8.3.33)$$

Since both $\mathbf{D}$ and $\mathbf{H}$ are divergence-free, we have two ways of choosing the vector potential:

$$\nabla \cdot \mathbf{D} = 0 \Rightarrow \mathbf{D} = \nabla \times \mathbf{C} \qquad\qquad (8.3.34)$$

or

$$\nabla \cdot \mathbf{H} = 0 \Rightarrow \mathbf{H} = \nabla \times \mathbf{A}' \qquad\qquad (8.3.35)$$

For convenience in considering dielectric problems with the source $\mathbf{j}_M$, let us use the vector potential $\mathbf{C}$ with

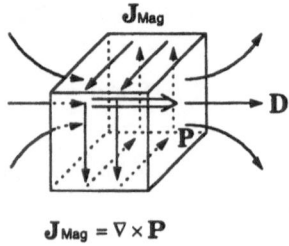

$$\mathbf{J}_{Mag} = \nabla \times \mathbf{P}$$

Figure 8.8.   The electric displacement $\mathbf{D}$ and induced polarization field $\mathbf{p}$ of a small dielectric object can be described in a unified form by using an equivalent source of magnetic current $\mathbf{j}_{Mag}$ running on the boundary surface. The magnetic current behaves as the source of electric displacement $\mathbf{D}$ in a similar way in which an electric current produces magnetic induction $\mathbf{B}$.

$$H = \frac{\partial C}{\partial t}, \quad \nabla \cdot C = 0 \tag{8.3.36}$$

Because of the divergence-free nature of the magnetic current it is straightforward to obtain the transverse wave equation for C:

$$\nabla \times \nabla \times C + \frac{1}{c^2} \frac{\partial^2 C}{\partial t^2} = j_M \tag{8.3.37}$$

We then see that C corresponds to the retarded potential due to the source j_M. Note that this wave equation holds only for nonmagnetic systems.

The wave equation for D is given by

$$\nabla \times \nabla \times D + \frac{1}{c^2} \frac{\partial^2 D}{\partial t^2} = \nabla \times j_M = \nabla \times \nabla \times [V(\mathbf{r})D] \tag{8.3.38}$$

where a scattering potential $V(\mathbf{r})$ is introduced:

$$V(\mathbf{r}) = \left(1 - \frac{\varepsilon_0}{\varepsilon(\mathbf{r})} \right) \tag{8.3.39}$$

The vector potential C is compared directly with the electric Hertz vector or the polarization potential Π^E [2] as

$$C = \nabla \times \Pi^E \tag{8.3.40}$$

since for dielectric problems the electric induction field D is produced by the Hertz vector as

$$D = \nabla \times \nabla \times \Pi^E \tag{8.3.41}$$

This can be easily understood by recalling that the Hertz vector represents the transverse retarded potential due to polarization p, whereas the potential field C is due to the magnetic current $j_M = \nabla \times p$.

8.3.2.2. Small Parameters in Scattering Problems and Near-Field Criterion

Let us study the time-independent scattering theory for the Fourier component of the vector potential, $C(\mathbf{r}, t) = C(\mathbf{r}, \omega) \exp(-i\omega t)$,

$$\left(\nabla^2 + K^2 \right) C(\mathbf{r}, \omega) = \left[-\nabla \left(\ln \varepsilon(\mathbf{r}) \right) \times \nabla \times + K^2 \left(1 - \frac{\varepsilon(\mathbf{r})}{\varepsilon_0} \right) \right] C \tag{8.3.42}$$

$$= \left(\hat{V}_S + \hat{V}_V \right) C(\mathbf{r}, \omega) \tag{8.3.43}$$

This can be derived from the d'Alembert equation for $C(\mathbf{r}, t)$ with $K = \omega/c$. Here $\hat{V}_S$ and $\hat{V}_V$ correspond, respectively, to the surface and bulk terms of the scattering potential defined by

$$\hat{V}_S = -\nabla (\ln \varepsilon(\mathbf{r})) \times \nabla \times \qquad (8.3.44)$$

$$\hat{V}_V = -K^2 \left(1 - \frac{\varepsilon(\mathbf{r})}{\varepsilon_0} \right) \qquad (8.3.45)$$

This equation is equivalent to the vector Schrödinger equation, and the discontinuity in the boundary condition which appeared in the case of the vector potential $\mathbf{A}$ is absorbed into the surface scattering potential (cf. Section 8.2.4.3). In other word, the surface potential $\hat{V}_S$ is an alternative representation of the electromagnetic boundary condition.

We can write down the solution of the Helmholtz equation for $C(\mathbf{r}, \omega)$ in the form of the Lippmann–Schwinger equation as

$$C(\mathbf{r}, \omega) = C^0(\mathbf{r}, \omega) + \int G^T(\mathbf{r}, \mathbf{r}', \omega) \left(\hat{V}_S + \hat{V}_V \right) C(\mathbf{r}', \omega) \, d^3 r' \qquad (8.3.46)$$

where $G^T(\mathbf{r}, \mathbf{r}', \omega)$ is the transverse Green's dyadic for the system. It is informative to investigate the magnitude of the scattering potentials in this perturbation series. When the spatial dimension of the scatterer is specified by a size parameter a, the radius of a dielectric sphere, for instance, we have a dimensionless parameter Ka, with $K = \omega/c$, which characterizes the magnitude of the scattering potential.

For a small-scatterer regime with $Ka \ll 1$ the surface and bulk terms in the perturbation series are, respectively, on the order of Ka and $(Ka)^2$, so that we get the relation

$$1 \gg \underbrace{\left| Ka \left(\frac{\varepsilon(\mathbf{r}) - \varepsilon_0}{\varepsilon_0} \right) \right|}_{\text{surface potential}} \gg \underbrace{\left| (Ka)^2 \left(\frac{\varepsilon(\mathbf{r}) - \varepsilon_0}{\varepsilon_0} \right) \right|}_{\text{bulk potential}} \qquad (8.3.47)$$

Note that the evaluation of the surface terms requires careful consideration of the discontinuity at a steep dielectric surface [30]. This shows that the surface term dominates the scattering process in the near-field regime of a subwavelength-size scatterer. The surface potential is also a small perturbation, so that we can resort to the Born approximation in the single-small-scatterer case and replace the self-consistent field $C(\mathbf{r}, \omega)$ in the integral by the incident one, $C^0(\mathbf{r}, \omega)$. These two criteria make the near-field problems extremely simple.

Let us investigate typical values for which this small-scatterer regime holds. Consider an incident light wave of wavelength $\lambda = 2\pi/K = 500$ nm and $\varepsilon(\mathbf{r})/\varepsilon_0$ of the order of unity; the size parameter $a = 1$–10 nm gives the characteristic parameter $Ka = 0.01$–0.1, which satisfies the small-scatterer condition. In contrast, when 100 nm is assumed for the value of a, which is close to the diffraction limit $a \sim \lambda/4$, Ka becomes on the order of unity and the scattering problem becomes much more complicated.

It is also important in considering the NOM process to find the criterion for the near-field measurement of a scattered field from a small object. This is equivalent to a potential barrier problem. Also in this case the near-field criterion applies, in which the dimensionless parameters in the scattering matrix are also Ka and Va^2 with the barrier height V and width a.

It should be noted that the discussion of the near-field criterion above applies only for small three-dimensional objects. When we consider a small two-dimensional scatterer, such as a thin cylindrical object lying on a substrate, the contribution from the bulk term is of the same order as the surface term with respect to Ka. Two- and three-dimensional scattering problems are basically different especially in the near-field regime. In addition, it is even more complicated when we observe a two-dimensionally localized object using a near-field probing technique such as NOM. That is, we need to consider the sensitivity of the probe tip to the longer range correlation due to the bulk effect of a long, thin object. The overall character of such sample–probe scattering demonstrates the important problem of screening or spatial frequency filtering by the NOM probe for a long-range interaction, which will be discussed later in this chapter.

8.3.2.3. Magnetic-Current-Coil Model for Estimating the Optical Near-Field

As an application of the theoretical description of the field $\mathbf{C}$, a very effective and informative representation of the optical near-field regime is available [30]. This is done by replacing the dielectric boundary with the streamline of a surface magnetic current, as long as the surface magnetic current is divergence-free and reproduces the induced polarization on the object on average. The procedure is similar to considering the magnetic flux produced by a coil of electric current. Using some geometrical intuition, it is not difficult to imagine the shape of the lines of magnetic flux for a given coil for the electric current. Such a replacement holds in the near-field regime discussed in the previous section.

Let us consider a subwavelength-size homogeneous dielectric object irradiated by a uniform incident electric field. The electric induction field $\mathbf{D}$ then satisfies the equation

$$\nabla \times \mathbf{D} = \mathbf{j}_M = \nabla \times \mathbf{p} \sim \nabla \left[\ln \varepsilon(\mathbf{r}) \right] \times \mathbf{D} \tag{8.3.48}$$

if we assume the near-field regime and neglect the bulk term in $\mathbf{j}_M$. This relation tells us that the spatial distribution of the $\mathbf{D}$ field for a given shape of dielectric object is formally equivalent to a distribution of magnetic fields produced by a solenoidal current running on the surface, where $\nabla[\ln \varepsilon(\mathbf{r})]$ has nonzero value. As the $\mathbf{D}$ field contains the polarization field $\mathbf{p}$ in it, they should be determined in a self-consistent manner in a rigorous treatment. However, by virtue of the Born approximation available in the near-field regime and of the solenoidal (nondivergent) nature of the magnetic current, what one can do is just wind a coil of magnetic current on the surface of a small object so as to reproduce the mean polarization field induced by an incident electric field. As long as the object's shape is not very complicated, the self-consistent field may not be very different from the approximation.

Let us investigate the distribution of the field $\mathbf{D}$ for rectangular, disk-shaped, and spherical objects of subwavelength size with homogeneous dielectric constant

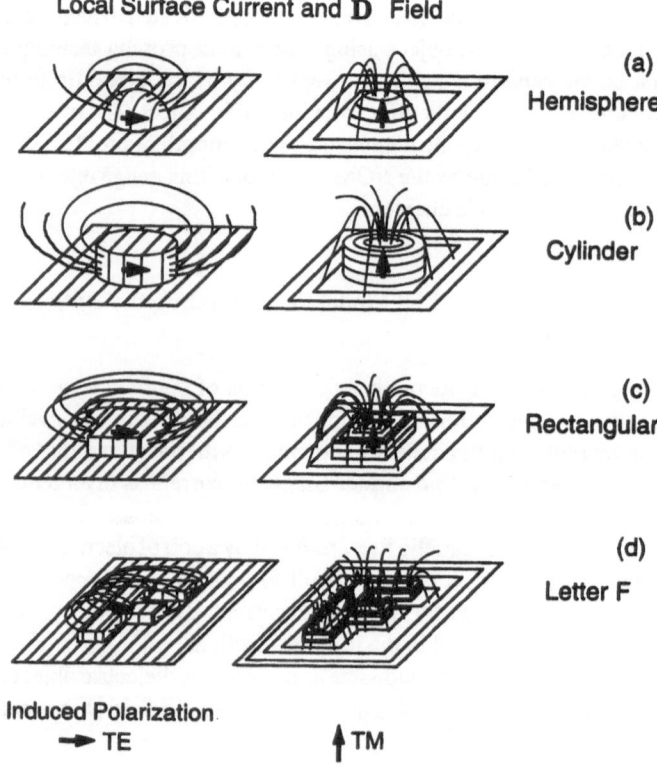

Figure 8.9. The magnetic-current-coil model is convenient for estimating the optical near-field of small dielectric protrusions of several different shapes. By assuming coils of magnetic current wound so as to reproduce the induced polarizations, we can easily estimate the optical near-field to a good approximation. In each figure the direction of induced polarization is indicated and coils are wound so as to rotate around it.

and an incident uniform electric field with well-defined polarization, as illustrated in Fig. 8.9. From the incident polarization we can estimate the mean polarization induced on the object and then wind a magnetic current coil on its surface. Then we can imagine the spatial distribution of the D field by analogy with the magnetic field produced by an electric current of the same spatial arrangement.

Next consider a rectangular protrusion fabricated on a flat and uniform dielectric surface and irradiated by an evanescent wave at total internal reflection. Let us evaluate the field distribution separately for the perfect plane and the protrusion by imposing the condition that the magnetic current distributions cancel on the surface at which we cut off the protrusion. In this way the continuity in the system is maintained.

The intensity distribution of D for a rectangular protrusion is numerically calculated by assuming magnetic-current coils as shown in Fig. 8.10. We first estimate a perturbation field D_{Coil} due to the magnetic-current coil in the same manner in which we calculate the magnetic field around an electric-current coil. Next, we add the applied illumination field $D_{Illumination}$ which is inducing the magnetic current on the rectangular protrusion, and obtain the total field $D_{Coil} + D_{Illumination}$. Because of the vector nature of the electric displacement we need to evaluate the effect of the protrusion by the absolute square of the total field, $|D_{Coil} + D_{Illumination}|^2$. The results are shown in Figs. 8.11 and 8.12 as the intensity distribution on a horizontal observation plane at distance d above the protrusion (see Fig. 8.10c). Figure 8.11 corresponds to the coil shown in Fig. 8.10a, and Fig. 8.12 to that in Fig. 8.10b, which correspond respectively to TM and TE illumination. In order to emphasize the importance of considering the vector nature of the process, we also indicate the intensity of the coil field alone, $|D_{Coil}|^2$.

We find the following features. First, imagine that an incident field on the system produces a mean polarization parallel to the surface of the substrate (usually referred to as s-polarization or TE polarization with respect to the plane of incidence). We can imagine that the perturbation of the D field near the protrusion is due to a coil of magnetic current wound with its solenoidal axis directed along the mean induced polarization (Fig. 8.10a). We can see that D is strong near both edges of the solenoid and weak above the solenoid (Fig. 8.11a). In the s-polarization case the perturbation field exhibits inverse polarity above the coil. Since the field D is equal to the electric field E coming out from the matter, we can see that the D due to the protrusion cancels the background above it and adds to the intensity at the edge of the solenoid. This results in a depression of the total field above the protrusion. As the observation plane becomes far from the protrusion, the intensity distribution smears out (Figs. 8.11b–8.11d). Since the field D is equal to the electric field E coming out from the matter, we can see that the D due to the protrusion cancels the background above it and adds to the intensity at the edge of the solenoid.

On the other hand, when it is illuminated by an incident field so as to produce a mean polarization perpendicular to the surface (usually referred to as p-polarization

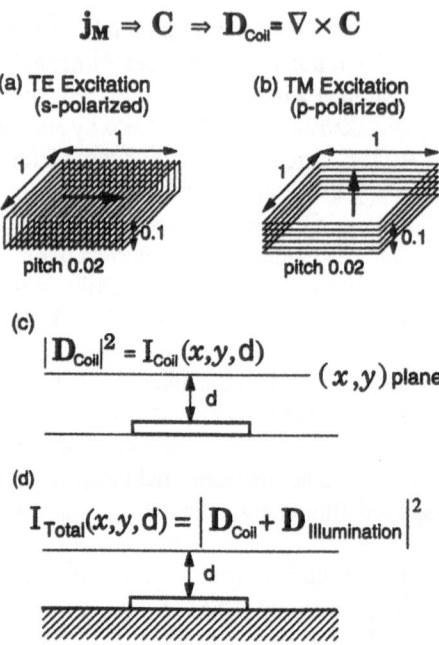

$$\mathbf{j_M} \Rightarrow \mathbf{C} \Rightarrow \mathbf{D_{Coil}} = \nabla \times \mathbf{C}$$

(a) TE Excitation (s-polarized)

(b) TM Excitation (p-polarized)

pitch 0.02

pitch 0.02

(c)

$$\frac{|\mathbf{D_{Coil}}|^2 = I_{Coil}(x,y,d)}{} \quad (x,y)\,\text{plane}$$

(d)

$$I_{Total}(x,y,d) = \left| \mathbf{D_{Coil}} + \mathbf{D}_{Illumination} \right|^2$$

Figure 8.10. Example of an optical near-field calculation by using magnetic-current coils which are equivalent to a dielectric object or protrusion of rectangular shape illuminated respectively by (a) *TE*-polarized and (b) *TM*-polarized incident fields. (c) The scattered amplitude is given by $|D_{Coil}|^2$, whereas (d) there arises an interference between the coil field and the illuminating field $|D_{Coil} + D_{Illumination}|^2$. The intensity distributions for both cases in a plane at a distance *d* above the object and protrusion are shown in Figs. 8.11 and 8.12, respectively.

or *TM* polarization with respect to the plane of incidence), we can consider a magnetic solenoid directed normal to the surface (Fig. 8.10b). In this case we can see that **D** is strong just above the edge of the protrusion and cancels the background field at its side. If one considers a protrusion of relatively large area one may find that the field enhancement is strong at the side and slightly less just above the center. As the observation plane becomes far from the protrusion, the intensity distribution smears out (Figs. 8.12b–8.12d).

In a similar way, in the near-field limit, we can estimate the field distribution around a small protrusion of arbitrary, but simple shape put on a dielectric substrate. The semiquantitative discussions above hold not only for the surface of homogeneous dielectrics, but also for general dielectric boundaries such as a buried or gradually varying dielectric boundary. These estimated results can be compared with numerical results [25, 31].

8.3.2.4. Model of Probes and Interpretation of NOM Images

We can derive a very important feature of the polarization dependence of NOM images and evaluate the role of the probe tip and its coupling to a waveguide using the magnetic-current-coil picture. Let us consider an intuitive model describing an NOM probe tip as an application of the surface magnetic current theory. Suppose a conical probe with a hemispherical probe tip is located on one end as illustrated in Fig. 8.13. The other end is considered to be coupled with a signal transfer scheme such as an optical waveguide connected to a photodetector.

It is convenient to separate the function of this probe according to the mean polarization induced on its coupling end to the waveguide. Let us consider two typical cases, say "s-preferred" and "p-preferred" probes, with regard to the polarization induced on the coupling end. According to the polarization to be expected at one end and considering the solenoidal nature of the surface magnetic current, we can model these probes by using magnetic current coils as shown in Fig. 8.13. In order to maintain the near-field and small object criteria, the probe considered here is still subwavelength in size.

Suppose that these so-called s-preferred and p-preferred probes are scanned in a plane above a small object represented by a spherical solenoid with direction depending on the polarization of illumination. As is shown in Fig. 8.14, the induced polarization depends on the position of the probe and also on the polarization of the sample object.

The induced polarization for an s-preferred probe takes its maximum where the $\mathbf{D}$ field parallel to the substrate is intense. This occurs just above the s-polarized sample (Fig. 8.14a) or at both sides of the p-polarized sample (Fig. 8.14c). In the former case the spatial distribution of the pickup efficiency is single peaked and in the latter it is double peaked. Here it should be stressed that in the s-polarized case the polarity of the field is opposite to the illuminating background, so that the observed signal shows a single dip with two additional bumps at both sides of the signal background.

In contrast, the induced polarization for a p-preferred probe takes its maximum where the $\mathbf{D}$ normal to the substrate is intense. This occurs at both sides of the s-polarized sample (Fig. 8.14b) or just above the p-polarized sample (Fig. 8.14d). In the former case the spatial distribution of the pickup efficiency is single peaked and in the latter case it is double peaked.

As is seen above, when we observe the NOM signal as the total intensity scattered from the probe we need to add up contributions from both the s-preferred and p-preferred aspects of the probe. However, in general, light emission from an oscillating polarization exhibits strong directionality, so that the signal intensity depends largely on the position of the photodetector.

Finally, let us consider the coupling of the probe system to an optical waveguide, as shown in Fig. 8.15. Usually an optical waveguide exhibits a strong

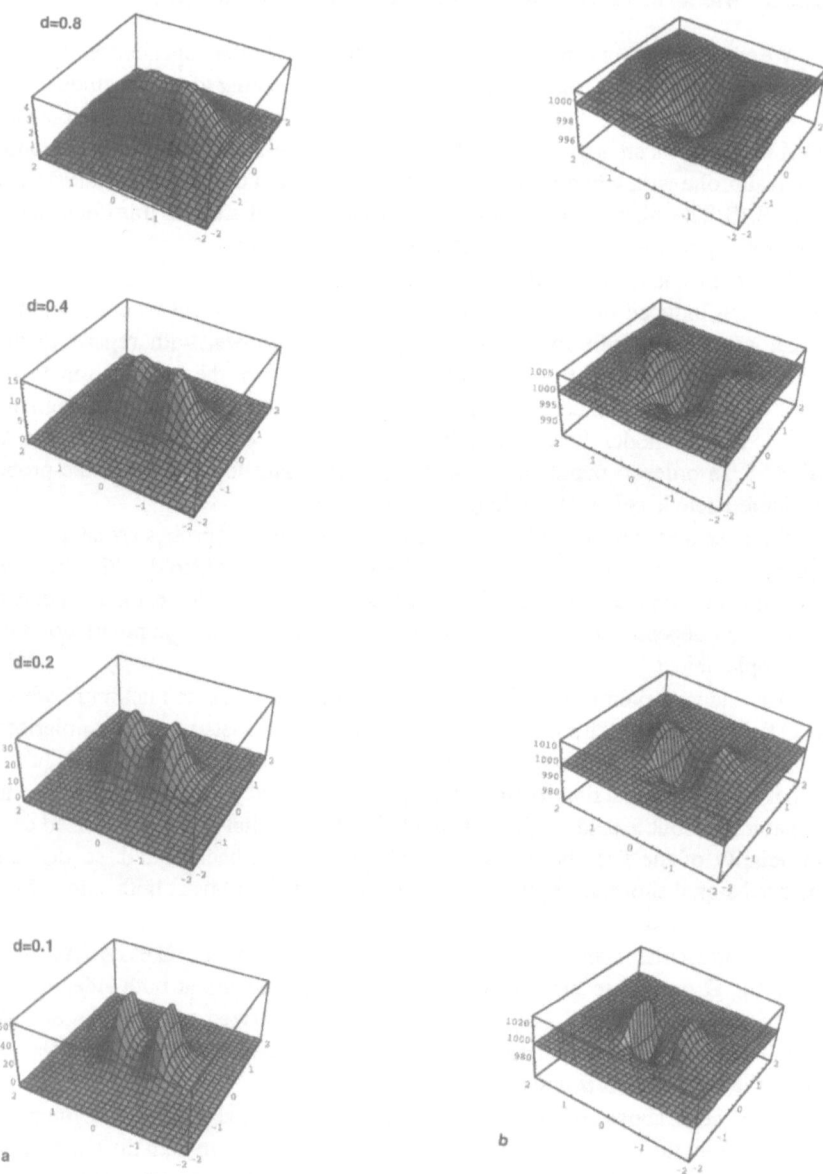

Figure 8.11. Intensity distribution with *TE* illumination field in a plane at a distance *d* above a rectangular dielectric protrusion, (a) scattered field alone and (b) total field: calculations are made using the magnetic-current-coil model shown in Fig. 8.10.

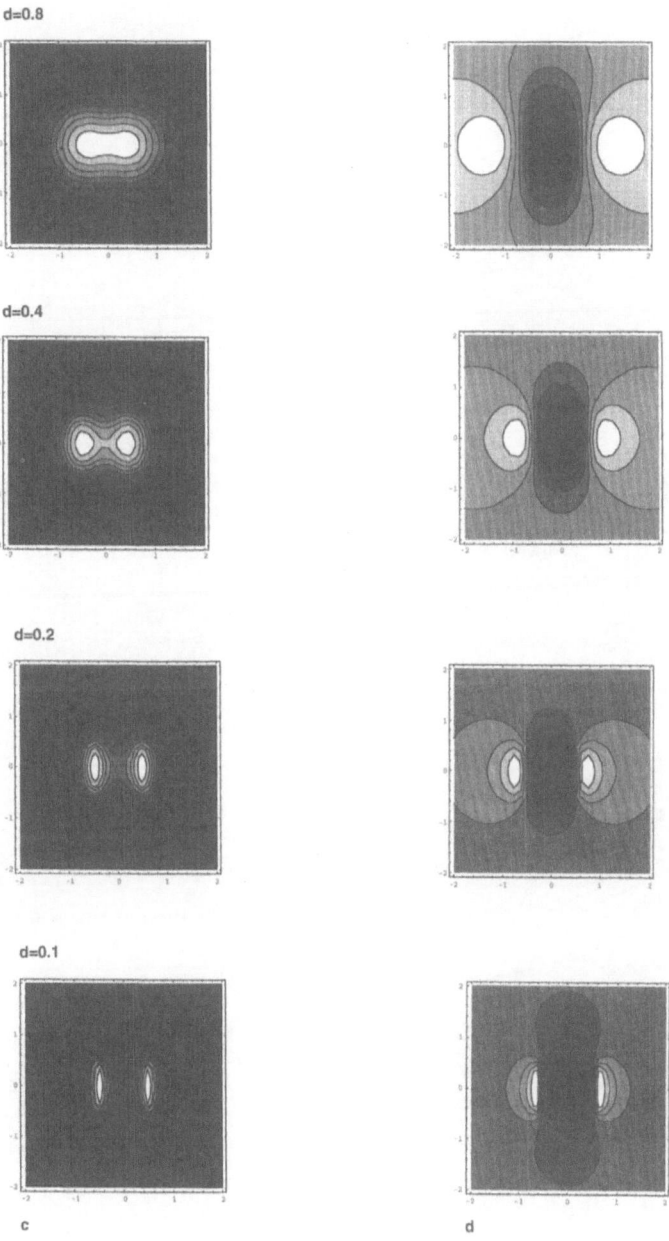

c

d

Fig. 8.11 Continued.

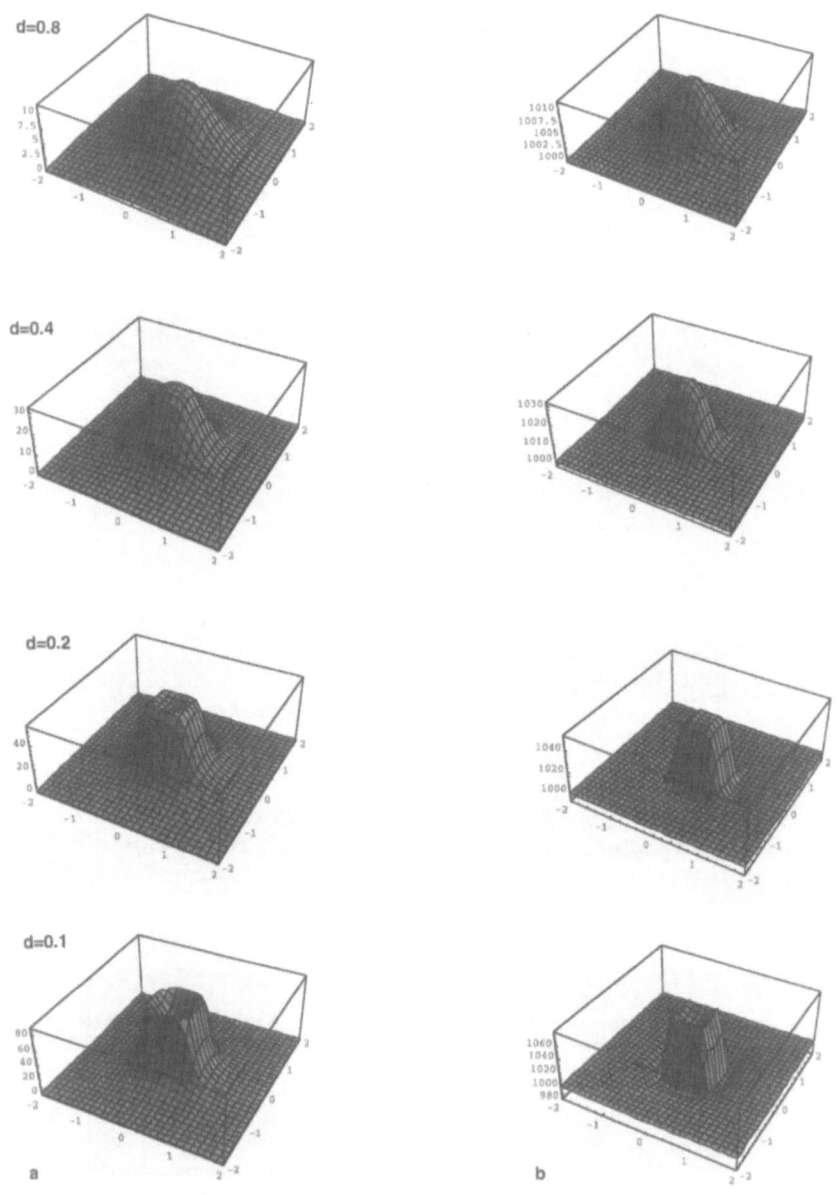

Figure 8.12. Intensity distribution with *TM* illumination field in a plane at a distance *d* above a rectangular dielectric protrusion, (a) scattered field alone and (b) total field: calculations are made using the magnetic-current-coil model shown in Fig. 8.10.

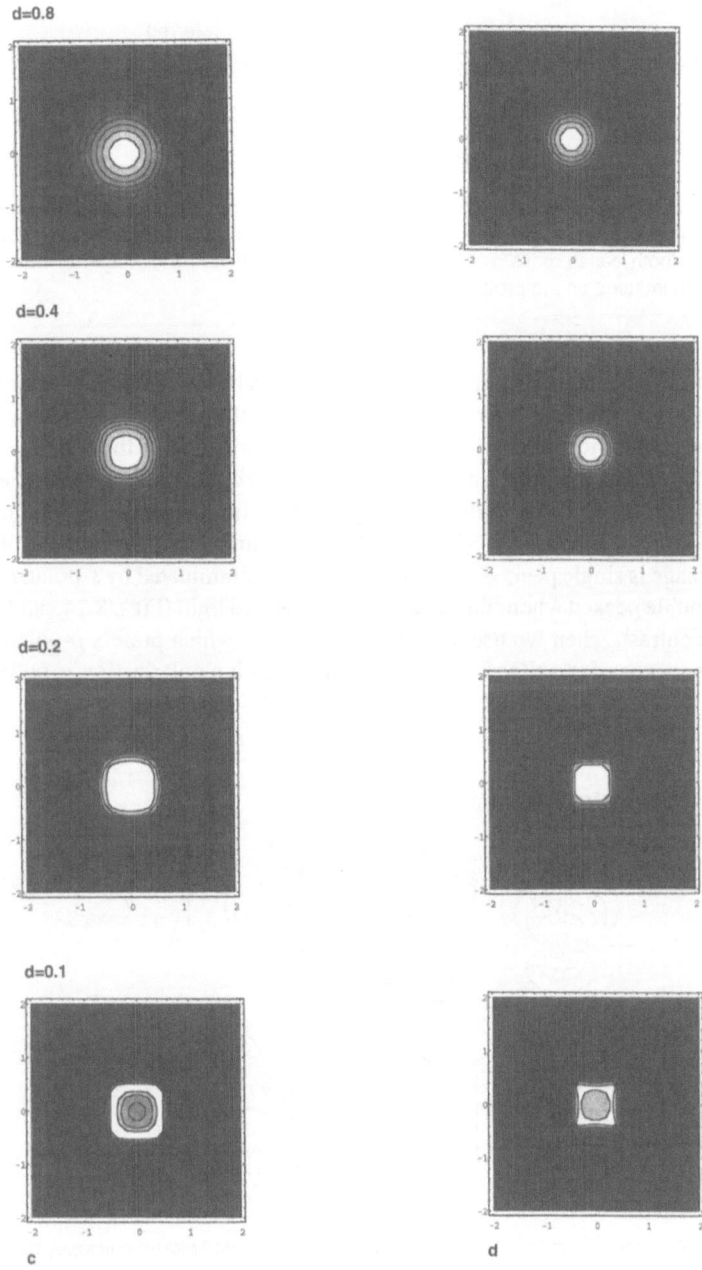

Fig. 8.12 Continued.

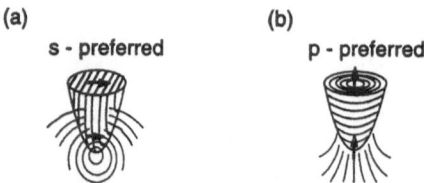

(a)
s - preferred

(b)
p - preferred

Figure 8.13. Magnetic-current-coil model of the NOM probe tip. The probe-tip function is classified into two types, s-preferred and p-preferred, depending on its connection to the far-field. In both cases, coils are wound so as to reproduce the direction of macroscopic polarization induced on the probe tip.

polarization dependence for the light waves that can be transmitted in it. In the case of a single-mode optical fiber used as a waveguide, the propagating light wave has a transverse electric field. In this case only the s-polarization on the end of the probe efficiently excites the light wave propagating in the optical fiber (Fig. 8.15a). This fact provides us the interpretation of the polarization dependence of the NOM images. When we use an optical fiber probe with a sharpened probe tip, the obtained NOM image is single peaked when the specimen is illuminated by s-polarized light and is double peaked when illuminated by p-polarized light (Figs. 8.14a and 8.14c).

In contrast, when we use an optical waveguide which prefers p-polarization, the dependence of the NOM image on the illumination polarization is opposite.

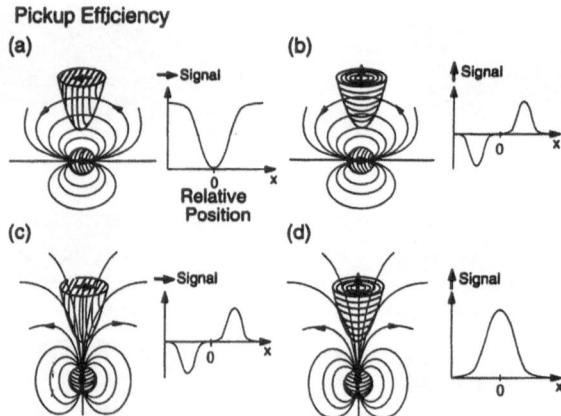

Figure 8.14. Intuitive understanding of the position-dependent pickup efficiency of the probe tip. The sample object is assumed to be a TE- or TM-illuminated dielectric sphere. Both the amplitude and polarity in reference to the direction of the sample polarizations are indicated for each case.

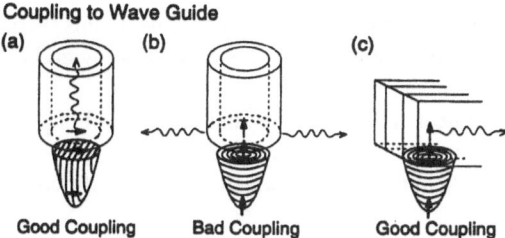

Coupling to Wave Guide

(a)   (b)   (c)

Good Coupling   Bad Coupling   Good Coupling

Figure 8.15. Intuitive understanding of the coupling efficiency of probe-tip polarizations and waveguides. The overall character of the NOM probe is determined by both the pickup efficiency indicated in Fig. 8.14 and the coupling efficiency of induced polarization to waveguides.

One can compare these intuitive interpretations with experimental results shown in the preceding chapters as well as with numerical calculations based on self-consistent solutions of the optical near-field regime [25, 31].

8.4. DIFFRACTION THEORY IN NEAR-FIELD OPTICS

8.4.1. Diffraction of Light from Subwavelength Aperture

In this section, we present a brief review of diffraction theory and discuss its relation to near-field optics. Of main concern is the diffraction of light through a subwavelength-size aperture in a thin conducting plane. This problem is important and instructive not only because it is related to the scanning aperture NOM originally proposed by Synge [32] and demonstrated by Pohl et al. [33], but also because the so-called "exact" solution derived by Leviatan [23] shows an important feature of the optical near-field regime, namely that the diffracted field exhibits a remarkable intensity distribution in the proximity of the aperture which is scaled not by the optical wavelength, but by the size of the aperture itself. The size-dependent localization is one of the most significant features of optical near-field problems. The problem was first set by Bethe [21], and the several different approaches it has inspired are summarized by Bowkamp [22]. For the small-aperture problem there are a number of useful theoretical frameworks, including a variational treatment [34]. Starting from Huygens, Young, and Fresnel and leading to modern quantum field theory such as that of Dyson and Wick, the treatment of diffraction theory has been among the most fundamental parts of theoretical physics.

Diffraction theory was originally developed as a far-field theory in which geometrical optics is modified due to the short optical wavelength; thus, conventional diffraction theory fails in the near-field regime. In this section we first review conventional diffraction theory based on the Kirchhoff integral and discuss both its applicability and limitations from the viewpoint of near-field optics. This is important because optical near-field problems and related techniques involve not only near-field effects, but also their measurement in the far-field region. Then we proceed to the theoretical treatment of small-aperture diffraction.

8.4.2. Kirchhoff's Diffraction Integral and Far-Field Theory

The scattering problem involves a system separated by a boundary between the source side and the detector side, with part of the boundary having an aperture or opaque region which diffracts the incident optical field. Using Green's theorem and several approximations, Kirchhoff formulated the diffracted field as an integral of the field variables only over the aperture region [1, 2].

Let us limit our study to scalar diffraction theory. Assume that the boundary consists of a screen part with an aperture in it and a remaining part lying infinitely far from the aperture. Kirchhoff assumed the following:

- Both the value and the normal derivative of the scalar potential vanish on the screen and are given simply by the incident field in the aperture region
- The diffracted field is described using the free-space Green's function $G(\mathbf{r}, \mathbf{r}') = \exp(ik|\mathbf{r} - \mathbf{r}'|)/4\pi|\mathbf{r} - \mathbf{r}'|$
- The field satisfies the radiation condition $\psi(\mathbf{r}) = \varphi(\theta, \phi) \exp(ikr)/r$

Here we assume monochromatic temporal variation $\exp(-i\omega t)$ and therefore the Helmholtz equation $(\nabla^2 + k^2)\psi(\mathbf{r}) = 0$.

Green's theorem with these approximation gives the Kirchhoff integral:

$$\psi(\mathbf{r}) = -\frac{1}{4\pi} \int_A \frac{e^{ikR}}{R} \hat{\mathbf{n}}' \cdot \left[\nabla'\psi + ik\left(1 + \frac{i}{kR}\right)\frac{\mathbf{R}}{R}\psi \right] da' \qquad (8.4.1)$$

where $\mathbf{R} = \mathbf{r} - \mathbf{r}'$, $R = |\mathbf{R}|$, and the integration is over the aperture region A with infinitesimal area da'. It is known that these assumptions are inconsistent both mathematically and physically; however, this integral works fine for the case of large aperture, $a > \lambda$, in the far-field regime, $R \gg \lambda$ [1]. When only the Dirichlet or Neumann condition is considered on the screen, the mathematical deficiencies are relaxed and one obtains the generalized Kirchhoff integral [1]. In the case of a planar conducting boundary, both the Dirichlet and Neumann Green functions can be derived simply by using the electromagnetic image method.

Even though the validity of this approach is limited, the Kirchhoff integral and its generalized versions are still available when considering the diffracted far-field due to the near-field of an object placed very close to the aperture. In fact the size effect in the NOM process introduced in Chapter 2 has been derived on the basis of the Kirchhoff integral [35], where the scattered optical near-field by a subwavelength-size dielectric sphere is integrated over a probing aperture region to estimate the far-field intensity of the diffracted field.

8.4.3. Small-Aperture Diffraction and Equivalent Problem

Small-aperture diffraction involves a mixed boundary problem in which the value of the potential is given on the conducting surface while its normal derivatives, i.e., the field continuity relation, is given in the aperture region. In this case, it is not straightforward to introduce a Green's function appropriate for the given boundary conditions. That is, if one insists on the original problem, one needs to start with an approximate form of the Green's function and introduce a self-consistent relation with respect to the Green's function regarding the mixed boundary conditions. Then by means of an iterative method one determines the self-consistent Green's function to evaluate the diffracted field at an arbitrary point in the half-space on the detector side. In this approach, one needs to resort to a numerical approach in evaluating the iterative procedures. The field in the proximity of the aperture should be modified significantly from the incident field. This is why ordinary diffraction theory fails in the near-field regime and long-wavelength limit.

On the other hand, it is possible to replace the problem by an equivalent one in which a well-defined Green's function is available. For aperture problems an equivalent problem is obtained by closing the aperture to form a perfect-conducting plane and then replacing the effect of the aperture by postulated sources attached to either side of the filled aperture. One can reproduce the original field which is consistent both with the illuminating field and the continuity relation on the aperture [21, 22]. In this case we can separate the field variable, the electric field, for example, into three components: an incident source field $\mathbf{E}^{(IN)}$, a reflected field by the perfect boundary plane $\mathbf{E}^{(REF)}$, and a field due to the sources induced in the aperture region, $\mathbf{E}^{(S)}$. Then the total field in the half-space of the light-source side (1) is given by $\mathbf{E}^{(1)} = \mathbf{E}^{(IN)} + \mathbf{E}^{(REF)} - \mathbf{E}^{(S)}$, and that in the diffracted half-space (2) is $\mathbf{E}^{(2)} = \mathbf{E}^{(S)}$. It can be shown that the exact diffracted field is given in terms of the half-space Green's function by

$$\mathbf{E}^{(S)} = \frac{1}{2\pi} \nabla \times \int_A (\hat{\mathbf{n}} \times \mathbf{E}) \frac{e^{ikR}}{R} \qquad (8.4.2)$$

Corresponding to this relation, we can assume effective magnetic and electric moments just next to the filled aperture. Only two magnetic moments tangent to the aperture surface and one electric moment normal to it are allowed to be introduced [1].

Let us now study details of Leviatan's theory. Since the results derived by Leviatan are based on this exact formula, the results derived are referred as "exact" in contrast to the approximate ones.

8.4.4. Magnetic Current Distribution and Self-Consistency

Consider a circular hole with radius a much smaller than the incident and diffracted optical wavelength λ. Since the field modification is significant in the near zone of the aperture, a quasistatic approach is valid. The theoretical treatment developed by Leviatan is pictured in Fig. 8.16.

Due to the small-aperture nature of the problem, the source field is due to induced polarization on the edge of the aperture, which is eliminated by filling the aperture to form a perfect-conducting plane. Then one replaces the induced polarization by a magnetic current running on either side of the filled aperture satisfying $\mathbf{j}_M = \nabla \times \mathbf{p}$ (see Fig. 8.16). Since the conductor shorts the tangential electric field applied originally in the aperture, the magnetic current is tangential to the filled aperture and represented by, with a unit normal vector $\hat{\mathbf{n}}$,

$$\mathbf{j}_M = \nabla \times \mathbf{p} = \mathbf{E} \times \hat{\mathbf{n}} \tag{8.4.3}$$

The corresponding magnetic field is produced both by the temporal derivative of the electric field and the quasistatic magnetic charge. Let the x–y plane lie in the conducting surface and z axis be normal to it. The magnetic currents in the x and y directions then give rise to the magnetic charges according to the conservation law

$$\frac{\partial \rho_{M\mu}}{\partial t} = -i\omega\rho_{M\mu} = -\nabla \cdot \mathbf{j}_{M\mu} \tag{8.4.4}$$

where $\mu = 1, 2$ stand for two tangential components in the x–y plane. In addition to these, we assume a solenoidal component which satisfies

$$\nabla \cdot \mathbf{j}_{M3} = 0 \tag{8.4.5}$$

This term gives rise to an electric polarization normal to the aperture (see Fig. 8.14). Leviatan described these three components of magnetic current as

$$\mathbf{j}_M = V_1\omega\, \mathbf{j}_{M1} + V_2\omega\mathbf{j}_{M2} + V_3\, \mathbf{j}_{M3} \tag{8.4.6}$$

The continuity relations are given for the tangential component of the magnetic field resulting from the magnetic current induced in the aperture:

$$\mathbf{H}_t^{(1)}(-\mathbf{j}_M) - \mathbf{H}_t^{(2)}(\mathbf{j}_M) = -\mathbf{H}_t^{(IN)} \tag{8.4.7}$$

where the regions denoted by (1) and (2) are the light source and detector sides of the conducting plane, respectively. The problem is then to determine the coefficients V_i ($i = 1, 2, 3$) consistent with the continuity relation imposed on $\mathbf{H}_t$.

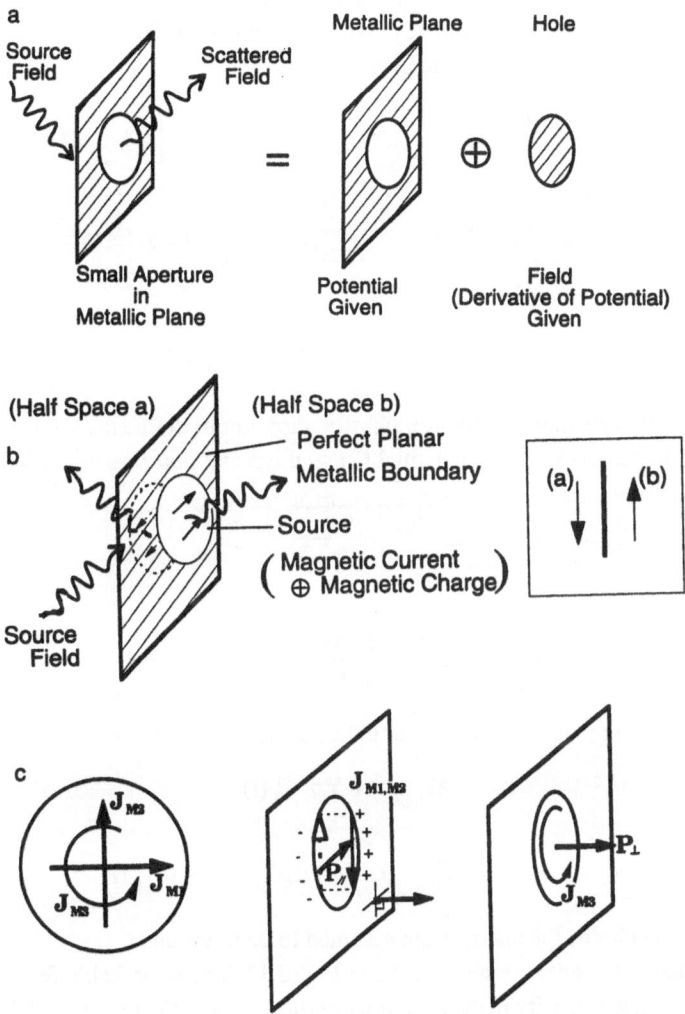

Figure 8.16. Theoretical treatments of small-aperture diffraction. (a) Small-aperture and relevant boundary conditions, (b) equivalent problem with a perfect planar metallic plate plus sources assumed in the aperture region, (c) induced polarizations in the aperture and three corresponding magnetic currents.

8.4.5. Leviatan's "Exact" Solutions for the Aperture Problem

Fortunately, an analytic solution can be found for the case of a circular aperture by virtue of its cylindrical symmetry. With respect to cylindrical coordinates (r, φ) and $\xi = \sqrt{a^2 - r^2}$ the solutions are given by

$$j_{M1r}(r, \varphi) = \frac{1}{\pi r^3} \xi \cos \varphi, \qquad j_{M1\varphi}(r, \varphi) = -\frac{1}{\pi r^3}\left(\xi + \frac{r^2}{2\xi}\right) \sin \varphi \qquad (8.4.8)$$

$$j_{M2r}(r, \varphi) = \frac{1}{\pi r^3} \xi \sin \varphi, \qquad j_{M2\varphi}(r, \varphi) = \frac{1}{\pi r^3}\left(\xi + \frac{r^2}{2\xi}\right) \cos \varphi \qquad (8.4.9)$$

$$j_{M3r}(r, \varphi) = 0, \quad j_{M3\varphi}(r, \varphi) = \frac{3}{2\pi r^3} \frac{r}{\xi} \qquad (8.4.10)$$

The field calculation is then straightforward. Let us define the Fourier components of the magnetic vector potential $\mathbf{C}$ and the scalar potential ϕ_M as follows:

$$\mathbf{C}_\mu(\mathbf{r}) = \frac{1}{2}\pi \int_{aperture} ds' \frac{\mathbf{j}_{M\mu}(\mathbf{r'}) \, e^{ik|\mathbf{r}-\mathbf{r'}|}}{|\mathbf{r}-\mathbf{r'}|} \qquad (8.4.11)$$

$$\phi_{M\mu}(\mathbf{r}) = \frac{1}{2\pi} \int_{aperture} ds' \frac{-\nabla \cdot \mathbf{j}_{M\mu}(\mathbf{r'}) \, e^{ik|\mathbf{r}-\mathbf{r'}|}}{i\omega|\mathbf{r}-\mathbf{r'}|} \qquad (\mu = 1, 2, 3) \qquad (8.4.12)$$

The electric and magnetic fields due to the source $\mathbf{j}_{M\mu}$ are then calculated by

$$\mathbf{E}_{(\mathbf{j}_{M\mu})}(\mathbf{r}) = \nabla \times \mathbf{C}_\mu(\mathbf{r}) \qquad (8.4.13)$$

$$\mathbf{H}_{(\mathbf{j}_{M\mu})}(\mathbf{r}) = -i\omega\varepsilon_0 \mathbf{C}_\mu(\mathbf{r}) - \nabla \phi_{M\mu}(\mathbf{r}) \qquad (8.4.14)$$

where both sides of the aperture are assumed to be in vacuum.

Numerical results are reproduced in Fig. 8.17. The remarkable feature is the almost flat field intensity in the proximity of the aperture; the field is well localized in the range determined just by the size of the aperture, except for the radiation part, which decays with $1/z$. These results, demonstrated first by Leviatan, represent one of the most instructive steps for understanding the basic characteristics of near-field optics.

It is possible to extend the results discussed above. For example, for the case of a relatively thick aperture, the distribution of the magnetic current needs to be modified [36].

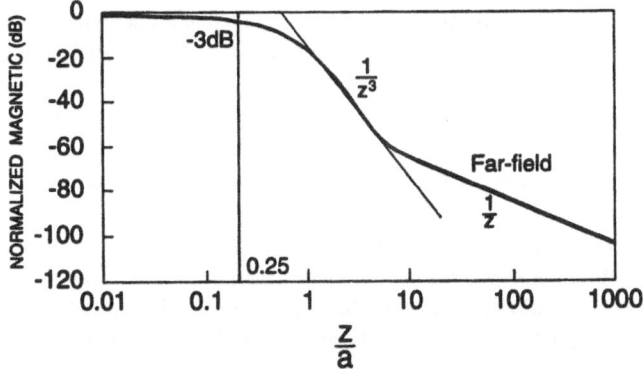

Figure 8.17. Schematic graph showing several characteristics of the so-called "exact" solution given by Leviatan [23] for the near-field intensity on the symmetry axis of a subwavelength-size aperture.

8.5. INTUITIVE MODEL OF OPTICAL NEAR-FIELD PROCESSES

8.5.1. Short-Range Quasistatic Nature of Optical Near-Field Processes

Leviatan's solution for the small-aperture problem demonstrated that the optical near-field of a small scatterer is highly localized (confined) in a spatial extent determined not by the incident optical wavelength, but by the size of the scattering object, i.e., the aperture [23]. Such a scaling character holds for a system characterized by a dimensionless small parameter $a/\lambda \ll 1$ (ratio of the size of the scatterer a to the incident and outgoing optical wavelength λ) and for an observation of the system at a distance r which satisfies the near-field criterion, $r/\lambda \ll 1$. In other words, there exists a highly localized effective field in small-aperture scattering phenomena which is relevant to the internal interactions of the small subsystem. Of course, the near-field component serves as the source of light waves propagating into the far field and cannot be observed unless they are scattered by another perturber (probe tip) placed in its proximity. It should be stressed that the localization of the optical near-field is not in the entire space, but in a half-space, as discussed in the context of the angular spectrum representation.

This size-dependent localization of the optical near-field is the foundation of NOM processes with resolution far beyond the diffraction limit. In general, Heisenberg's uncertainty principle tells us that the resolution of a microscope is determined by the shortest wavelength, or highest wavenumber, available in a specific process of measurement. In this sense, electromagnetic fields of extremely large wavenumber should contribute to NOM. It is therefore of great interest to investigate the near-field part of a scattered field, which is separated from the far-field part on the basis of spatial Fourier analysis. That is, the study of the spatial frequency spectrum of the optical near-field, which involves complex wavenumbers, provides a measure of the NOM resolution as well as a way of interpreting NOM images.

As discussed in Chapter 1, localized (confined) fields are relevant to short-range correlations between induced material excitations, or screened interactions in matter due to an external excitation field. Since we are interested only in the local disturbance involved in scattered fields, we need to find a way to extract short-range components out of the entire electromagnetic process to a good approximation. Short-range interactions in NOM processes also exhibit a quasistatic nature in the sense that propagation or retardation effects are negligible. For example, the traversal time of a light wave (photons) across a 10-nm gap is only of the order of 10^{-17} sec. As a consequence, understanding the quasistatic and short-range nature of optical near-field processes is a key to contructing an empirical model of NOM.

Let us study an intuitive model of NOM, which will help illustrate the physics involved in general optical near-field problems. Before proceeding, we emphasize that the materials in the following part of this section are not intended to introduce a well-established theory, but to suggest the possible future development of optical near-field theory.

8.5.2. Intuitive Model Based on Yukawa-Type Screened Potential

Let us start with an analysis of the localization of the optical near-field in the case of small-aperture scattering. First we will study how we can represent the short-range quasistatic nature of the electromagnetic interaction involved in small-aperture scattering [37]. Then we will see how to eliminate long-range fields. The introduction of screening is one of the most important ideas involved in NOM systems, as discussed in Chapter 1.

One of the simplest ways to describe a quasistatic short-range interaction is to employ a Yukawa-type screened potential defined by

$$\phi(r) \propto \frac{e^{-r/\bar{\lambda}_c}}{r} \tag{8.5.1}$$

Here $\bar{\lambda}_c$ is the screening length, or the electromagnetic correlation length, which characterizes the relevant effective fields.

In general, when one considers an effective field relevant to an interaction of finite range, one can assume that it is mediated by a quasiparticle with an effective mass $m_{eff} = \hbar/\overline{\lambda}_C c$, where $\hbar$ is the Planck constant divided by 2π, and c is the velocity of light in vacuum. In this sense $\overline{\lambda}_C$ can be referred to as the Compton wavelength (divided by 2π) of the effective field. In analogy to elementary excitations in a solid, such as exciton polaritons, let us call the quasiparticles relevant to general optical near-field processes virtual photons, which describe the excitations of coupled modes of photons with matter. The use of quasiparticles does not immediately mean that we need field quantization. Field quantization in a confined space, or in the presence of a material boundary, involves a number of delicate problems and requires very careful consideration [38, 14]. This is beyond the scope of this book, although the quantum behavior of the optical near-field regime is quite interesting and important for the further development of near-field photonics, which applies in the weak-field limit or narrow-space limit, as in the case of single-exciton tunneling.

The idea of the effective mass of a quasiparticle is an alternative to the short-range internal interaction in a many-body system and serves as a measure of the system response to an external field [39]. In other words, this implies that complicated microscopic processes inside the many-body system are renormalized, or already included, into an effective mass in order to separate the macroscopic response of the system to an additional external field. It is also worth noting that the idea of the virtual photon is an alternative to the evanescent field, and vice versa [40].

The Yukawa-type screened potential corresponds to a general solution of the modified Helmholtz equation (Klein–Gordon equation) for a scalar field $\phi(\mathbf{r})$ defined by

$$\left(\nabla^2 - \overline{\lambda}_C^{-2}\right)\phi(\mathbf{r}) = 0 \tag{8.5.2}$$

Here the minus sign in front of the square of the wavenumber, $\overline{\lambda}_C^{-2}$, results from the renormalization of the induced source fields, which are usually placed on the right-hand side of the ordinary Helmholtz equation. That is, the local coupling of the electromagnetic field (or photon) with matter is described in terms of an effective field of quasistatic short-range nature which is characterized by $\overline{\lambda}_C$. Unfortunately, such a renormalization procedure is difficult for a rigorous theoretical treatment in the case of a complicated many-body system with no specific resonance behavior. Here, let us try to see how this assumption fits a simple optical near-field problem.

Before proceeding, it is instructive to consider the meaning of the modified Helmholtz equation and its relation to optical near-field phenomena. Let us consider the dispersion relation of light waves (or photons)

$$\left(\frac{\omega}{c}\right)^2 - k_{\parallel}^2 - k_{\perp}^2 = 0 \tag{8.5.3}$$

where ω is the temporal frequency and $k_\parallel$ and $k_\perp$ are, respectively, wavenumbers parallel and perpendicular to an assumed planar boundary. If we assume a light field confined in a small region on the assumed boundary plane, it should contain spatial Fourier components with $k_\parallel$ as large as the inverse of the dimensions of the localization. Let the size of confinement be $\bar{\lambda}_C$. Then the confined field contains waves lying on the section of the dispersion relation at $k_\parallel \sim 1/\bar{\lambda}_C$, as shown in Fig. 8.18. Since the section corresponds to a parabolic function with respect to $k_\perp$, the confined field behaves like a wave of a massive particle when one considers its motion perpendicular to the assumed boundary plane, in the sense that it obeys a dispersion relation described by

$$\left(\frac{\hbar\omega}{c}\right)^2 - (\hbar k_\perp)^2 = m_{\text{eff}}^2 c^2, \qquad \frac{m_{\text{eff}}^2 c^2}{\hbar^2} = k_\parallel^2 \tag{8.5.4}$$

A corresponding quasiparticle with $k_\parallel > \omega/c$ is considered to be "virtual" since its energy $\hbar\omega$ lies below the vertex of the parabola, $\hbar\,k_\parallel c$ or $\hbar c/\bar{\lambda}_C$, as shown in Fig. 8.18. In this case the dispersion relation can be rewritten as

$$\left\{ m_{\text{eff}}^2 c^2 - \left(\frac{\hbar\omega}{c}\right)^2 \right\} + (\hbar k_\perp)^2 = 0, \qquad \frac{m_{\text{eff}}^2 c^2}{\hbar^2} > \frac{\hbar^2\omega^2}{c^2} \tag{8.5.5}$$

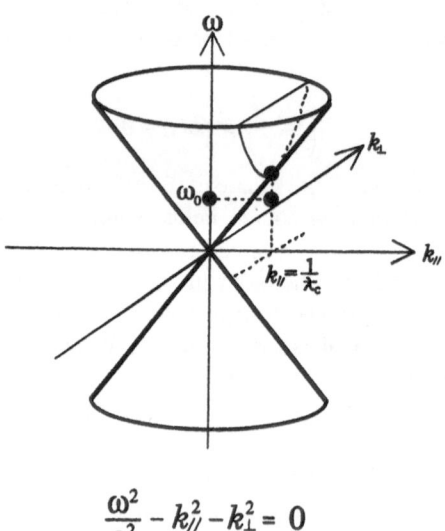

$$\frac{\omega^2}{c^2} - k_{//}^2 - k_\perp^2 = 0$$

Figure 8.18. Dispersion relation of light waves (photons) and confined electromagnetic field. A section of the dispersion curve at a $k_\parallel$ shows a parabolic shape. This corresponds to a reduced one-dimensional dispersion relation for confined fields.

and this results in a modified Helmholtz equation which corresponds to an analytic continuation of $k_\perp$ into the imaginary region. If we consider the near-field limit $\omega/c \ll k_\parallel$ we obtain Eq. (5.3.2). Then the virtual photon can appear only in the context of the Heisenberg uncertainty relation, $\Delta t \Delta E \sim \hbar$. This implies that we can obtain such a localized field only as an effective field excited in the narrow vicinity of a material surface which serves as the source of such a highly localized field. Therefore, in general, the virtual photon description holds only in relation to half-space problems regardless of the shape of the assumed boundary.

8.5.3. Application of Virtual Photon Model for Diffraction from a Small Aperture

Let us apply the intuitive Yukawa-potential model to evaluate the localization of the optical near-field of a subwavelength-size circular aperture. Assume that the incident light wavelength λ is much larger than the aperture radius a, i.e., $a/\lambda \ll 1$. The only parameter we can adjust is the screening distance $\overline{\lambda}_C$.

Let us first assume that $\overline{\lambda}_C$ is equal to the scale of the localization a, i.e., the aperture radius, and consider that contributions from each infinitesimal surface element dS' lying at a position $\mathbf{r}'$ in the aperture region are described by using a Yukawa-type screened potential. Then the total potential field around the aperture is obtained by summing up all the contributions from surface elements in the aperture area A (see Fig. 8.19a):

$$\psi(\mathbf{r}) \propto \int_A \frac{e^{-|\mathbf{r}-\mathbf{r}'|/\overline{\lambda}_C}}{|\mathbf{r}-\mathbf{r}'|} dS' \tag{8.5.6}$$

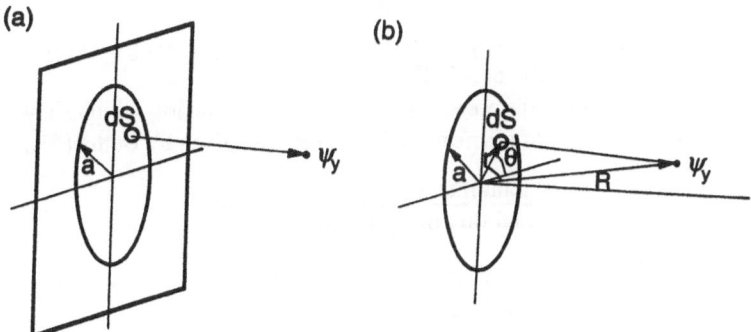

Figure 8.19. Coordinate system used for evaluation of the small-aperture near-field using a Yukawa-type screened potential.

In general, the Hamiltonian density for a Yukawa-type screened field is given by the sum of a so-called gradient term $|\nabla\psi|^2$ and a mass term $|\psi|^2/\bar{\lambda}_C^2$:

$$\mathcal{H}_y = |\nabla\psi|^2 + \frac{1}{\bar{\lambda}_C^2} |\psi|^2 \tag{8.5.7}$$

It is then straightforward to calculate the field intensity distribution near the aperture if we use this Hamiltonian density.

As an example, we can calculate the potential on the symmetry axis (z axis) of the aperture as

$$\psi(\zeta) = \bar{\lambda}_C \left[\exp\left(-\frac{a}{\bar{\lambda}_C}\zeta\right) - \exp\left(-\frac{a}{\bar{\lambda}_C}\sqrt{\zeta^2+1}\right) \right] \tag{8.5.8}$$

where $\zeta = z/a$ is the normalized distance from the aperture plane and the screening distance $\bar{\lambda}_C$. In virtue of the cylindrical symmetry, its gradient is simply given by differentiation with respect to the direction normal to the aperture plane,

$$\nabla\psi(\zeta) = \left[\frac{\zeta}{\sqrt{\zeta^2+1}} \exp\left(-\frac{a}{\bar{\lambda}_C}\sqrt{\zeta^2+1}\right) - \exp\left(-\frac{a}{\bar{\lambda}_C}\zeta\right) \right]\hat{\mathbf{z}} \tag{8.5.9}$$

where $\hat{\mathbf{z}}$ is the unit vector along the z axis. A three-dimensional calculation in the half-space separated by the aperture plane is also straightforward. With the coordinate system shown in Fig. 8.19b, we obtain

$$\phi(R, \varphi) = \frac{1}{\pi a^2} \int_0^a r\, dr \int_{-\pi}^{\pi} d\theta\, \frac{\exp\left[-(1/\bar{\lambda}_C)\sqrt{R^2+r^2-2Rr\cos\theta\cos\varphi}\right]}{\sqrt{R^2+r^2-2Rr\cos\theta\cos\varphi}} \tag{8.5.10}$$

Figure 8.20 shows numerical results for the intensity distribution $\mathcal{H}_y$ obtained from the model calculation for the case of $a = \lambda/50$ with the screening length assumed to be $\bar{\lambda}_C = a$. The near-field behavior simulated by the Yukawa potential fits very well with that given by the so-called exact calculation given by Leviatan [23], which is shown by the dots in Fig. 8.20.

We see that the only difference is in the far-field behavior, which shows a $1/r$ decrease. From the ratio of the near- and far-field flux given by Bethe [21],

$$\frac{\text{total flux at far field}}{\text{total flux in proximity}} = \frac{24}{(\lambda/a)^2} \tag{8.5.11}$$

we can easily estimate the far-field component. This simulates very well the entire behavior of the "exact" calculation, as shown in Fig. 8.20.

Figure 8.21 shows a three-dimensional intensity distribution and its cross section at several values of the distance z in the x–z plane. The results represent even the details of the behavior found by Leviatan [23]. In Fig. 8.21d it is important

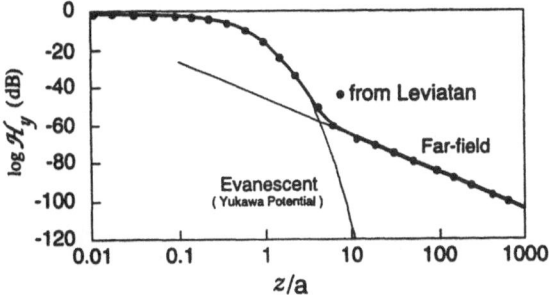

Figure 8.20. Localization of the optical near-field of a small circular aperture. An aperture radius of $\lambda/50$ is assumed. The solid line indicates the near-field intensity on the symmetry axis of the aperture estimated by using a Yukawa-type screened potential and the far-field scattered intensity given by Bethe [21]. The thin line shows the screened potential alone. Dots are the "exact" calculation given by Leviatan [23]. It can be seen that the locality of the small-aperture near-field is equal to that of the effective field with screening length $\bar{\lambda}_C = a$, the radius of the aperture.

to note the qualitative difference in the intensity distribution around the aperture edge in the case of the cross section at z very close to the aperture: Leviatan's result shows a steep decrease at the edge for very small z, whereas the result in Fig. 8.21d exhibits a relatively slow decay. This is because we only took account of the effective field with Compton wavelength $\bar{\lambda}_C = a$. In fact, if we investigate the effective field in the narrow vicinity of the aperture edge we observe shorter range effective fields due to the steepness of the boundary at the edge of aperture. This difference is very important when we study the relation between the effective field and the range of correlation as well as the distance between the object and the probe point.

In the above, we found that we can reproduce the near-field behavior of small-aperture diffraction by assuming a short-range quasistatic interaction and by using a screened potential scaled by the dimension of the scatterer, $\bar{\lambda}_C = a$. Although it is not derived directly from a rigorous theory of the electromagnetic interaction of matter, the virtual-photon picture introduced here provides an intuitive understanding of optical near-field phenomena. Of course, we need to take the vector nature of electromagnetic interactions into account. However, this point is not clear for the virtual photon model at present. However, we note that a scalar model often provides a good approximation of a fully vector description when the system considered has a certain symmetry—cylindrical in the case of the small-aperture problem. Let us look at some extensions of the model before we give a brief consideration of the physical meaning underlying the virtual photon picture.

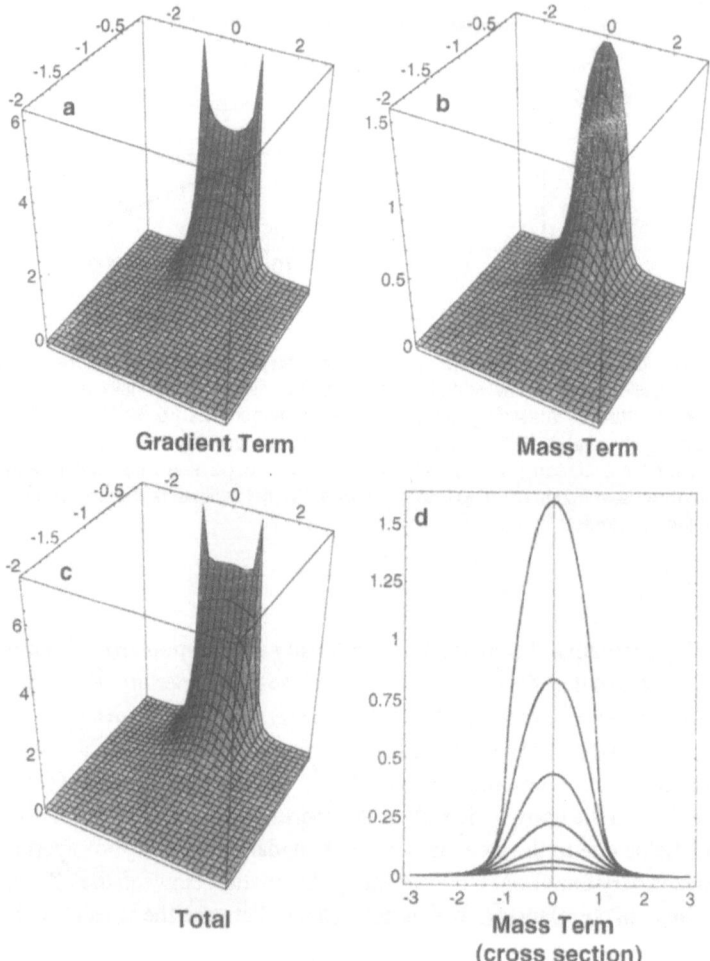

Figure 8.21. Three-dimensional intensity distribution of Yukawa-type screened field for a source distributed uniformly in a circular aperture: (a) gradient term, (b) mass term, (c) total intensity. (d) Cross section of panel (c). The intensity distribution in panel (c) reproduces the near-field behavior of the small-aperture near-field given by Leviatan's result.

8.5.4. Virtual Photon Model of NOM

Let us consider an intuitive model of the sample–probe interaction in NOM. Here we assume that the fundamental process can be described as a short-range quasistatic interaction via virtual photons between two closely spaced small apertures or dielectric spheres corresponding to an object and a probe tip. We consider that both the object and probe tip are coupled by some means to incoming and outgoing propagating photons. This model requires an extension of the optical near-field interaction to those of closely spaced disconnected material objects. That is, we need to consider the interaction propagator connecting the electric polarizations on two closely spaced scatterers. Note that solutions of Maxwell's equations with disconnected boundaries are in general very difficult to obtain.

It is interesting to compare NOM interactions with van der Waals interactions. According to field theory, the origin of the van der Waals force is the exchange force of virtual photons connecting electric polarizations induced on two closely spaced dielectric objects due to vacuum fluctuations [41]. Since electric polarizations are induced by incident light in the NOM process, the range of optical near-field interactions should be longer than in the van der Waals case. NOM processes, in general, involve destructive measurements in the optical near-field regime, and therefore we should concentrate on evaluating optical near-field interactions in the object–probe system rather than on calculating the optical near-field intensity itself.

As we found that the screening length $\overline{\lambda}_C$ of the optical near-field regime is the counterpart of the lateral spatial frequency $k_\parallel$, we can assume a resonance character with respect to the spatial frequencies of the object and probe near-fields: the magnitude of the interaction, or coupling efficiency, between sample and probe tip takes its maximum when the screening distances $\overline{\lambda}_C$ of the object and probe tip coincide. This can be regarded as a momentum conservation rule in relation to the exchange processes of virtual photons. Although orthogonal relations for virtual photons cannot be strict, such a resonance assumption is considered to give a good approximation.

Once virtual photons with short penetration depths are assumed, the problem we need to solve is similar to electron tunneling. This is because the process is described by a modified Helmholtz equation. One might introduce the tunneling current of virtual photons similar to Bardeen's tunneling current for electrons [42]. The resonance assumption provides an intuitive way to understand for the size dependence in NOM processes.

8.5.4.1. Resolution and Signal Transfer Function of NOM

This simple picture of NOM leads to a natural definition of NOM resolution. Let us consider a small dielectric sphere of radius a placed on a surface. The surface itself can be described by dielectric spheres with a radius equal to the penetration depth λ_{pen} ($>>a$) of the illuminating evanescent wave. We consider a conical probe described by a set of dielectric spheres, as shown in Fig. 8.22. According to the resonance assumption, the highest NOM resolution is determined by the radius of the sphere at the apex of the conical probe. Another important feature determining the NOM resolution is the filtering of the illuminating field. Let us investigate how these filtering characteristics can be described. Consider that the apex of the probe

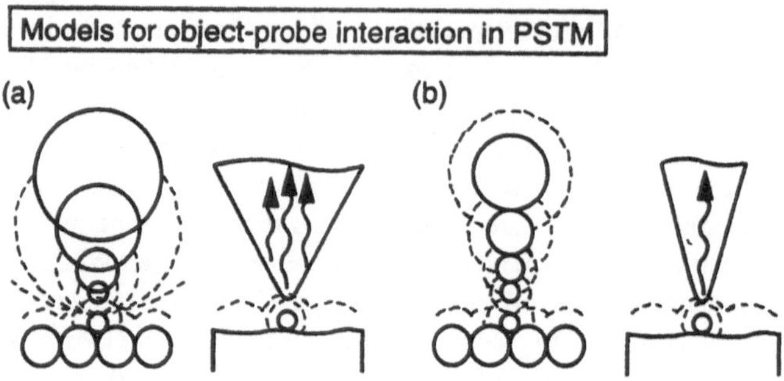

Figure 8.22.   Optical near-field interactions of conical bare optical fiber probe tips with obtuse and acute conical angles for photon scanning tunneling microscopy (PSTM).

is positioned at a distance equal to its radius of curvature a from the microscopic object. In this case, the tunneling probability of the evanescent field with screening distance $\overline{\lambda}_C = a$ is of the order of $\sim e^{-1}$, and evanescent fields with shorter screening distances due to small bumps on the object are filtered out since their interaction ranges are much shorter than the probe–object distance. Therefore, the short-range character of the probe is determined by the radius of curvature at the apex and the sample–probe distance.

Next, let us consider the long-range characteristics. When the probe tip has an obtuse conical angle (Fig. 8.22a), evanescent fields having penetration depth larger than a have a relatively large tunneling probability, and consequently the ratio of the picked-up power from the object and that from the illumination background is less than unity. This results in a poor contrast of NOM images. On the other hand, when the probe tip has an acute conical angle (Fig. 8.22b), the evanescent fields with penetration depth larger than a decay very rapidly and then they are filtered since the corresponding spheres recede rapidly as their radii increase.

As a consequence, the NOM process can be viewed as a kind of spatial frequency bandpass filter for the optical near-field.

8.5.4.2.   Comparison of Model Description with Experiments

The validity of the resonance assumption is tested for the simple case of an evanescent field at total internal reflection with incident angle θ_{in}, refractive index n, and optical wavelength λ. The penetration depth of the illuminating evanescent wave is given by

$$\overline{\lambda}_C = \lambda / \{2\pi(n^2\sin^2\theta_{in} - 1)^{1/2}\} \qquad (8.5.12)$$

Let us evaluate the optical power picked up from the illuminating evanescent wave by bare optical fiber probes of different conical angles. The evanescent field with penetration depth $\overline{\lambda}_C$ is assumed to couple with a sphere of radius $\overline{\lambda}_C$ comprising the probe.

A simple calculation shows that when the apex of the probe is held at a constant height from the surface, the pickup power from the evanescent wave is given by

$$\exp[-2\alpha/\sin(\theta/2)] \qquad (8.5.13)$$

where θ is the conical angle of the probe. Here α is a form parameter that is dependent on the shape of the probe, conical in the present case.

This conical angle dependence of the pickup power is compared with experimental data showing a decay length of 350 nm, which is in good agreement with the screening distance $\overline{\lambda}_C$ for $\theta_{in} = \pi/4$, $n = 1.5$, and $\lambda = 780$ nm. The solid line in Fig. 8.23 shows the theoretical prediction with a form parameter $\alpha = \pi^{-1}$, which is given by the calculation of the overlap integral of the whole probe–object Yukawa field. Figure 8.23 shows good agreement of the prediction with experiments.

Because of the spatial filtering character of the bare optical fiber probe, the pickup power of ~1 pW due to interaction between the nanometric object and the sphere at the extremity of the probe tip becomes comparable to the evanescent background when the conical angle of the probe tip is decreased to ~25 deg.

As shown in the example above, the assumption of size resonance allows us to calculate the pickup power for object–probe systems of arbitrary shape represented as a set of dielectric spheres. This is an essential feature of the local theory we have sought.

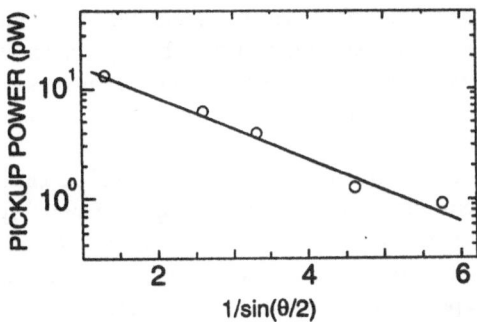

Figure 8.23. Screening characteristics of bare optical fiber probes with different conical angles. The picked-up optical power from an evanescent wave on a planar dielectric surface is indicated as a function of the conical angle of bare optical fiber probes with apex held 5 nm above the surface. A penetration depth of $\overline{\lambda}_C = 350$ nm is assumed for the evanescent wave at total internal reflection. The solid line estimated by using the screened-potential model with a form factor $\alpha = \pi^{-1}$ is in good agreement with the experimental results shown by the circles.

8.5.5. Meaning of the Screened Potential Model and Physical Meaning of the Virtual Photon

Here we consider the physical meaning of the intuitive model developed in the above sections. Recall that the origin of the difficulty in handling optical near-field problems is that the electromagnetic interactions, or induced sources, play fundamental roles. In order to avoid an explicit treatment of the interactions or sources one needs in general to resort to the idea of the macroscopic response of matter, as is discussed in Section 8.2. Otherwise, one has to solve complicated inhomogeneous Helmholtz equations. Furthermore, the material response itself is a function of field variables, so that self-consistent considerations are required.

One way to avoid explicit use of sources is to replace the microscopic interactions by macroscopic dielectric functions in matter and by a set of boundary conditions at the material surface. In contrast, we also saw how an assumed source distribution can replace boundary conditions, when we studied the assumed magnetic current in Section 8.3.2.

Another way to replace microscopic source fields is to introduce the effective mass of an effective field. This type of replacement is popular in quantum field theory and its application to solid-state physics. For example, the behavior of electrons under the influence of the environment, such as the quantum vacuum or a crystalline field, is described in terms of the effective mass of the electron, which serves as a measure of the electron motion in response to the external field. In relation to the electromagnetic field, an important example is the case of a superconducting surface where a photon acquires an effective mass and exhibits a shallow penetration. This result is due to the direct proportionality of the screening current to the vector potential, known as the London equation [43, 44]. A simpler example is the case of a metallic surface which screens the penetration of an incident electromagnetic field by a conducting current. It might be an interesting problem to investigate the analogy between the conducting current on a metallic surface and the assumed magnetic current on a dielectric surface as sources which screen electromagnetic interactions.

The problem we need to consider first is how to find an appropriate description of the electromagnetic interaction of matter for a given material system and incident field as well as observed outgoing field. It is emphasized that the appropriateness of the theoretical description depends on both the spatial and temporal scales concerned with the problem. For instance, the characteristic parameter is the ratio of the object size and the optical wavelength, which is equivalent to the product of the traversal time of a photon across the object and the optical frequency. When we consider a short-range correlation it is naturally quasistatic.

The screened-potential model discussed above can serve as an empirical model appropriate to describe the near-field behavior of an electromagnetic problem characterized in the small-parameter limit.

8.5.5.1. Quantization of Evanescent Waves and Comparison with Quasiparticle Model

We can compare the virtual photons associated with the screened potential with the surface plasmons used in the description of specific types of near-field optical microscopes for metallic surfaces which utilize a metallic probe tip [45–47]. Although a strict resonance behavior is utilized both in the excitation and detection of illuminating and scattered plasmon fields, the specimen–probe interaction is due to a localized electronic correlation (Coulomb interaction) with a certain screening length. The term "localized plasmon" is commonly used in reference to such a localized interaction via electronic correlation in a sea of positive charge. In a similar way, the virtual photon description for the dielectric case can be related to a surface excitation which in turn may be related to exciton polaritons representing an electromagnetic field coupled to a dielectric. In this case an illuminating field is often provided in the form of evanescent waves at total internal reflection, which exhibits a broadened excitation spectrum in resonance with an incident propagating field.

A more careful consideration is required when one applies the idea of fundamental excitations for a localized excitation in matter, even in the case of the popular plasmon. In contrast to usual fundamental excitations, no rigid dispersion relation can be assumed when the interaction takes place in a space-time volume much smaller than the periods of spatial and temporal oscillation of the incident polarization wave. The fundamental excitation is due to a collective motion or long-range correlation of a material excitation obtained as a result of tracing out the short-range correlations or fluctuations. On the other hand, a local interaction is often relevant to the short-range behavior of material excitations depending on the nature of the specific material and the excitation frequencies.

It should also be noted that there arises a difficulty in applying second quantization of the optical field in a highly confined space-time smaller than the optical wavelength. That is, the optical near-field component here represents only the tail of the mode function defined in a macroscopic volume λ^3 insofar as one considers the problem on the basis of Fourier analysis with a specific wavelength λ [38]. In fact when we treat the electromagnetic field defined at each point of space-time, we need to assume nonlocality in the material response or modified boundary conditions as a natural consequence [9–12].

Accordingly, the picture of a pointlike virtual particle propagating along a path might be a convenient way to treat near-field problems where the evanescent field is described in terms of the electromagnetic propagator connecting material objects placed in a space-time volume smaller than λ^4/c. In this sense, no free photons can be considered in the treatment of the evanescent electromagnetic field in this section [38]. The microscopic nature of the evanescent field is considered to consist in the quasistatic (Coulomb-like) interaction between electric polarizations induced on the material surface by incident and scattered light. In this sense, the evanescent field correlates a pair of incident and reflected waves via induced polarizations $\mu[\phi_{in}(x_1); \phi_{ref}(x_1)]$ at a position x_1 with a pair $\mu[\phi_{in}(x_2); \phi_{ref}(x_2)]$ at a closely spaced position x_2 on the material surface. This implies that the evanescent field represents the propagator similar to the two-particle Green's function, which integrates the many-body interaction on the material surface. This picture is consistent with the quantum theory of the evanescent electromagnetic field near a flat dielectric surface developed by Carniglia and Mandel [40].

Here it is instructive to compare the quantized evanescent field with the exciton polariton description developed first by Hopfield [8]. The exciton polariton corresponds to the quasiparticle representation of the space-time correlation between atomic excitations in a crystal. By means of Fourier analysis, coupled modes of light waves with an induced polarization field are formulated. These modes correspond to eigenstates of Maxwell's equations coupled to the constitutive equation

$$\frac{1}{\omega_0^2} \frac{\partial^2 \mathbf{p}}{\partial t^2} + \mathbf{p} = \beta \mathbf{E} \tag{8.5.14}$$

where ω_0 is the oscillation frequency of the polarization, and a linear optical response is assumed as $\beta \mathbf{E}$. One can find in the original article of Hopfield that the polariton part of the Hamiltonian consists of the polariton number operator, the polariton–photon interaction, and the self-coupling of polariton modes (see, e.g., Eq. (27) in Hopfield [8]). On the other hand, in the quantization of evanescent waves presented by Carniglia and Mandel, a coupled mode consists of incident, reflected, and evanescent waves, which are connected at the planar dielectric boundary by the Fresnel relation, and are described on the basis of Fourier analysis due to translational symmetry along the boundary plane. Also, in the quantized evanescent waves, the interaction processes of the induced polarization field are described by both the interaction with incident and reflected waves and self-coupling of induced polarizations. Here, instead of using constitutive equations, a set of boundary conditions was imposed since the material response is explicitly known at the planar dielectric–air boundary.

Although the similarity between quantized evanescent waves and exciton polaritons is quite intuitive, it provides some hints on how to proceed further in developing optical near-field theory. In fact, the idea of a quasiparticle on a material surface itself is an important problem in surface physics. A comprehensive quantum model describing general optical near-field problems is still under development.

In addition, besides near-field optics, it is instructive to note that a Yukawa-type screened potential plays an important role in the Fourier expansion of the Coulomb-like $1/r$ interaction propagator; a Yukawa potential instead of a Coulomb potential is Fourier transformed, and then the screening distance $\overline{\lambda}_C$ is extended to infinity:

Fourier transform of $\left[\dfrac{1}{r}\right]$

$$= \lim_{\lambda_c \to \infty} \int_0^\infty d^3r \frac{e^{-\frac{|\mathbf{r}|}{\lambda_c}}}{|\mathbf{r}|} e^{-i\mathbf{q}\cdot\mathbf{r}} = \lim_{\lambda_c \to \infty} \frac{4\pi}{|\mathbf{q}|^2 + \bar{\lambda}_C^{-2}} = \frac{4\pi}{|\mathbf{q}|^2} \qquad (8.5.15)$$

It would be interesting to consider the relation between the extraction of a short-range component relevant to a small interaction volume and the cutoff of the long-range terms in the integration in the Fourier transform.

8.5.5.2. Coarse Graining of Microscopic Interactions in a Many-Body System and Meaning of the Screened Potential

It is instructive for further development of a phenomenological theory to consider the statistical characteristics of microscopic electromagnetic interactions taking place in optical near-field processes. It is interesting to consider the so-called "coarse graining" viewpoint adopted in relation to the many-body process. Optical near-field processes and measurements are concerned with the extraction of information about electromagnetic correlations between induced polarizations at a certain coarse-grained scale lying between the optical wavelength and the atomic size. By means of coarse graining one can extract longer range electromagnetic correlations which result in optical responses viewed at that scale. On the other hand, if one views interaction processes in matter in detail, one finds microscopic interactions which are screened on the larger scale.

Landau introduced a phenomenological model of phase transitions based on an effective Hamiltonian [48, 49]:

$$\mathcal{H}_L = \int \left[|\nabla\phi(\mathbf{r})|^2 + \mathcal{R}(L, T)|\phi(\mathbf{r})|^2 + \mathcal{U}(L, T)|\phi(\mathbf{r})|^4 - H(\mathbf{r})\phi(\mathbf{r}) \right] d^d r \qquad (8.5.16)$$

where $\phi(\mathbf{r})$ is an order parameter describing the state at position $\mathbf{r}$, L is the size of the unit cell considered, $\mathcal{R}(L, T)$ is a characteristic function of L and the temperature T corresponding to [correlation length]$^{-2}$, $\mathcal{U}(L, T)$ is the coefficient of interactions between cells, and d is the dimension of the system. The last term in the integral represents the interaction with an applied external field H. For a spin system, the order parameter represents magnetization. Let $\mathcal{U}(L, T) = \xi_L^{-2}$, where ξ_L is the screening length when we view the interaction process in terms of the coarse-graining parameter L. When the internal interactions $\mathcal{U}(L, T)|\phi(\mathbf{r})|^4$ and external field $H(\mathbf{r})$ are omitted, the averaged value of the order parameter $<\phi(\mathbf{r})>$ is given by a screened function:

$$<\phi(\mathbf{r})> \sim H \frac{1}{|\mathbf{r}|^{d-2}} \exp\left(-\frac{|\mathbf{r}|}{\xi_L}\right) \qquad (8.5.17)$$

By using this phenomenological Hamiltonian and the theory of scaling based on renormalization group theory, a number of important critical behaviors of the phase transition have been discussed mainly in relation to spin systems.

It is interesting to investigate the relation between the coarse graining of a system and the optical properties of matter, though no critical behavior would be expected to arise in the case of an ordinary dielectric system. One might expect some interesting features when one investigates light scattering by a random surface. In fact the screened correlation in Eq. (8.5.17) provides an important measure by which to evaluate a random process. An introduction to these theoretical treatments is, however, beyond the scope of this book. Here, we employ a theoretical description which shows the relation between the macroscopic polarization and its local disorder. Instead of spins in a magnetic medium, let us consider induced polarizations in a dielectric medium. As discussed in Section 8.3 in this chapter, the Hertz vector representation yields important characteristics of the local fluctuations and the macroscopic average of the induced polarization. The Hertz vector in a nonmagnetic medium represents the retardation potential produced by a polarization field $\mathbf{p}$:

$$\mathbf{\Pi}^E(\mathbf{r}, t) = \int_{V_1} \frac{d\mathbf{p}\left(\mathbf{r}_1, t - \dfrac{|\mathbf{r} - \mathbf{r}_1|}{c}\right)}{4\pi|\mathbf{r} - \mathbf{r}_1|} \tag{8.5.18}$$

Let us consider a nonuniform dielectric medium; we have a wave equation for the electric field assuming a monochromatic response $\exp(-i\omega t)$:

$$\nabla^2 \mathbf{E}_\omega + K^2 \mathbf{E}_\omega - \nabla(\nabla \cdot \mathbf{E}_\omega) = -\frac{K^2}{\varepsilon_0} \mathbf{p}_\omega \tag{8.5.19}$$

where the term $\nabla(\nabla \cdot \mathbf{E}_\omega)$ is due to the nonuniformity of the induced polarizations. Using the relation $\mathbf{E}_\omega = \nabla(\nabla \cdot \mathbf{\Pi}_\omega^E) + K^2 \mathbf{\Pi}_\omega^E$, we obtain

$$\nabla^2 \mathbf{\Pi}_\omega^E + K^2 \mathbf{\Pi}_\omega^E = -\frac{1}{\varepsilon_0} [\mathbf{p}_\omega - (\varepsilon - \varepsilon_0)\mathbf{E}_\omega] \tag{8.5.20}$$

where the right-hand side represents the local fluctuations of the induced polarization. This is clear when we take the average of the induced polarization over a macroscopic volume,

$$\langle \mathbf{p}_\omega \rangle = (\varepsilon - \varepsilon_0)\langle \mathbf{E}_\omega \rangle \qquad (\nabla^2 \mathbf{\Pi}_\omega^E + K^2 \mathbf{\Pi}_\omega^E = 0) \tag{8.5.21}$$

It would be of further interest to consider the meaning of coarse graining in a dielectric medium and its relation to the effective field described by a Yukawa-type screened potential on the bases of the Hertz vector representation or the vector potential $\mathbf{C}$ developed in Section 8.3.

8.6. REFERENCES

1. J. D. Jackson, *Classical Electrodynamics*, 2nd ed., Wiley, New York, 1975.
2. M. Born and E. Wolf, *Principles of Optics*, 3rd ed., Pergamon Press, Oxford, 1965.
3. A. Sommerfeld, *Partial Differential Equations in Physics*, Academic Press, New York, 1949.
4. P. M. Morse and H. Feshbach, *Methods of Theoretical Physics, Part 1*, McGraw-Hill, New York, 1953.
5. P. M. Morse and H. Feshbach, *Methods of Theoretical Physics, Part 2*, McGraw-Hill, New York, 1953.
6. J. J. Sakurai, *Modern Quantum Mechanics*, Addison-Wesley, Redwood City, California, 1985.
7. C. Kittel, *Quantum Theory of Solids*, Wiley, New York, 1963.
8. J. J. Hopfield, Theory of the contribution of excitons to the complex dielectric constant of crystals, *Phys. Rev.* **112**: 1555–1567 (1958).
9. R. Fuchs and F. Claro, Multipolar response of small metallic spheres: Nonlocal theory, *Phys. Rev. B* **35**: 3722–3727 (1987).
10. F. Hache, D. Richard, and C. Girard, Optical nonlinear response of small metal particles: A self-consistent calculation, *Phys. Rev. B* **38**: 7990–7996 (1988).
11. K. Cho, Nonlocal theory of radiation–matter interaction: Boundary-condition-less treatment of Maxwell equations, *Progr. Theor. Phys. Suppl.* **106**: 225–233 (1991).
12. K. Cho, Y. Ohfuti, and K. Arima, Study of scanning near-field optical microscopy (SNOM) by nonlocal response theory, *Jpn. J. Appl. Phys.* **34**: 267–270 (1994).
13. C. Kittel, *Introduction to Solid State Physics*, 6th ed., Wiley, New York, 1986.
14. G. S. Agarwal, Quantum electrodynamics in the presence of dielectrics and conductors. I. Electromagnetic-field response functions and blackbody fluctuations in finite geometries, *Phys. Rev. A* **11**: 230–242 (1975).
15. G. S. Agarwal, Quantum electrodynamics in the presence of dielectrics and conductors. II. Theory of dispersion forces, *Phys. Rev. A* **11**: 243–252 (1975).
16. G. S. Agarwal, Quantum electrodynamics in the presence of dielectrics and conductors. III. Relations among one-photon transition probabilities in stationary and non stationary fields, density of states, the field-correlation functions, and surface-dependent response functions, *Phys. Rev. A* **11**: 253–264 (1975).
17. E. Yablonovitch and T. J. Gmitter, Photonic band structure: The face-centered-cubic case, *Phys. Rev. Lett.* **63**: 1950–1953 (1989).
18. S. John and J. Wang, Quantum electrodynamics near a photonic band gap: Photon bound states and dressed atoms, *Phys. Rev. Lett.* **64**: 2418–2421 (1990).
19. Z. Zhang and S. Satpathy, Electromagnetic wave propagation in periodic structure: Bloch wave solution of Maxwell's equation, *Phys. Rev. Lett.* **65**: 2650–2653 (1990).
20. R. P. Feynman, R. B. Leighton, and M. Sands, *The Feynman Lectures on Physics, Vol. 2*, Addison-Wesley, Reading, Massachusetts, 1964.
21. H. A. Bethe, Theory of diffraction by small holes, *Phys. Rev.* **66**: 163–182 (1944).
22. C. J. Bowkamp, Diffraction theory, *Rep. Progr. Phys.* **17**: 35–100 (1954).
23. Y. Leviatan, Study of near-zone fields of a small aperture, *J. Appl. Phys.* **60**: 1577–1583 (1986).
24. I. Banno, to be published.
25. O. J. F. Martin, C. Girard, and A. Dereux, Generalized field propagator for electromagnetic scattering and light confinement, *Phys. Rev. Lett.* **74**: 526–529 (1995).
26. C. Girard and C. Girardet, Self-consistent interaction potential for a molecule absorbed on a dielectric surface: A symmetric top molecule on an ionic crystal, *J. Chem. Phys.* **86**: 6531–6539 (1987).
27. C. Girard, Theoretical atomic-force-microscopy study of a stepped surface: Nonlocal effects in the probe, *Phys. Rev. B* **43**: 8822–8828 (1991).

28. C. Girard and M. Spajer, Model for reflection near field optical microscopy, *Appl. Opt.* **29:** 3726–3733 (1990).

29. I. Banno, H. Hori, and T. Inoue, An Ampere-like law for displacement vector field and near-field optical microscopy, *Opt. Rev.* **3:** 454–457 (1996).

30. I. Banno and H. Hori, to be published.

31. O. J. F. Martin, C. Girard, and A. Dreux, Dielectric versus topographic contrast in near-field, *J. Opt. Soc. Am. A* **13:** 1801–1808 (1996).

32. E. A. Synge, A suggested method for extending microscopic resolution into the ultra-microscopic region, *Phil. Mag.* **6:** 356–362 (1928).

33. D. W. Pohl, W. Denk, and M. Lanz, Optical stethoscopy: Image recording with resolution $\lambda/20$, *Appl. Phys. Lett.* **44:** 651–653 (1984).

34. H. Levine and J. Schwinger, On the theory of diffraction by an aperture in an infinite plane screen. I, *Phys. Rev.* **74:** 958–974 (1948).

35. K. Jang and W. Jhe, Nonglobal model for a near-field scanning optical microscope using diffraction of the optical near field, *Opt. Lett.* **21:** 236–238 (1996).

36. A. Roberts, Small-hole coupling of radiation into a near-field probe, *J. Appl. Phys.* **70:** 4045–4049 (1991).

37. H. Hori, Quantum optical picture of photon STM and proposal of single atom manipulation, in *Near Field Optics*, D. W. Pohl and D. Courjon, eds., Kluwer, Dordrecht, 1993, pp. 105–114.

38. L. Mandel, Configuration-space photon number operators in quantum optics, *Phys. Rev.* **144:** 1071–1077 (1966).

39. G. E. Brown, *Many-Body Problems*, North-Holland, Amsterdam, 1972.

40. C. K. Carniglia and L. Mandel, Quantization of evanescent electromagnetic waves, *Phys. Rev. D* **3:** 280–296 (1971).

41. C. Itzykson and J.-B. Zuber, *Quantum Field Theory*, McGraw-Hill, New York, 1980, Chapter 3.

42. J. Bardeen, Tunneling from a many-particle point of view, *Phys. Rev. Lett.* **6:** 57–59 (1961).

43. C. P. Enz, *A Course on Many-Body Theory Applied to Solid-State Physics*, World Scientific, Singapore, 1992.

44. I. J. R. Aitchson and A. J. G. Hey, *Gauge Theories in Particle Physics*, 2nd ed., Adam Higler, Bristol, England, 1989.

45. U. Ch. Fischer and D. W. Pohl, Observation of single-particle plasmon by near-field optical microscopy, *Phys. Rev. Lett.* **62:** 458–461 (1989).

46. M. Specht, J. D. Pedaring, W. M. Heckl, and T. W. Hänsch, Scanning plasmon near-field microscope, *Phys. Rev. Lett.* **68:** 476–479 (1992).

47. P. Dawson, F. de Fornel, and J.-P. Goudonnet, Imaging of surface plasmon propagation and edge interaction using a photon scanning tunneling microscope, *Phys. Rev. Lett.* **72:** 2927–2930 (1994).

48. L. D. Landau and E. M. Lifshitz, *Statistical Physics, Part 1*, 3rd ed., Pergamon Press, Oxford, 1977.

49. Y. M. Ivanchenko and A. A. Lisyansky, *Physics of Critical Fluctuations*, Springer-Verlag, New York, 1995.

NINE

THEORETICAL DESCRIPTION OF NEAR-FIELD OPTICAL MICROSCOPE

In this chapter, we study the basic electromagnetic processes involved in near-field optical microscopes (NOM) from the theoretical viewpoint of the multiple scattering of a vector field. The purpose is to examine the localization of the near-field interaction between small dielectric objects. Such a process is relevant to the near-field detection of near-field phenomena. We will show how we can obtain information about localized electromagnetic interactions between small objects.

We first prepare a mathematical foundation on which to describe vector fields in terms of spherical, planar, and cylindrical vector mode functions. These vector modes are related, respectively, to light scattering, to the interaction between scatterers, and to the coupling of a scattered field to macroscopic material systems. Here the scatterers correspond to the sample object and the probe tip, and the macroscopic material systems correspond to the dielectric substrate and optical waveguide in the case of NOM.

We next study transformations between these vector mode functions on the basis of rotation group theory with analytic continuation. The vector nature of the electromagnetic field is defined by using plane waves of well-defined polarizations and is extended to spherical and cylindrical mode functions by means of rotational transforms. This manner of representation provides us a clear geometrical understanding of the vector nature of the electromagnetic field as well as a good basis to describe and understand NOM processes.

By means of analytic continuation of the rotation angle into the complex region we obtain evanescent vector mode functions which correspond to the angular spectrum representation of the scattered fields. As discussed in Chapter 1, the angular spectrum representation provides a very useful way to evaluate the range and localization relevant to optical near-field processes.

The mathematical foundation developed in this chapter can serve as a powerful tool in handling general optical near-field problems, including both diagnosis and manipulation of material objects utilizing optical near-field techniques.

9.1. ELECTROMAGNETIC PROCESSES INVOLVED IN THE NEAR-FIELD OPTICAL MICROSCOPE

As discussed in Chapter 2, the NOM consists of several subsystems with different characteristic scales. We describe these subsystems theoretically in terms of a nonglobal treatment which is illustrated in Fig. 9.1a. Interaction between the specimen and the probe tip in the near-field regime is the most important part of the NOM, which is described in terms of the multiple scattering of a vector field in a subwavelength-size material system. The coupling of the local fields of the sample and the probe tip to the macroscopic field of the substrate and an optical waveguide leads to the observation of the near-field process from the far-field.

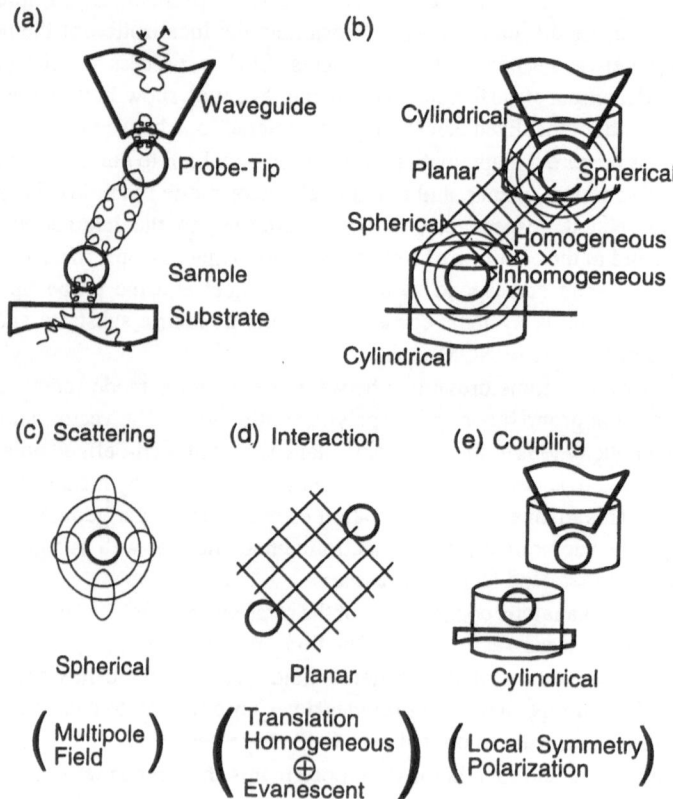

Figure 9.1. Fundamental processes involved in the near-field optical microscope and relevant electromagnetic modes.

Figure 9.1b shows the relation of these fundamental processes to our theoretical description of electromagnetic mode functions. Three fundamental processes arise.

First, the scattering of a vector field in the near-field regime is the most important part of our theoretical study of the NOM. As shown in Fig. 9.1b, a local description of the scattering process involves the vector spherical mode functions (Fig. 9.1c). In order to provide a foundation for the theoretical treatment, we start with a brief review of the spherical representation of a vector field and then introduce the idea of the interaction propagator or Green's dyadic in Section 9.2. The reader should refer to standard textbooks for details [1–5]. We introduce mathematical formulations convenient for optical near-field problems. The vector nature of the electromagnetic field is described in Section 9.3 on the basis of well-defined polarization states of plane waves and their extension by means of rotational transforms. An intuitive and clear geometrical understanding of the vector nature of spherical and cylindrical electromagnetic modes is obtained. We show several graphs describing the results of numerical calculations for the scattered field from vector electric and magnetic multipoles. The near-field distributions and far-field radiation patterns of the scattered field are also compared.

Second, the interaction between subwavelength-size material objects is developed in the near-field regime as a basis to describe the specimen–probe-tip interactions of the NOM. The interaction is formulated in Section 9.4 in terms of the spatial displacement of electromagnetic states on the basis of a vector plane-wave expansion of scattered fields, as illustrated in Fig. 9.1d. In the near-field regime, evanescent waves or inhomogeneous plane waves play the most important role. We formulate the transformations from vector spherical and cylindrical modes to plane waves and vice versa in terms of the angular spectrum representation of vector multipole modes. These formulas provide the basis to describe the interaction propagator or Green's dyadic. As a result we obtain the angular spectrum representation of interaction propagators, which provides the basis to study various remarkable features involved in optical near-field phenomena, of which the idea of the penetration depth, or range of interaction, and its localization are the most important features both in understanding and analyzing NOM processes.

Here and throughout this chapter we use the term "penetration depth" to describe the localization of interaction in a half-space demarcated by an arbitrary planar boundary in contact with a scattering object, which is introduced as the frame of the angular spectrum representation. The term "localization" is used to describe the two-dimensional localization of the interaction in a plane parallel to the assumed planar boundary. Therefore, we indicate "two plus one-half space localization" in three-dimensional space by using the expression "finite penetration depth plus localization of the interaction."

Third, the coupling of the scattered vector fields with macroscopic systems, such as a planar substrate and an optical waveguide, is relevant to the NOM process as providing the connection between a near-field event and its far-field measurement. As shown in Fig. 9.1e, the local description of coupling processes involves vector cylindrical mode functions. The vector Bessel mode functions are introduced in Section 9.2 to formulate the vector spherical mode functions by means of the rotational transform of the vector plane waves. The transformations between vector spherical and cylindrical mode functions are derived in Section 9.4. In order to describe the evanescent wave of Fresnel, which is often used as an illumination field in the collection-mode operation of NOM, we will present the spherical expansion of the evanescent mode functions given by Carniglia and Mandel as the basis of the quantization of evanescent waves [6].

Finally, based on the vector mode functions and interaction propagators, we give in Section 9.5a nonglobal theoretical description of NOM processes. We discuss the penetration depth and localization of the electromagnetic interaction between a specimen and a probe tip in terms of the angular spectrum representation [7] of the interaction propagator. We will discuss the meaning of the near-field regime in which the interaction process is characterized not by the optical wavelength, but by the size of the scatterers and the distance between scatterers. As a practical example, the localization of the dielectric sphere–surface interaction in the near-field regime is studied in an analytical form which allows us to recognize the role of the surface field or surface excitation.

The theoretical treatment of optical near-field problems using rotation operator representations and angular spectrum expansions is based mainly on the work of Inoue and Hori [8, 9] which should be consulted for further theoretical background.

9.2. REPRESENTATION OF THE ELECTROMAGNETIC FIELD AND THE INTERACTION PROPAGATOR

9.2.1. Spherical Representation of Scalar Waves

In this section, we recall briefly the theoretical treatment of scalar spherical waves as a foundation for our study of vector spherical waves [1].

9.2.1.1. Helmholtz Equation in Spherical Coordinates

Let us consider scalar waves in terms of a temporal Fourier analysis,

$$\psi(\mathbf{r}, t) = \int_{-\infty}^{\infty} \psi(\mathbf{r}, \omega)e^{-i\omega t}\, d\omega \qquad (9.2.1)$$

The wave equation for $\psi(\mathbf{r}, t)$ is reduced to a Helmholtz equation for each Fourier component $\psi(\mathbf{r}, \omega)$:

$$[\nabla^2 + K^2]\,\psi(\mathbf{r}, \omega) = 0 \tag{9.2.2}$$

where $K = \omega/c$, and c is the velocity of light in vacuum.

It is useful to study the solution of the Helmholtz equation in terms of a spherical harmonic expansion given by

$$\psi(\mathbf{r}, \omega) = \sum_{l,m_l} f_l(r) Y_l^{m_l}(\theta, \varphi) \tag{9.2.3}$$

where we use the spherical coordinate system $\mathbf{r} = (r\sin\theta\cos\varphi,\, r\sin\theta\sin\varphi,\, r\cos\theta)$, and l is a positive integer including $l = 0$, and m_l is an integer in the range $-l \le m_l \le +l$. With the use of a normalized radial distance and radial wave function defined by

$$\rho = Kr, \qquad f_l(r) = \sqrt{\frac{\pi}{2\rho}}\, u_l(\rho) \tag{9.2.4}$$

the radial and angular parts are separated into two differential equations:

$$\left[\frac{d^2}{d\rho^2} + \frac{1}{\rho}\frac{d}{d\rho} + \left(1 - \frac{\nu^2}{\rho^2}\right)\right] u_l(\rho) = 0 \qquad \left(\nu = l + \frac{1}{2}\right) \tag{9.2.5}$$

$$-\left[\frac{1}{\sin\theta}\frac{\partial}{\partial\theta}\left(\sin\theta\frac{\partial}{\partial\theta}\right) + \frac{1}{\sin\theta^2}\frac{\partial^2}{\partial\varphi^2}\right] Y_l^{m_l} = l(l+1) Y_l^{m_l} \tag{9.2.6}$$

9.2.1.2. Spherical Cylindrical Functions

The first differential equation is of cylindrical type with basic solutions of the Bessel function $u_l(\rho) = J_\nu(\rho)$ and Neumann function $u_l(\rho) = N_\nu(\rho)$ of fractional order $\nu = l + 1/2$. The radial wave function $f_l(r)$ is given by

$$f_l(r) = \begin{cases} j_l(Kr) = \sqrt{\dfrac{\pi}{2Kr}}\, J_{l+1/2}(Kr) \\[2mm] n_l(Kr) = \sqrt{\dfrac{\pi}{2Kr}}\, N_{l+1/2}(Kr) \end{cases} \tag{9.2.7}$$

where $j_l(Kr)$ and $n_l(Kr)$ are spherical Bessel and Neumann functions, respectively. Their polynomial expressions are given by

$$j_l(\rho) = (-\rho)^l \left(\frac{1}{\rho}\frac{d}{d\rho}\right)^l \left(\frac{\sin\rho}{\rho}\right) \tag{9.2.8}$$

$$n_l(\rho) = -(-\rho)^l \left(\frac{1}{\rho}\frac{d}{d\rho}\right)^l \left(\frac{\cos\rho}{\rho}\right) \tag{9.2.9}$$

For $Kr \to 0$, $j_l(Kr)$ converges to zero, whereas $n_l(Kr)$ diverges to $-\infty$. The asymptotic behaviors for $Kr \gg 1$ are, respectively,

$$j_l(Kr) \sim \frac{1}{Kr} \sin\left(Kr - \frac{l\pi}{2}\right) \tag{9.2.10}$$

$$n_l(Kr) \sim -\frac{1}{Kr} \cos\left(Kr - \frac{l\pi}{2}\right) \tag{9.2.11}$$

According to the asymptotic behavior, we can compose another set of solutions for $f_l(r)$, which behave for $Kr \gg 1$ as outgoing spherical waves and their complex conjugates. They are spherical Hankel functions of first and second kinds defined, respectively, by

$$h_l^{(1)}(Kr) = j_l(Kr) + in_l(Kr) \tag{9.2.12}$$

$$h_l^{(2)}(Kr) = \left[h_l^{(1)}(Kr)\right]^* \tag{9.2.13}$$

In fact these functions exhibit asymptotic behaviors as outgoing spherical waves:

$$h_l^{(1)}(Kr) \sim (-i)^{l+1} \frac{e^{iKr}}{Kr} \qquad \text{for} \qquad Kr \gg 1 \tag{9.2.14}$$

The explicit forms for lower values of l are

$$h_0^{(1)}(Kr) = \frac{e^{iKr}}{iKr} \tag{9.2.15}$$

$$h_1^{(1)}(Kr) = -\frac{e^{iKr}}{Kr}\left(1 + \frac{i}{Kr}\right) \tag{9.2.16}$$

$$h_2^{(1)}(Kr) = i\frac{e^{iKr}}{Kr}\left(1 + \frac{3i}{Kr} - \frac{3}{(Kr)^2}\right) \tag{9.2.17}$$

These appear in the expressions for dipole radiation discussed later.

9.2.1.3. Spherical Harmonics and Polynomials

The second differential equation (9.2.6) gives the spherical harmonics defined in normalized form by

$$Y_l^{m_l}(\theta, \varphi) = \sqrt{\frac{2l+1}{4\pi} \frac{(l-m_l)!}{(l+m_l)!}} \, P_l^{m_l}(\cos\theta)e^{im_l\varphi} \tag{9.2.18}$$

$$Y_l^{-m_l}(\theta, \varphi) = (-1)^{m_l} [Y_l^{m_l}(\theta, \varphi)]^* \tag{9.2.19}$$

with the associated Legendre functions

$$P_l^{m_l}(x) = \begin{cases} \dfrac{(-1)^{m_l}}{2^l l!} (1-x^2)^{m_l/2} \dfrac{d^{l+m_l}}{dx^{l+m_l}} (x^2-1)^l & \text{for} \quad m_l \geq 0 \\[3mm] (-1)^{m_l} \dfrac{(l-m_l)!}{(l+m_l)!} P_l^{m_l}(x) & \text{for} \quad m_l \leq 0 \end{cases} \tag{9.2.20}$$

The spherical harmonics defined here are orthonormal:

$$\int \left[Y_{l'}^{m'_l}(\theta, \varphi) \right]^* Y_l^{m_l}(\theta, \varphi) \, d\Omega = \delta_{l,l'} \delta_{m_l, m'_l} \tag{9.2.21}$$

$$\sum_{l=0}^{\infty} \sum_{m_l=-l}^{+l} [Y_l^{m_l}(\theta', \varphi')]^* Y_l^{m_l}(\theta, \varphi) = \delta(\cos\theta - \cos\theta') \, \delta(\varphi - \varphi') \tag{9.2.22}$$

where $d\Omega = \sin\theta \, d\theta \, d\varphi$ represents the infinitesimal solid angle.

9.2.1.4. Spherical Waves and Free-Space Green's Function

The general solution of the Helmholtz equation (9.2.2) is represented in terms of spherical harmonics and spherical Hankel functions by

$$\psi(\mathbf{r}, \omega) = \sum_{l, m_l} \left[A_{\omega, lm_l}^{(1)} h_l^{(1)}(Kr) + A_{\omega, lm_l}^{(2)} h_l^{(2)}(Kr) \right] Y_l^{m_l}(\theta, \varphi) \tag{9.2.23}$$

When a set of boundary conditions is given, the coefficients $A_{\omega, lm_l}^{(1,2)}$ are determined. We can generate a free-space Green's function by using spherical functions.

The free-space Green's function corresponds to a solution of the Helmholtz equation with a point source,

$$\left[\nabla^2 + K^2 \right] G(\mathbf{r}, \mathbf{r}', \omega) = -\delta(\mathbf{r} - \mathbf{r}') \tag{9.2.24}$$

and is given explicitly by

$$G(\mathbf{r}, \mathbf{r}', \omega) = \frac{e^{iK|\mathbf{r}-\mathbf{r}'|}}{4\pi|\mathbf{r}-\mathbf{r}'|} \tag{9.2.25}$$

When we consider the spherical expansion of $G(\mathbf{r}, \mathbf{r}', \omega)$ as

$$G(\mathbf{r}, \mathbf{r}', \omega) = \sum_{l, m_l} g_l(r, r', \omega) \, [Y_l^{m_l}(\theta', \varphi')]^* Y_l^{m_l}(\theta, \varphi) \tag{9.2.26}$$

with $r = |\mathbf{r}|$ and $r' = |\mathbf{r}'|$, the radial part $g_l(r, r', \omega)$ satisfies a cylindrical equation similar to the $f_l(r)$ case with a point source term:

$$\left[\frac{d^2}{dr^2} + \frac{2}{r}\frac{d}{dr} + \left\{K^2 - \frac{l(l+1)}{r^2}\right\}\right] g_l(r, r', \omega) = -\frac{1}{r^2}\delta(r - r') \qquad (9.2.27)$$

Here the cylindrical differential equation is written with respect to r.

The solution for $g_l(r, r', \omega)$ should be finite at the origin and exhibit the asymptotic behavior of an outgoing spherical wave at infinity, so that $g_l(r, r', \omega)$ can be expressed as a composition of spherical Bessel and spherical Hankel functions continued on a certain spherical boundary Σ lying in the middle of r and r'. Let r_{in} correspond to either of r or r' which is closer to the origin and therefore lies inside of Σ, and let r_{out} be that lying outside of Σ. Then $g_l(r_{in}, r_{out}, \omega)$ can be written as

$$g_l(r_{in}, r_{out}, \omega) = iKj_l(Kr_{in})h_l^{(1)}(Kr_{out}) \qquad (9.2.28)$$

where the coefficient iK is chosen with regard to continuity on the boundary surface. The spherical representation of the free-space Green's function is then given by

$$G(\mathbf{r}, \mathbf{r}', \omega) = \sum_{l, m_l} iKj_l(Kr_{in})h_l^{(1)}(Kr_{out}) [Y_l^{m_l}(\theta', \varphi')]^* Y_l^{m_l}(\theta, \varphi) \qquad (9.2.29)$$

This multipolar representation of the free-space Green's function provides a very useful and informative basis for describing electromagnetic interactions in material systems. According to this feature, we also refer to the Green's function as the interaction propagator. For instance, one can describe the electromagnetic field associated with a material system in terms of the free-space Green's function together with the material response functions. We can obtain the whole electromagnetic process involved in the system by accounting for microscopic interactions and resulting modifications of the field in a self-consistent manner. Such a multiple scattering treatment provides a practical approach to boundary value problems.

Before entering into these theories, we need to study the vector nature of the electromagnetic field. Whereas certain aspects of electromagnetic processes can be described on the basis of scalar field theory, the vector nature of the field plays a very important role in most electromagnetic problems. A vector field theory needs to be established since no general criterion is available for identifying when the vector nature of the field is significant for an electromagnetic problem.

9.2.1.5. Multipole Expansion of Plane Waves

In addition to the theory of the scalar multipole field, it is useful to study the asymptotic behavior of the free-space Green's function in the limit of $r_{out} \to \infty$. This results in a useful formula: the multipole expansion of plane waves.

Let $\mathbf{r}' = r'\hat{\mathbf{n}}$ and consider $r' \to \infty$; we can approximate the distance $|\mathbf{r} - \mathbf{r}'|$ and the spherical Hankel function as

$$|\mathbf{r} - \mathbf{r}'| \approx r' - \hat{\mathbf{n}} \cdot \mathbf{r}, \qquad h_l^{(1)}(Kr') = (-i)^{l+1} \frac{e^{iKr'}}{Kr'} \qquad (9.2.30)$$

In this limit the free-space Green's function behaves as

$$G(\mathbf{r}, \mathbf{r}', \omega) = \frac{e^{iK|\mathbf{r}-\mathbf{r}'|}}{4\pi|\mathbf{r}-\mathbf{r}'|} \sim \frac{e^{iKr'}}{4\pi r'} e^{-iK\hat{\mathbf{n}}\cdot\mathbf{r}} \qquad (9.2.31)$$

$$= iK \frac{e^{iKr'}}{Kr'} \sum_{l,m_l} (-i)^{l+1} j_l(Kr) \, [Y_l^{m_l}(\theta', \varphi')]^* \, Y_l^{m_l}(\theta, \varphi) \qquad (9.2.32)$$

With this approximation we get the multipole expansion of plane waves as

$$e^{-i\mathbf{K}\cdot\mathbf{r}} \equiv e^{-iK\hat{\mathbf{n}}\cdot\mathbf{r}} \qquad (9.2.33)$$

$$= 4\pi i \sum_{l,m_l} (-i)^{l+1} j_l(Kr) \, [Y_l^{m_l}(\theta', \varphi')]^* \, Y_l^{m_l}(\theta, \varphi) \qquad (9.2.34)$$

With the angle γ between $\mathbf{K}$ and $\mathbf{r}$, i.e., $\mathbf{K} \cdot \mathbf{r} = Kr \cos \gamma$, we get the following well-known formula of the spherical representation of plane waves:

$$e^{-iKr \cos \gamma} = \sum_l (-i)^l (2l + 1) j_l(Kr) P_l(\cos \gamma) \qquad (9.2.35)$$

$$= \sum_l (-i)^l \sqrt{4\pi(2l + 1)} \, j_l(Kr) Y_l^0(\gamma) \qquad (9.2.36)$$

9.2.2. Vector Nature of the Electromagnetic Field

9.2.2.1. Multipole Solutions and Vector Spherical Harmonics

Let us consider Maxwell's equations in terms of a Fourier analysis with respect to time. The electromagnetic field equations with time evolution $\exp(-i\omega t)$ lead to

$$\nabla \times \mathbf{E} = i\omega \mathbf{B}, \qquad c^2 \nabla \times \mathbf{B} = -i\omega \mathbf{E} \qquad (9.2.37)$$

$$\nabla \cdot \mathbf{E} = 0, \qquad \nabla \cdot \mathbf{B} = 0 \qquad (9.2.38)$$

The first two are wave equations for $\mathbf{E}$ and $\mathbf{B}$, and the last two correspond to gauge conditions.

According to the twofold gauge selection, we have two alternative representations of the electromagnetic field [1], one for the **B**-field wave equation with defined **E**,

$$\left[\nabla^2 + K^2\right] \mathbf{B} = 0 \qquad (9.2.39)$$

with

$$\mathbf{E} = i\frac{c}{K}\nabla \times \mathbf{B} \qquad (9.2.40)$$

where $K = \omega/c$, and the other for the **E**-field wave equation with defined **B**,

$$\left[\nabla^2 + K^2\right] \mathbf{E} = 0 \qquad (9.2.41)$$

with

$$\mathbf{B} = -\frac{i}{cK}\nabla \times \mathbf{E} \qquad (9.2.42)$$

It is convenient to use multipolar solutions for the wave equations:

$$\mathbf{B} \text{ or } \mathbf{E} = \sum_{l,m_l}\left[\mathbf{A}_{lm_l}^{(1)}h_l^{(1)}(Kr) + \mathbf{A}_{lm_l}^{(2)}h_l^{(2)}(Kr)\right]Y_l^{m_l}(\theta, \varphi) \qquad (9.2.43)$$

with

$$\nabla \cdot \mathbf{B} = 0 \qquad \text{or} \qquad \nabla \cdot \mathbf{E} = 0 \qquad (9.2.44)$$

This type of solution extends to spherical waves in the long-distance limit. Multipolar expansions of incoming waves, which have no singularity at the origin, can be obtained simply by replacing the spherical Hankel functions by spherical Bessel functions, $j_l(Kr)$.

According to the type of gauge condition, the solution representing the transverse magnetic field is generally referred to as the E-mode, and the one representing the transverse electric field as the M-mode. Since the vector nature of the field is characterized by an angular momentum corresponding to spin 1, it is convenient to represent the field with a spherical basis [10–12]. The transverse conditions for the spherical waves are given by

$$\hat{\mathbf{r}} \cdot \mathbf{A}_{lm_l}^{(1,2)} \, Y_l^{m_l}(\theta, \varphi) = 0 \qquad (9.2.45)$$

where $\hat{\mathbf{r}}$ represents the radial unit vector defined by

$$\hat{\mathbf{r}} \equiv (\sin\theta\cos\varphi, \sin\theta\sin\varphi, \cos\theta) \qquad (9.2.46)$$

Then we can use the following convenient expression for the vector coefficients $\mathbf{A}_{lm_l}^{(1)}$ and $\mathbf{A}_{lm_l}^{(2)}$ with the angular momentum operator $\hat{\mathbf{L}}$:

$$\mathbf{A}_{lm_l}^{(1,2)} = a_{lm_l}^{(1,2)}\hat{\mathbf{L}} \tag{9.2.47}$$

where $\hat{\mathbf{L}}$ is given by

$$\hat{\mathbf{L}} = -i(r\hat{\mathbf{r}} \times \nabla); \quad \hat{\mathbf{r}} \cdot \hat{\mathbf{L}} = 0, \quad \hat{\mathbf{r}} \cdot (\hat{\mathbf{L}} \times \hat{\mathbf{L}}) = \hat{\mathbf{r}} \cdot (i\hat{\mathbf{L}}) = 0 \tag{9.2.48}$$

An orthonormal set of this type of vector mode function corresponds to vector spherical harmonics defined by

$$\mathbf{Y}_l^{m_l}(\theta, \varphi) = \frac{1}{\sqrt{l(l+1)}} \hat{\mathbf{L}} Y_l^{m_l}(\theta, \varphi) \tag{9.2.49}$$

$$\int \mathbf{Y}_{l'}^{m_l'}(\theta, \varphi) \mathbf{Y}_l^{m_l}(\theta, \varphi) \, d\Omega = \delta_{l,l'} \delta_{m_l,m_{l'}'} \tag{9.2.50}$$

Here $\mathbf{Y}_{00}(\theta, \varphi) = 0$ by definition.

By using vector spherical harmonics, we can obtain a set of basis mode functions corresponding to multipolar solutions of the vector wave equations: for transverse $\mathbf{B}^{(E)}$ waves we obtain

$$\mathbf{B}^{(E)} = \sum_{l,m_l} f_l(Kr) \mathbf{Y}_l^{m_l}(\theta, \varphi) \tag{9.2.51}$$

with

$$\mathbf{E}^{(E)} = i\frac{c}{K} \nabla \times \mathbf{B}^{(E)} \tag{9.2.52}$$

and for transverse $\mathbf{E}^{(M)}$ waves we get

$$\mathbf{E}^{(M)} = \sum_{l,m_l} g_l(Kr) \mathbf{Y}_l^{m_l}(\theta, \varphi) \tag{9.2.53}$$

with

$$\mathbf{B}^{(M)} = -\frac{i}{cK} \nabla \times \mathbf{E}^{(M)} \tag{9.2.54}$$

The radial functions are represented as functions of the normalized radial distance $Kr = (\omega/c)r$. Then we can describe an outgoing or scattered wave in terms of the spherical Hankel functions as

$$f_l(Kr) \quad \text{or} \quad g_l(Kr) = A_l^{(1)} h_l^{(1)}(Kr) + A_l^{(2)} h_l^{(2)}(Kr) \tag{9.2.55}$$

Finally we obtain the general solution for the free-space Maxwell equations described in terms of multipolar expansions:

$$\mathbf{B}(Kr, \theta, \varphi) = \sum_{l,m_l} \left[a_{lm_l}^{(E)} f_l(Kr) Y_l^{m_l}(\theta, \varphi) - \frac{i}{cK} a_{lm_l}^{(M)} \nabla \times g_l(Kr) Y_l^{m_l}(\theta, \varphi) \right] \quad (9.2.56)$$

$$\mathbf{E}(Kr, \theta, \varphi) = \sum_{l,m_l} \left[\frac{ic}{K} a_{lm_l}^{(E)} \nabla \times f_l(Kr) Y_l^{m_l}(\theta, \varphi) + a_{lm_l}^{(M)} g_l(Kr) Y_l^{m_l}(\theta, \varphi) \right] \quad (9.2.57)$$

The vector spherical harmonics make up the angular distribution $Y_{lm_l}(\theta, \varphi)$ with vector operator $\hat{\mathbf{L}}$. It is convenient to represent the vector $\hat{\mathbf{L}}$ with a set of circular basis vectors which are defined with Cartesian unit vectors $\hat{\mathbf{x}}, \hat{\mathbf{y}}$, and $\hat{\mathbf{z}}$ by

$$\hat{\mathbf{e}}_0 \equiv \hat{\mathbf{z}}, \qquad \hat{\mathbf{e}}_{\pm 1} \equiv \mp \frac{1}{\sqrt{2}} (\hat{\mathbf{x}} \pm i\hat{\mathbf{y}}) \quad (9.2.58)$$

Then the vector spherical harmonics are represented as three types of basis vector functions

$$\hat{\mathbf{e}}_\mu Y_l^{m_l}(\theta, \varphi) \qquad (\mu = 0, \pm 1) \quad (9.2.59)$$

The meaning of these vector functions is now clear: the basis vector functions correspond to waves with angular momentum l and its z-projection m_l, i.e., the state $|l, m_l\rangle$, made up of the vector space with spin $s = 1$ and z-projection $\mu = 0, \pm 1$, i.e., the state $|1, \mu\rangle$. Here we have utilized the quantum mechanical expression of states in the description of electromagnetic fields [10, 13, 14].

Let us examine the nature of the electromagnetic field described by the spherical vector harmonics $\hat{\mathbf{e}}_\mu Y_l^{m_l}(\theta, \varphi)$. This corresponds to the state vector produced by the direct product of spherical harmonics and a vector,

$$|l, m_l\rangle \otimes |s = 1, \mu\rangle \equiv |l, s = 1; m_l, \mu\rangle \quad (9.2.60)$$

Then we can expand this state in terms of states $|J, M\rangle$ which represent vector fields with total angular momentum J and its z-projection M,

$$\mathbf{Y}_{J,l,(s=1),M}(\theta, \varphi) = \sum_\mu Y_l^{M-\mu} \hat{\mathbf{e}}_\mu \langle l, s = 1; M - \mu, \mu | J, M \rangle \qquad (l = J, J \pm 1) \quad (9.2.61)$$

Here only three configurations, $l = J, J \pm 1$, are allowed since the total angular momentum is the composition of l and spin $s = 1$ (vector). These states consist of the bases of the $2J$ multipolar fields. The explicit form of the vector spherical harmonics of rank J with z-projection M is given by

$$\mathbf{Y}_{J,l,(s=1),M}(\theta, \varphi) = \begin{cases} \dfrac{1}{\sqrt{J(2J+1)}} (r\nabla + J\hat{\mathbf{r}}) Y_J^M(\theta, \varphi), & l = J - 1 \\[2ex] \dfrac{1}{\sqrt{J(J+1)}} \hat{\mathbf{L}} Y_J^M(\theta, \varphi), & l = J \\[2ex] \dfrac{1}{\sqrt{(J+1)(2J+1)}} (r\nabla - (J+1)\hat{\mathbf{r}}) Y_J^M(\theta, \varphi), & l = J + 1 \end{cases}$$

$$(9.2.62)$$

Usually the subscript ($s = 1$), being trivial for the vector field, is omitted. Henceforth we use the following notations:

$$\mathbf{Y}_{J,l,M}(\hat{\mathbf{r}}) \equiv \mathbf{Y}_{J,l,M}(\theta, \varphi) \equiv \mathbf{Y}_{J,l,(s=1),M}(\theta, \varphi) \tag{9.2.63}$$

Here $\hat{\mathbf{r}}$ stands for the directional unit vector, which is used instead of (θ, φ) for simplicity. The field components $\mathbf{E}^{(M)}$ and $\mathbf{B}^{(E)}$ being solutions of the Helmholtz equations and expanded into series of $Y_l^{m_l} \propto \hat{L} Y_l^{m_l}$, are therefore relevant to the component $\mathbf{Y}_{J,J,M}$ of the vector $2J$ multipolar field. The associated field components $\mathbf{E}^{(E)}$ and $\mathbf{B}^{(M)}$ derived by $\nabla \times \mathbf{B}^{(E)}$ and $\nabla \times \mathbf{E}^{(M)}$, respectively, are expressed in terms of the superposition of $\mathbf{Y}_{J,(J-1),M}$ and $\mathbf{Y}_{J,(J+1),M}$.

9.2.2.2. Complex Vector Mode Functions of Magnetic, Electric, and Longitudinal Types

Instead of defining the basis set of vector spherical expressions for each field component, it is convenient to define potentials in terms of vector spherical functions. Let us adopt the Coulomb gauge and set the scalar potentials to be zero, and define the basis of the complex multipolar vector potentials $\mathbf{A}_{K,J,M}^{(M,SC)}$ and $\mathbf{A}_{K,J,M}^{(E,SC)}$, which extend to outgoing spherical waves, as

$$\mathbf{A}_{K,J,M}^{(M,SC)}(\mathbf{r}) = i^J h_J^{(1)}(Kr)\mathbf{Y}_{J,J,M}(\hat{\mathbf{r}}) \tag{9.2.64}$$

$$\Phi_{K,J,M}^{(M)}(\mathbf{r}) = 0 \tag{9.2.65}$$

$$\mathbf{A}_{K,J,M}^{(E,SC)}(\mathbf{r}) = i^{J+1} \sqrt{\frac{J}{2J+1}}\, h_{J+1}^{(1)}(Kr)\mathbf{Y}_{J,J+1,M}(\hat{\mathbf{r}})$$
$$+ i^{J-1} \sqrt{\frac{J+1}{2J+1}}\, h_{J-1}^{(1)}(Kr)\mathbf{Y}_{J,J-1,M}(\hat{\mathbf{r}}) \tag{9.2.66}$$

$$\Phi_{K,J,M}^{(E)}(\mathbf{r}) = 0 \tag{9.2.67}$$

Here we omit the complex conjugate $h_{l=J,J\pm1}^{(2)}(Kr)$ for simplicity, and therefore $\mathbf{A}_{K,J,M}^{(M/E,SC)}$ represent complex vector fields. For the multipolar expansions of incoming waves, which have no singularity at the origin, the spherical Hankel functions should be replaced by spherical Bessel functions, $j_l(Kr)$. Then complex vector potentials corresponding to incoming waves are defined by

$$\mathbf{A}_{K,J,M}^{(M,IN)}(\mathbf{r}) = i^J j_J(Kr)\mathbf{Y}_{J,J,M}(\hat{\mathbf{r}}) \tag{9.2.68}$$

$$A_{K,J,M}^{(E,IN)}(\mathbf{r}) = i^{J+1}\sqrt{\frac{J}{2J+1}}\, j_{J+1}(Kr)\mathbf{Y}_{J,J+1,M}(\hat{\mathbf{r}})$$
$$+ i^{J-1}\sqrt{\frac{J+1}{2J+1}}\, j_{J-1}(Kr)\mathbf{Y}_{J,J-1,M}(\hat{\mathbf{r}}) \qquad (9.2.69)$$

Each of the magnetic and electric vector potentials is a solution of a Helmholtz equation,

$$\left(\nabla^2 + K^2\right)A_{K,J,M}^{(M)}(\mathbf{r}) = 0 \qquad (9.2.70)$$

$$\left(\nabla^2 + K^2\right)A_{K,J,M}^{(E)}(\mathbf{r}) = 0 \qquad (9.2.71)$$

which stand for both the incident and scattered fields. The field components $\mathbf{E}^{(M)}$ and $\mathbf{B}^{(E)}$ are then given by

$$\mathbf{E}_{K,J,m}^{(M)}(\mathbf{r}) = iKc A_{K,J,M}^{(M)}(\mathbf{r}) \qquad (9.2.72)$$

$$\mathbf{B}_{K,J,M}^{(E)}(\mathbf{r}) = \nabla \times A_{K,J,M}^{(E)}(\mathbf{r}) = -K A_{K,J,M}^{(M)}(\mathbf{r}) \qquad (9.2.73)$$

The associated field components are given, respectively, in similar form by

$$\mathbf{B}_{K,J,M}^{(M)}(\mathbf{r}) = \nabla \times A_{K,J,M}^{(M)}(\mathbf{r}) = -K A_{K,J,M}^{(E)}(\mathbf{r}) \qquad (9.2.74)$$

$$\mathbf{E}_{K,J,M}^{(E)}(\mathbf{r}) = iKc A_{K,J,M}^{(M)}(\mathbf{r}) \qquad (9.2.75)$$

As a set of linear combinations of vector spherical harmonics, we can construct a convenient basis for describing the electromagnetic field in terms of *magnetic*, *electric*, and *longitudinal* type vector fields, respectively:

$$\mathbf{Y}_{J,M}^{(M)}(\hat{\mathbf{r}}) = \mathbf{Y}_{J,J,M}(\hat{\mathbf{r}}) \qquad (9.2.76)$$

$$\mathbf{Y}_{J,M}^{(E)}(\hat{\mathbf{r}}) = \sqrt{\frac{J}{2J+1}}\, \mathbf{Y}_{J,J+1,M}(\hat{\mathbf{r}}) + \sqrt{\frac{J+1}{2J+1}}\, \mathbf{Y}_{J,J-1,M}(\hat{\mathbf{r}}) \qquad (9.2.77)$$

$$\mathbf{Y}_{J,M}^{(L)}(\hat{\mathbf{r}}) = \sqrt{\frac{J}{2J+1}}\, \mathbf{Y}_{J,J-1,M}(\hat{\mathbf{r}}) - \sqrt{\frac{J+1}{2J+1}}\, \mathbf{Y}_{J,J+1,M}(\hat{\mathbf{r}}) \qquad (9.2.78)$$

Using the inverted definitions

$$\mathbf{Y}_{J,J,M}(\hat{\mathbf{r}}) = \mathbf{Y}_{J,M}^{(M)}(\hat{\mathbf{r}}) \qquad (9.2.79)$$

$$\mathbf{Y}_{J,J+1,M}(\hat{\mathbf{r}}) = \sqrt{\frac{J}{2J+1}}\, \mathbf{Y}_{J,M}^{(E)}(\hat{\mathbf{r}}) - \sqrt{\frac{J+1}{2J+1}}\, \mathbf{Y}_{J,M}^{(L)}(\hat{\mathbf{r}}) \qquad (9.2.80)$$

$$\mathbf{Y}_{J,J-1,M}(\hat{\mathbf{r}}) = \sqrt{\frac{J+1}{2J+1}}\, \mathbf{Y}_{J,M}^{(E)}(\hat{\mathbf{r}}) + \sqrt{\frac{J}{2J+1}}\, \mathbf{Y}_{J,M}^{(L)}(\hat{\mathbf{r}}) \qquad (9.2.81)$$

we can rewrite the complex vector multipole potentials as

$$\mathbf{A}_{K,J,M}^{(M,SC)}(\mathbf{r}) = i^J h_J^{(1)}(Kr) \mathbf{Y}_{J,M}^{(M)}(\hat{\mathbf{r}}) \qquad (9.2.82)$$

$$\mathbf{A}_{K,J,M}^{(E,SC)}(\mathbf{r}) = \frac{1}{2J+1}\left[i^{J+1} J h_{J+1}^{(1)}(Kr) + i^{J-1}(J+1)h_{J-1}^{(1)}(Kr) \right] \mathbf{Y}_{J,M}^{(E)}(\hat{\mathbf{r}})$$

$$- \frac{\sqrt{J(J+1)}}{2J+1}\left[i^{J+1} h_{J+1}^{(1)}(Kr) - i^{J-1} h_{J-1}^{(1)}(Kr) \right] \mathbf{Y}_{J,M}^{(L)}(\hat{\mathbf{r}}) \quad (9.2.83)$$

Here, it is convenient to note formulas for the complex conjugate of complex potentials. Since

$$[Y_l^{m_l}(\theta, \varphi)]^* = (-1)^{m_l} Y_l^{m_l}(\theta, \varphi) \qquad (9.2.84)$$

and

$$\langle l, s = 1; M - \mu, \mu | J, M \rangle = (-1)^{(J+l+1)}\, \langle l, s = 1; -M + \mu, -\mu | J, -M \rangle \qquad (9.2.85)$$

we obtain

$$\left[\mathbf{Y}_{J,l,M}(\hat{\mathbf{r}}) \right]^* = (-1)^{(J+l+M+1)} \mathbf{Y}_{J,l,-M}(\hat{\mathbf{r}}) \qquad (9.2.86)$$

$$\left[\mathbf{A}_{K,J,M}^{(E,IN)}(\mathbf{r}) \right]^* = (-1)^{(M+1)} \mathbf{A}_{K,J,-M}^{(E,IN)}(\mathbf{r}) \qquad (9.2.87)$$

$$\left[\mathbf{A}_{K,J,M}^{(M,IN)}(\mathbf{r}) \right]^* = (-1)^{(M+1)} \mathbf{A}_{K,J,-M}^{(M,IN)}(\mathbf{r}) \qquad (9.2.88)$$

Note that these relations hold only for incoming fields which consist of spherical Bessel functions. No such simple relation holds for scattered vector multipole fields containing spherical Hankel functions, which exhibit singularities at the origin.

9.2.2.3. Near- and Far-Field Behaviors of Vector Multipole Fields

Here it is informative to study the asymptotic behaviors of the vector multipole potentials both in the far-field ($Kr \gg 1$) and near-field ($Kr \ll 1$) limits.

In the far-field limit the spherical Hankel function is approximated by an outgoing spherical wave as

$$i^J h_J^{(1)}(Kr) \sim \frac{e^{iKr}}{iKr} \qquad (Kr \gg 1) \tag{9.2.89}$$

and the asymptotic behaviors of the vector multipole potentials are

$$A_{K,J,M}^{(M,SC)}(\mathbf{r}) \sim \frac{e^{iKr}}{iKr} Y_{J,M}^{(M)}(\hat{\mathbf{r}}) \tag{9.2.90}$$

$$A_{K,J,M}^{(E,SC)}(\mathbf{r}) \sim \frac{1}{2J+1}\left[J \frac{e^{iKr}}{iKr} + (J+1)\frac{e^{iKr}}{iKr}\right] Y_{J,M}^{(E)}(\hat{\mathbf{r}})$$

$$- \frac{\sqrt{J(J+1)}}{2J+1}\left[\frac{e^{iKr}}{iKr} - \frac{e^{iKr}}{iKr}\right] Y_{J,M}^{(L)}(\hat{\mathbf{r}})$$

$$= \frac{e^{iKr}}{iKr} Y_{J,M}^{(E)}(\hat{\mathbf{r}}) \tag{9.2.91}$$

Note that the longitudinal fields cancel in the far-field limit and both of the vector multipole potentials give rise to outgoing vector spherical waves.

In the near-field limit, the Hankel function is approximated by

$$h_J^{(1)}(Kr) \sim \frac{-i(2J-1)!!}{(Kr)^{J+1}} \qquad (Kr \ll 1) \tag{9.2.92}$$

and the vector multipole potentials then exhibit the asymptotic behaviors

$$A_{K,J,M}^{(M,SC)}(\mathbf{r}) \sim -i^{J+1}\frac{(2J-1)!!}{(Kr)^{J+1}} Y_{J,M}^{(M)}(\hat{\mathbf{r}}) \tag{9.2.93}$$

$$A_{K,J,M}^{(E,SC)}(\mathbf{r}) \sim i^J \frac{(2J-1)!!}{(Kr)^{J+1}}\left[\frac{J}{2J+1} Y_{J,M}^{(E)}(\hat{\mathbf{r}}) - \frac{\sqrt{J(J+1)}}{2J+1} Y_{J,M}^{(L)}(\hat{\mathbf{r}})\right] \tag{9.2.94}$$

Here the potentials involve the longitudinal component, so that the leading order is the component with highest J. Therefore the vector multipole components $A_{K,J,M}^{(E)}(\mathbf{r})$ and the associated $\mathbf{E}_{K,J,M}^{(E)}(\mathbf{r})$ and $\mathbf{B}_{K,J,M}^{(M)}(\mathbf{r})$ play the dominant roles in the near-field regime. For example, the electric dipole field in the near-field regime is written as

$$A_{K,1,M}^{(E,SC)}(\mathbf{r}) = -\sqrt{\frac{1}{3}}\, h_2^{(1)}(Kr)Y_{1,2,M}(\hat{\mathbf{r}}) + \sqrt{\frac{2}{3}}\, h_0^{(1)}(Kr)Y_{1,0,M}(\hat{\mathbf{r}}) \tag{9.2.95}$$

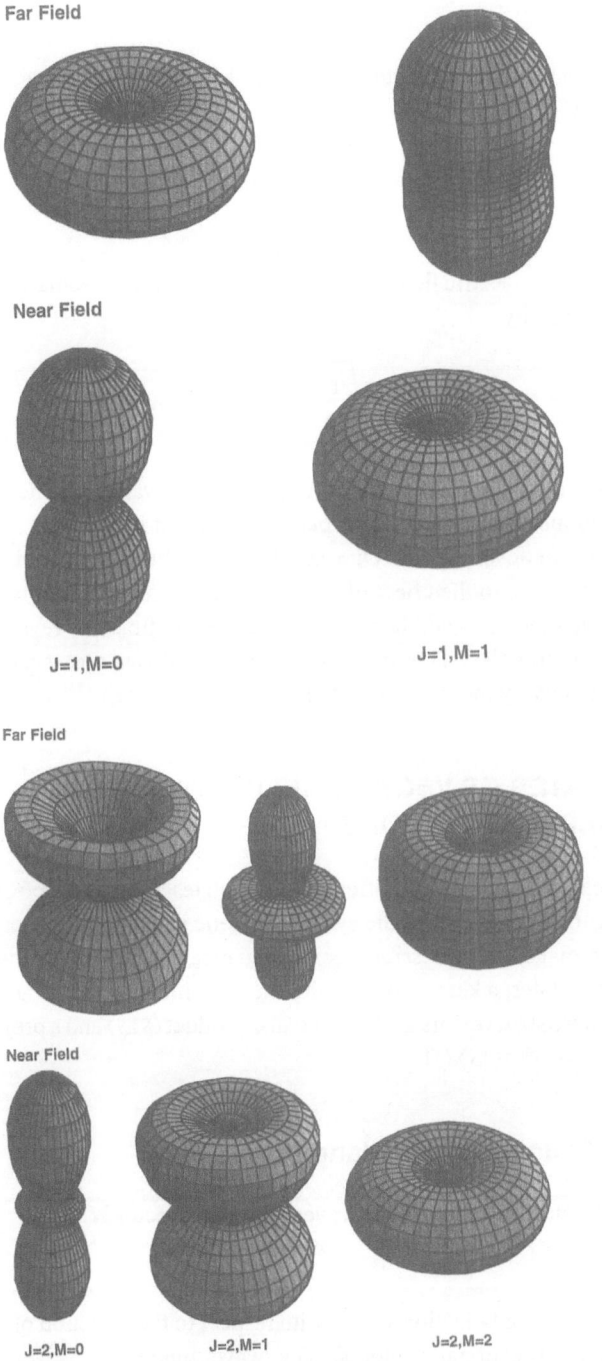

Far Field

Near Field

J=1,M=0

J=1,M=1

Far Field

Near Field

J=2,M=0

J=2,M=1

J=2,M=2

Figure 9.2. Radial intensity distributions of multipole fields for near-field and far-field regimes.

The electric and magnetic multipoles are described using the polarization vector $\hat{e}_m$ by

$$p = \sum_{M=-1}^{+1} (-1)^M P_{-M} \hat{e}_M \tag{9.2.96}$$

For example, the electric field associated with a multipole oscillating at a frequency $\omega = Kc$ is given by

$$E(r) = iK^3 \sqrt{\frac{1}{6\pi}} \sum_{M=-1}^{+1} (-1)^M P_{-M} A_{K,1,M}^{(E,SC)}(r) \tag{9.2.97}$$

Substituting the asymptotic forms of $A_{K,1,M}^{(E,SC)}(r)$, we evaluate the near-field intensity distribution and far-field radiation pattern as shown in Fig. 9.2; we can also find these for magnetic multipoles. One can clearly see the differences between the near and far fields, due to the effect of the longitudinal mode functions. A remarkable feature is that the near- and far-field angular distributions are reversed for the case of electric multipoles. This implies that the significance of near-field effects is observable only by means of near-field detection.

9.3. STATES OF VECTOR FIELDS AND THEIR REPRESENTATIONS

We can utilize several different basis states for the electromagnetic field according to the character of the electromagnetic interactions under consideration. In order to make the expressions simple and clear, let us use quantum mechanical notation for states, a ket vector $|S\rangle$ with the state index S and its adjoint bra vector $\langle S|$ [15]. These state vectors generate a scalar product $\langle S|S\rangle$ and a projection operator onto the state $|S\rangle$ as $|S\rangle\langle S|$.

9.3.1. State of Vector Plane Waves

Let the state of vector plane waves with wavevector $\mathbf{K}$ be

$$|\mathbf{K}; \hat{e}(\hat{s})\rangle, \qquad \mathbf{K} = K\hat{s} \tag{9.3.1}$$

where $\hat{e}(\hat{s})$ is the polarization vector with respect to the direction of the wavevector $\mathbf{K}$ with unit directional vector $\hat{s}$. The wave function is given by projecting $|\mathbf{K}; \hat{e}(\hat{s})\rangle$ onto the coordinate basis state $|r\rangle$ as

$$|\mathbf{r}\rangle \langle \mathbf{r}|K; \hat{\mathbf{e}}(\hat{\mathbf{s}})\rangle = \frac{1}{\sqrt{(2\pi)^3}} \hat{\mathbf{e}}(\hat{\mathbf{s}})e^{i\mathbf{K}\cdot\mathbf{r}} |\mathbf{r}\rangle \qquad (9.3.2)$$

where the coordinate basis states are orthonormal,

$$\langle \mathbf{r}|\mathbf{r}'\rangle = \int d^3\mathbf{K} \langle \mathbf{r}|\mathbf{K}\rangle \langle \mathbf{K}|\mathbf{r}'\rangle = \delta(\mathbf{r} - \mathbf{r}') \qquad (9.3.3)$$

The vector plane waves are synthesized from scalar plane waves, $|K; l, m_l\rangle$, by a direct product with a vector, $|s = 1, \mu = 0, \pm 1\rangle$, corresponding to the spin state $s = 1$ with z-projection $s_z = 0, \pm 1$. Due to the transverse nature of electromagnetic waves, two independent polarization states are enough to constitute a basis set. Let us start with the vector plane wave propagating in the z direction with well-defined polarization $s_z = \pm 1$: $|K\hat{\mathbf{z}}; \hat{\mathbf{e}}_{\pm 1}\rangle$. Then vector plane waves with arbitrary wavevector $\mathbf{K} = K\hat{\mathbf{s}}$ are generated by rotational transforms from the state $|K; \hat{\mathbf{e}}_{\pm 1}\rangle$ as

$$|K; \hat{\mathbf{e}}(\hat{\mathbf{s}})\rangle = \hat{\mathbf{D}}(\varphi, \theta, 0) | K\hat{\mathbf{z}}; \hat{\mathbf{e}}_{\pm 1}\rangle \qquad (9.3.4)$$

where $\hat{\mathbf{D}}(\varphi, \theta, 0)$ is the rotation operator with angular parameter corresponding to the rotated wavevector, $\hat{\mathbf{s}} = (\sin\theta \cos\varphi, \sin\theta \sin\varphi, \cos\theta)$. The rotation operator $\hat{\mathbf{D}}(\alpha, \beta, \gamma)$ transforms a state as successive rotations $\hat{\mathbf{R}}_z(\gamma)$, $\hat{\mathbf{R}}_y(\beta)$, and $\hat{\mathbf{R}}_z(\alpha)$ with respect to the fixed Cartesian coordinates, and φ and θ are the directional angles of the wavevector $\mathbf{K}$ (Fig. 9.3).

As is discussed later, however, the vector nature of the electromagnetic field is often described in terms of an orthogonal set of linearly polarized states. For further convenience, let us define the x-polarized and y-polarized plane waves $|K\hat{\mathbf{z}}; \hat{\mathbf{x}}\rangle$ and $|K\hat{\mathbf{z}}; \hat{\mathbf{y}}\rangle$, respectively, with Cartesian unit basis vectors

$$\hat{\mathbf{x}} = -\frac{1}{\sqrt{2}} (\hat{\mathbf{e}}_{+1} - \hat{\mathbf{e}}_{-1}) \qquad (9.3.5)$$

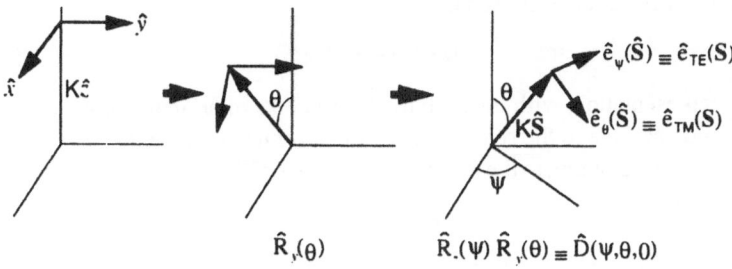

Figure 9.3. Vector plane waves and their polarization states described by rotational transforms of z-directed vector plane waves.

$$\hat{\mathbf{y}} = +\frac{i}{\sqrt{2}}(\hat{\mathbf{e}}_{+1} + \hat{\mathbf{e}}_{-1}) \tag{9.3.6}$$

The rotation transform generates an orthogonal set of φ-polarized and θ-polarized states given by [13]

$$\langle \mathbf{r}|K; \hat{\mathbf{e}}_\varphi(\hat{\mathbf{s}})\rangle = \langle \mathbf{r}|\hat{\mathbf{D}}(\varphi, \theta, 0)|K\hat{\mathbf{z}}; \hat{\mathbf{y}}\rangle \tag{9.3.7}$$

$$\langle \mathbf{r}|K; \hat{\mathbf{e}}_\theta(\hat{\mathbf{s}})\rangle = \langle \mathbf{r}|\hat{\mathbf{D}}(\varphi, \theta, 0)|K\hat{\mathbf{z}}; \hat{\mathbf{x}}\rangle \tag{9.3.8}$$

Here we assume an incident plane spanned by $\hat{\mathbf{z}}$ and $\hat{\mathbf{s}}$ and refer to the polarization state with $\hat{\mathbf{e}}_\varphi(\hat{\mathbf{s}}) \equiv \hat{\mathbf{e}}_{TE}(\hat{\mathbf{s}})$ as TE-polarized and that with $\hat{\mathbf{e}}_\theta(\hat{\mathbf{s}}) \equiv \hat{\mathbf{e}}_{TM}(\hat{\mathbf{s}})$ as TM-polarized, as shown in Fig. 9.3. This is because the directions of these polarization vectors are perpendicular and parallel to the assumed incident plane, respectively.

9.3.2. State of Vector Spherical Waves

The scalar spherical wave is described by the state vector $|K; l, m_l\rangle$ with K the wavenumber, l the orbital angular momentum, and m_l its z-projection. The scalar wave function corresponding to this state is obtained as the $r_{\text{out}} \to \infty$ limit of the Green's function with $r_{\text{in}} = r$,

$$\langle \mathbf{r}|K; l, m_l\rangle = \sqrt{\frac{2}{\pi}} \, K \, i^l j_l(Kr) Y_l^{m_l}(\hat{\mathbf{r}}) \tag{9.3.9}$$

The vector spherical waves are synthesized by the direct product of the scalar spherical wave $|K; l, m_l\rangle$ and the vector (spin 1 state) $|s = 1, \mu = 0, \pm1\rangle$. There are several different representations of vector spherical waves. One convenient basis state, $|K; l, s = 1; m_l, \mu\rangle$, is given by the direct product of these states,

$$|K; l, s = 1; m_l, \mu\rangle \equiv |K; l, m_l\rangle \otimes |s = 1, \mu = 0, \pm1\rangle \tag{9.3.10}$$

On the other hand, the state vector

$$|K; l, s = 1; J, M\rangle \tag{9.3.11}$$

is the representation with respect to the total angular momentum J with its z-projection M. Due to the selection rule (or sum rule) $J = l, l \pm 1$, the vector wave functions corresponding to these states are expressed in terms of the vector spherical harmonics by

$$\langle \mathbf{r}|K; l, s = 1; J, M\rangle = \sqrt{\frac{2}{\pi}} \, K i^l j_l(Kr) \mathbf{Y}_{J,l,M}(\hat{\mathbf{r}}) \qquad (l = J, J \pm 1) \tag{9.3.12}$$

This formula is derived later with the help of the wavenumber representation.

9.3.3. State of Vector Cylindrical Waves

Scalar cylindrical waves are described as the product of plane waves and radial waves, the former propagating along the symmetry axis (z axis) and the latter described by the radial coefficient of a Bessel function, J_M. Then we can describe a scalar cylindrical wave as

$$J_{K,\alpha,M}(\mathbf{r}) \equiv i^M J_M(Kr \sin \alpha) \, e^{iM\varphi} \, e^{iKz\cos\alpha} \qquad (9.3.13)$$

where we use the cylindrical coordinate system $\mathbf{r} = (r, \varphi, z)$. The state of the vector cylindrical wave

$$|K; \, TE \text{ or } TM; \, \alpha, M\rangle \qquad (9.3.14)$$

is synthesized from the product of scalar cylindrical waves and polarization vectors $\hat{\mathbf{e}}_0 \equiv \hat{\mathbf{z}}$ by means of rotational transforms by

$$\mathbf{J}_{K,\alpha,M}^{(TE)} \equiv \langle \mathbf{r}|K; \, TE; \, \alpha, M\rangle = \frac{i}{K\sin\alpha} \nabla \times \hat{\mathbf{e}}_0 J_{K,\alpha,M}(\mathbf{r}) \qquad (9.3.15)$$

$$\mathbf{J}_{K,\alpha,M}^{(TM)} \equiv \langle \mathbf{r}|K; \, TM; \, \alpha, M\rangle = \frac{-1}{K^2 \sin\alpha} \nabla \times \nabla \times \hat{\mathbf{e}}_0 J_{K,\alpha,M}(\mathbf{r}) \qquad (9.3.16)$$

These functions are often referred to as vector Bessel functions. On the basis of linearly polarized vector plane waves, the polarization vectors for the cylindrical waves are defined by

$$\hat{\mathbf{e}}_0 \times \hat{\mathbf{s}} = \hat{\mathbf{e}}_{TE}(\hat{\mathbf{s}}) \sin \alpha, \qquad \hat{\mathbf{e}}_{TE}(\hat{\mathbf{s}}) \times \hat{\mathbf{s}} = \hat{\mathbf{e}}_{TM}(\hat{\mathbf{s}}) \qquad (9.3.17)$$

and the rotational transform is described simply by

$$\mathbf{J}_{K,\alpha,M}^{(\mu)}(\mathbf{r}) = \frac{1}{2\pi} \int_{-\pi}^{+\pi} d\varphi \, \hat{\mathbf{e}}_\mu(\hat{\mathbf{s}}) \, e^{iM\varphi} \, e^{iK\hat{\mathbf{s}}\cdot\mathbf{r}} \qquad (\mu = TE, TM) \qquad (9.3.18)$$

We will see later in Section 9.4.5 a convenient description of $\mathbf{J}_{K,\alpha,M}^{(\mu)}(\mathbf{r})$ in terms of the reduced rotation matrix.

9.3.4. Spatial Fourier Representation of Electromagnetic Fields

The wavevector representation or spatial Fourier representation of electromagnetic fields is also very useful. The projection of a wavevector state $|\mathbf{k}\rangle$ onto $|\mathbf{r}\rangle$ gives a plane wave and is also expanded into spherical waves as

$$\langle \mathbf{r} | \mathbf{k} \rangle = \frac{1}{\sqrt{(2\pi)^3}} e^{i\mathbf{k}\cdot\mathbf{r}} \tag{9.3.19}$$

$$= \frac{1}{\sqrt{(2\pi)^3}} \sum_{l,m_l} (4\pi) i^l j_l(Kr) Y_l^{m_l}(\hat{\mathbf{r}}) \left[Y_l^{m_l}(\hat{\mathbf{k}}) \right]^* \tag{9.3.20}$$

which corresponds to the Green's function defined in Eq. (9.2.32) in the limit of $r_{out} \to \infty$ and $r_{in} = r$. Similar to the spatial Fourier representation of a scalar multipole field,

$$\langle \mathbf{k} | K; l, m_l \rangle = \frac{\delta(K-k)}{K} Y_l^{m_l}(\hat{\mathbf{k}}) \tag{9.3.21}$$

a vector multipolar field is described by

$$\langle \mathbf{k}; \sigma | K; l, s = 1; m_l, \mu \rangle = \langle \mathbf{k} | K; l, m_l \rangle \hat{\mathbf{e}}_\sigma^\dagger \cdot \hat{\mathbf{e}}_\mu \tag{9.3.22}$$

$$= \frac{\delta(K-k)}{K} Y_l^{m_l}(\hat{\mathbf{k}}) \delta_{\sigma\mu} \tag{9.3.23}$$

Dropping the scalar product with polarization vector $\hat{\mathbf{e}}_\sigma$, we obtain the vector multipole basis

$$\langle \mathbf{k} | K; l, s = 1; m_l, \mu \rangle = \langle \mathbf{k} | K; l, m_l \rangle \hat{\mathbf{e}}_\mu \tag{9.3.24}$$

and the spatial Fourier representation of vector spherical harmonics

$$\langle \mathbf{k} | K; l, s = 1; J, M \rangle = \frac{\delta(K-k)}{K} \sum_\mu \langle l, s = 1; M - \mu, \mu | J, M \rangle \hat{\mathbf{e}}_\mu Y_l^{M-\mu}(\hat{\mathbf{k}}) \tag{9.3.25}$$

By using $\mathbf{k} = k\hat{\mathbf{s}}$ and the definition of the vector spherical harmonics in Eq. (9.2.61) we can represent this relation as

$$\langle \mathbf{k} | K; l, s = 1; J, M \rangle = \frac{\delta(K-k)}{K} \sum_\mu \mathbf{Y}_{J,l,M}(\hat{\mathbf{s}}), \qquad (l = J, J \pm 1) \tag{9.3.26}$$

As is discussed in Section 9.2.2, we can define another convenient set of basis vector functions by

$$\mathbf{Y}_{J,M}^{(M)}(\hat{\mathbf{s}}) = \mathbf{Y}_{J,J,M}(\hat{\mathbf{s}})$$

$$\mathbf{Y}_{J,M}^{(E)}(\hat{\mathbf{s}}) = \sqrt{\frac{J}{2J+1}} \, \mathbf{Y}_{J,J+1,M}(\hat{\mathbf{s}}) + \sqrt{\frac{J+1}{2J+1}} \, \mathbf{Y}_{J,J-1,M}(\hat{\mathbf{s}})$$

$$\mathbf{Y}_{J,M}^{(L)}(\hat{\mathbf{s}}) = \sqrt{\frac{J}{2J+1}}\,\mathbf{Y}_{J,J-1,M}(\hat{\mathbf{s}}) - \sqrt{\frac{J+1}{2J+1}}\,\mathbf{Y}_{J,J+1,M}(\hat{\mathbf{s}})$$

which are referred to as magnetic (M), electric (E), and longitudinal (L) type vector fields.

9.3.5. Multipole Expansion of Vector Plane Waves

Let us study how vector spherical waves are related to vector plane waves with well-defined polarizations.

Consider the state of a vector plane wave specified by the state vector $|\mathbf{K};\hat{\mathbf{e}}(\hat{\mathbf{s}})\rangle$ with wavevector $\mathbf{K} = K\hat{\mathbf{s}}$ and polarization $\hat{\mathbf{e}}(\hat{\mathbf{s}})$ with respect to the directional unit vector of propagation $\hat{\mathbf{s}}$. The projection of the state onto the coordinate representation $|\mathbf{r}\rangle$ gives the wave function of the corresponding vector plane wave:

$$\langle\mathbf{r}|\mathbf{K};\hat{\mathbf{e}}(\hat{\mathbf{s}})\rangle = \frac{1}{\sqrt{(2\pi)^3}}\,\hat{\mathbf{e}}(\hat{\mathbf{s}})e^{i\mathbf{K}\cdot\mathbf{r}} \tag{9.3.27}$$

On the other hand, a good basis for spherical representations of a vector field is provided by the states $|l, s = 1; m_l, \mu\rangle$ of spin $s = 1$ (vector) and its z-projection $\mu = 0, \pm1$. For plane electromagnetic waves propagating in the z direction, $\mathbf{K} = K\hat{\mathbf{z}}$, the transversality condition allows only the superposition of the right/left-circularly polarized states, $|K\hat{\mathbf{z}};\hat{\mathbf{e}}(\hat{\mathbf{z}}) = \hat{\mathbf{e}}_{\pm1}\rangle$, with respect to the quantization axis z. Therefore the spherical representation of a z-directed plane wave becomes simple. The projection of the basis state $|l, s = 1; m_l, \mu = \pm1\rangle$ onto the right/left-circularly polarized plane waves $|K\hat{\mathbf{z}};\hat{\mathbf{e}}(\hat{\mathbf{z}}) = \hat{\mathbf{e}}_\mu\rangle$ or simply $|K\hat{\mathbf{z}};\mu\rangle$ is

$$\langle K\hat{\mathbf{z}};\mu \mid K'; l, s = 1; m_l, \sigma\rangle = \frac{\delta(K'-K)}{K'}\,Y_l^{m_l}(\hat{\mathbf{z}})\delta_{\mu\sigma} \tag{9.3.28}$$

Since only spherical harmonics with nonvanishing component in the z direction are relevant to states with $m_l = 0$,

$$Y_l^{m_l}(\hat{\mathbf{z}}) = Y_l^0(\hat{\mathbf{z}})\delta_{m_l 0} = \sqrt{\frac{2l+1}{4\pi}}\,\delta_{m_l 0} \tag{9.3.29}$$

the multipole expansion of $|K\hat{\mathbf{z}};\mu\rangle$ is described by

$$|K\hat{\mathbf{z}};\mu\rangle = \int dK' \sum_{l,m_l,\sigma} |K'; l, s = 1; m_l,\sigma\rangle$$

$$\times \langle K'; l, s = 1; m_l, \sigma \mid K\hat{\mathbf{z}};\mu\rangle \tag{9.3.30}$$

$$= \int dK' \sum_{l,m_l,\sigma} \frac{\delta(K'-K)}{K'} \sqrt{\frac{2l+1}{4\pi}} \, \delta_{m_l 0}\delta_{\mu\sigma} \, | \, K'; l, s = 1; m_l, \sigma\rangle \quad (9.3.31)$$

$$= \frac{1}{K} \sum_l \sqrt{\frac{2l+1}{4\pi}} \, | \, K; l, s = 1; 0, \mu\rangle \quad (9.3.32)$$

Since the state of vector plane waves with wavevector $\mathbf{K}$ is generated by rotational transform from the state $|\mathbf{K}; \hat{\mathbf{e}}_{\pm 1}\rangle$ as

$$|\mathbf{K}; \hat{\mathbf{e}}(\hat{\mathbf{s}})\rangle = \hat{\mathbf{D}}(\varphi, \theta, 0) \, | \, K\hat{\mathbf{z}};\hat{\mathbf{e}}_{\pm 1}\rangle \quad (9.3.33)$$

we get the vector spherical description of vector plane waves as

$$|\mathbf{K}; \hat{\mathbf{e}}(\hat{\mathbf{s}})\rangle = \frac{1}{K} \sum_l \sqrt{\frac{2l+1}{4\pi}} \, \hat{\mathbf{D}}(\varphi, \theta, 0) \, | \, K; l, s = 1; 0, \mu\rangle \quad (9.3.34)$$

A good basis for evaluating the rotation operator is the state $|K; l, s = 1; J, M\rangle$ with total angular momentum J and its z-projection M. The rotation matrix element $\mathcal{D}_{MM'}^{(J)}$ is then defined by

$$\mathcal{D}_{MM'}^{(J)} = \langle K; l, s = 1; J, M \, | \, \hat{\mathbf{D}}(\varphi, \theta, 0) \, | \, K; l, s = 1; J, M'\rangle \quad (9.3.35)$$

The multipole expansion of vector plane waves is derived by inserting the identity operator

$$1 = \sum_{J,l,M} | \, K; l, s = 1; J, M\rangle\langle K; l, s = 1; J, M | \quad (9.3.36)$$

and by using the relations given above:

$$\langle \mathbf{r} | \, \mathbf{K}; \hat{\mathbf{e}}(\hat{\mathbf{s}})\rangle = \frac{1}{\sqrt{(2\pi)^3}} \, \hat{\mathbf{e}}(\hat{\mathbf{s}}) e^{i\mathbf{K}\cdot\mathbf{r}} = \langle \mathbf{r} | \, \hat{\mathbf{D}}(\varphi, \theta, 0) \, | \, K\hat{\mathbf{z}}; \mu = \pm 1\rangle \quad (9.3.37)$$

$$= \langle \mathbf{r} | \, 1\hat{\mathbf{D}}(\varphi, \theta, 0)1 \, | \, K\hat{\mathbf{z}}; \mu = \pm 1\rangle \quad (9.3.38)$$

$$= \sum_J \sum_{l=J-1}^{J+1} \sum_M \frac{\sqrt{2l+1}}{\sqrt{2\pi}} \, i^l j_l(Kr) \mathbf{Y}_{J,l,M}(\hat{\mathbf{r}}) \mathcal{D}_{M,\pm 1}^{(J)} \langle l, 1; J, M \, | \, l, 1; 0, \pm 1\rangle$$

$$(9.3.39)$$

$$= \frac{1}{\sqrt{(2\pi)^3}} \sum_{J,M} \mathcal{D}_{M,\pm 1}^{(J)} (\varphi, \theta, 0) \sqrt{2\pi} \sqrt{2J+1}$$

$$\times \left[i^{J+1} \sqrt{\frac{J}{2J+1}} \, j_{J+1}(Kr) \mathbf{Y}_{J,J+1,M}(\hat{\mathbf{r}}) \right.$$

$$\left. + i^{J-1} \sqrt{\frac{J+1}{2J+1}} \, j_{J-1}(Kr) \mathbf{Y}_{J,J-1,M}(\hat{\mathbf{r}}) - (\pm 1) i^{J} j_{J}(Kr) \mathbf{Y}_{J,J,M}(\hat{\mathbf{r}}) \right] \quad (9.3.40)$$

$$= \frac{1}{\sqrt{(2\pi)^3}} \sum_{J,M} \mathcal{D}^{(J)}_{M,\pm 1}(\varphi, \theta, 0) \sqrt{2\pi} \sqrt{2J+1}$$

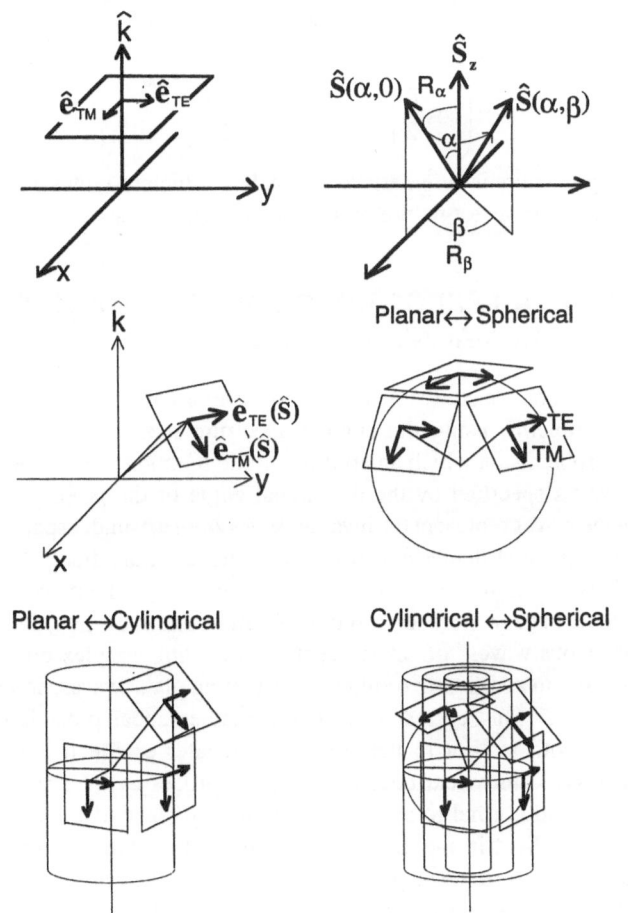

Figure 9.4. Transformations between planar, spherical, and cylindrical vector mode functions described by rotational transforms.

$$\times \left[A_{K,J,M}^{(E,IN)}(\hat{\mathbf{r}}) - (\pm 1)A_{K,J,M}^{(M,IN)}(\hat{\mathbf{r}}) \right] \tag{9.3.41}$$

where $\langle l, s; J, M \mid l, s; m_l, \mu \rangle$ is the Clebsch–Gordan coefficient, and $A_{K,J,M}^{(E,IN)}(\hat{\mathbf{r}})$ and $A_{K,J,M}^{(M,IN)}(\hat{\mathbf{r}})$ are incident complex vector multipole potentials of electric (E) and magnetic (M) types, respectively. The explicit form of the rotational matrix element is given by

$$\mathcal{D}_{M,M'}^{(J)}(\varphi,\theta,0) = e^{-iM\varphi} d_{M,M'}^{(J)}(\theta) \tag{9.3.42}$$

with the reduced rotation matrix elements given by (Wigner's formula) [10, 12,15]

$$d_{M,M'}^{(J)}(\theta) = \sum_x (-1)^x \frac{\sqrt{(J+M)!(J-M)!(J+M')!(J-M')!}}{(J+M-x)!(J-M'-x)!x!(x-M+M')!}$$

$$\times \left(\cos\frac{\theta}{2} \right)^{2J+M-M'-2x} \left(\sin\frac{\theta}{2} \right)^{2x-M+M'} \tag{9.3.43}$$

The above procedure is summarized in Fig. 9.4, which provides a simple geometrical picture of the vector nature of the electromagnetic field.

9.4. ANGULAR SPECTRUM REPRESENTATION OF ELECTROMAGNETIC INTERACTIONS

In this section we study one of the most important ways of describing optical near-field problems, using the angular spectrum representation of the field in scattering problems, in which an arbitrary mode function is expanded into a series of plane waves specified by the directional angle of the propagation vector. Of course, plane waves represent the invariance or symmetry under spatial translational transform or parallel displacement in space. An arbitrary field configuration, in general, does not exhibit such a type of invariance or symmetry. Nevertheless, it is possible to expand a field of arbitrary configuration by introducing so-called "inhomogeneous waves" or "evanescent waves" with complex directional angles of propagation, in addition to ordinary propagating plane waves, so-called "homogeneous waves." This corresponds to the mathematical procedure of analytic continuation with respect to the angular parameter extended into the complex region. By this means we can describe scattering processes in terms of a penetration depth of the field normal to an assumed boundary plane and a spatial frequency spectrum of the field in the direction parallel to the boundary. Investigating the inhomogeneous part of the angular spectrum, such as the central spatial frequency, at which the spectrum takes its maximum, the spectral intensity, the spectral width, and so on, we can evaluate the character of the scattered field, such as its range or effective distance as well as its localization. These features provide an understanding and theoretical description of the near-field optical microscope and of general optical near-field problems.

9.4.1. Angular Spectrum Representation of Scattering Problems

Let us first study the angular spectrum representation and its meaning by considering a scattering problem for a scalar field.

The scattering of a scalar field by a material object, or equivalently the scattering potential in general, is described in terms of a time-independent scattering problem based on the Helmholtz equation

$$\left[\nabla^2 + K^2\varepsilon(\mathbf{r})\right]\varphi(\mathbf{r}) = 0 \tag{9.4.1}$$

where $K = \omega/c$ and $\varepsilon(\mathbf{r})$ are the wavenumber in vacuum and dielectric function of the scattering system, respectively, and $V(\mathbf{r}) = K^2[1 - \varepsilon(\mathbf{r})]$ is interpreted as a scattering potential. The Helmholtz equation describes the interaction of incident and scattered fields with the induced source field on the material system.

Let us assume a scattering system localized in a finite domain D in space and assume also tentative planar boundaries Σ^+ and Σ^- attached at either end of the scatterer, as shown in Fig. 9.5. Then the whole space is separated into right, R^+, and left, R^-, half-spaces. The problem we consider is to describe the scattered scalar field φ in terms of plane waves in both the R^+ and R^- half-spaces separated by the assumed boundaries.

Even for a scattering system with no spatial translational symmetry, one can describe the scattered field in terms of plane waves if one resorts to analytic continuation of the directional angle of the wavevectors into the complex region [3, 4]. This is the angular spectrum representation of the scattered field,

$$\varphi^{(SC)}(\mathbf{r}) = -\frac{1}{4\pi}\int_D G_0(\mathbf{r}, \mathbf{r}', \omega)V(\mathbf{r}')\varphi(\mathbf{r}')\, d^3r' \tag{9.4.2}$$

based on the Weyl transform of the Green's function [2] given by

$$G_0(\mathbf{r}, \mathbf{r}', \omega) = \frac{e^{iK|\mathbf{r}-\mathbf{r}'|}}{4\pi|\mathbf{r}-\mathbf{r}'|} = \frac{iK}{2\pi}\int_{-\pi}^{\pi} d\beta \int_{c^{\pm}} d\alpha \sin\alpha\, e^{iK\hat{s}\cdot(\mathbf{r}-\mathbf{r}')} \tag{9.4.3}$$

Here, as shown in Fig. 9.6, the C^+ contour of integration runs over 0 to $\pi/2$ on the real axis and $\pi/2$ to $\pi/2 - i\infty$ on the imaginary path in the complex α plane, for the transmitting scattered wave into the right half-space, whereas, C^- runs over $\pi/2 + i\infty$ to $\pi/2$ on the complex region and $\pi/2$ to π on the real axis.

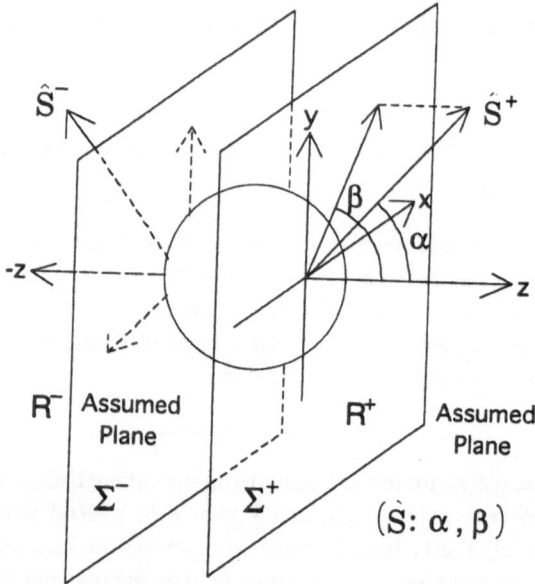

Figure 9.5. Assumed boundary planes and coordinate systems attached to an object for angular spectrum representation of the scattered field.

The analytic nature of the angular spectrum representation was studied extensively by Wolf and Niet-Vesperinas, who concluded that the angular spectrum is an entire function of α [7]. This results in an instructive theorem that any scattered field includes both propagating (homogeneous) and evanescent (inhomogeneous) waves at the same time.

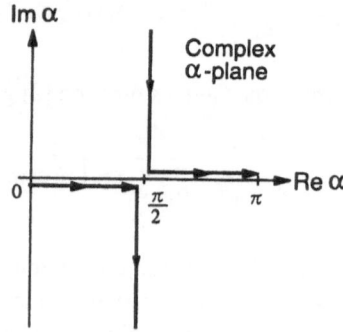

Figure 9.6. Contour of integration in the complex α plane.

For an evanescent wave, the real part of the complex wavenumber forms a vector which represents the symmetry under spatial translation in the direction parallel to the assumed boundary plane. On the other hand, the imaginary part represents an exponential decay of the field with respect to the direction normal to the assumed boundary plane. In the NOM process, the large wavevector of evanescent waves propagating parallel to the assumed boundary is the source of the ultrahigh spatial resolution beyond the diffraction limit.

9.4.2. Meaning of the Angular Spectrum Representation

We can see the meaning of the angular spectrum representation of scattered fields with the aid of some examples.

Let us examine the integral representation of the spherical Hankel function

$$h_l^{(1,2)}(Kr) = (-i)^l \int_{C_{1,2}} P_l(\xi) e^{iKr\xi} d\xi \qquad (9.4.4)$$

where the contour of integration is C_2 from -1 via 0 analytically continued to $0 + i\infty$ and C_1 from $0 + i\infty$ via 0 continued to 1, as shown in Fig. 9.7.

An outgoing spherical wave is described by

$$h_l^{(1)}(Kr) = \frac{e^{iKr}}{iKr} \qquad (9.4.5)$$

$$= \int_0^1 e^{iKr\xi} d\xi + \frac{1}{iKr} \qquad (9.4.6)$$

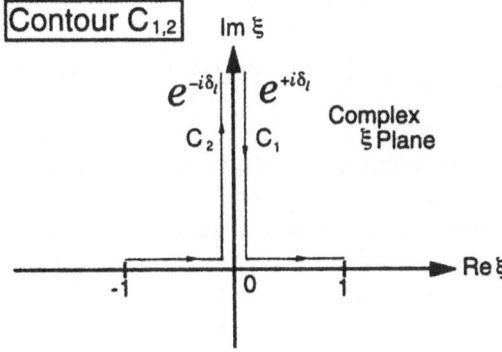

Figure 9.7.   Contours C_1 and C_2 in the complex ξ plane and associated phase shift due to a planar boundary.

$$= \int_0^1 e^{iKr\xi}\,d\xi + \int_{i\infty}^0 e^{iKr\xi}\,d\xi \qquad (9.4.7)$$

$$= \int_{C_1} e^{iKr\xi}d\xi \qquad (9.4.8)$$

Consider the scattering of a scalar spherical wave by a spherical potential described by the step function $V = V_0\theta(a - r)$ with the Helmholtz equation

$$\left[\nabla^2 + K^2\right]\varphi(\mathbf{r}) = V_0\theta(a - r) \qquad (9.4.9)$$

Here $\theta(x) = 1$ for $x > 0$ and $\theta(x) = 0$ for $x < 0$, and $r = |\mathbf{r}|$. The radial solutions can be described in the following forms:

$$\varphi_l^{(IN)}(\mathbf{r}) = C_l^{(IN)}j_l(K^{(IN)}r) \qquad (9.4.10)$$

$$\varphi_l^{(OUT)}(\mathbf{r}) = C_l^{(OUT)}\left[j_l(K^{(OUT)}r)\cos\delta_l - n_l(K^{(OUT)}r)\sin\delta_l\right] \qquad (9.4.11)$$

with the continuation relation at the boundary

$$\left.\frac{\partial\ln\varphi_l^{(IN)}}{\partial r}\right|_{r=a} = \left.\frac{\partial\ln\varphi_l^{(OUT)}}{\partial r}\right|_{r=a} \qquad (9.4.12)$$

By using the spherical Hankel functions we obtain

$$\varphi_l^{(OUT)}(Kr) = C_l^{(OUT)}\left[j_l(K^{(OUT)}r)\cos\delta_l - n_l(K^{(OUT)}r)\sin\delta_l\right] \qquad (9.4.13)$$

$$= \frac{1}{2}\left[h_l^{(1)}(Kr)e^{i\delta_l} + h_l^{(2)}(Kr)e^{-i\delta_l}\right] \qquad (9.4.14)$$

$$= \frac{1}{2}(-i)^l\left[\left(\int_{C_1} P_l(\xi)e^{iKr\xi}d\xi\right)e^{i\delta_l} + \left(\int_{C_2} P_l(\xi)e^{iKr\xi}d\xi\right)e^{-i\delta_l}\right] \qquad (9.4.15)$$

This implies that the effect of the scattering potential manifests itself as a phase shift $\pm\delta_l$ between waves relevant, respectively, to the contours C_1 and C_2 of integration in the angular spectrum representation. That is, for the case of nonzero scattering potential, $V \neq 0$, the contributions from the imaginary parts of the contours C_1 and C_2 do not cancel because of the phase shift $\exp[\pm i\delta_l]$. Then the scattered field involves evanescent waves. In contrast, for the case of $V = 0$ the contributions from the imaginary parts of the integration contours cancel and we obtain only homogeneous waves.

For example, the solution for $V = 0$ is

$$\varphi_l(\mathbf{r}) = j_l(Kr) = \frac{(-1)^l}{2} \int_{-1}^{1} P_l(\xi) \, e^{iKr\xi} d\xi \tag{9.4.16}$$

In order to verify this, let us multiply by $\Sigma_l(2l + 1)i^l P_l(\xi')$ and perform the integration. The result is

$$\sum_l (2l + 1)i^l j_l(Kr) P_l(\xi') = \int_{-1}^{1} \sum_l \frac{2l + 1}{2} P_l(\xi') P_l(\xi) e^{iKr\xi} d\xi \tag{9.4.17}$$

$$= \int_{-1}^{1} \delta(\xi - \xi') e^{iKr\xi} d\xi \tag{9.4.18}$$

$$= e^{iKr\xi'} \tag{9.4.19}$$

What we have obtained here is the spherical Bessel function expansion of plane waves.

As is shown above, the phase shift between the incoming and outgoing waves due to a scattering potential can be expressed in terms of inhomogeneous or evanescent waves appearing in the angular spectrum representation of the scattered field. Thus one sees the importance of the optical near-field.

9.4.3. Angular Spectrum Representation of Scalar Multipole Field and Propagator

9.4.3.1. Angular Spectrum Representation of a Scalar Multipole Field

In general, one can show that the scalar $2l$ multipole field

$$\varphi^{(SC)}(\mathbf{r}) = \langle \mathbf{r}|K; l, m_l \rangle = \sqrt{\frac{2}{\pi}} \, K \, i^l h_l^{(1)}(Kr) Y_l^{m_l}(\hat{\mathbf{r}}) \tag{9.4.20}$$

is described in terms of plane waves by using the angular spectrum representation of spherical waves:

$$i^l h_l^{(1)}(Kr) Y_l^{m_l}(\hat{\mathbf{r}}) = \frac{1}{2\pi} \int_{-\pi}^{\pi} d\beta \int_{C\pm} \sin\alpha \, d\alpha \, Y_l^{m_l}(\alpha, \beta) e^{iK\hat{s}\cdot\mathbf{r}} \tag{9.4.21}$$

$$= \frac{1}{2\pi} \int_{C} d\Omega_{\hat{s}} \, Y_l^{m_l}(\hat{s}) e^{iK\hat{s}\cdot\mathbf{r}} \tag{9.4.22}$$

Here $\hat{s}$ is the angular directional unit vector $\hat{s} = (\sin \alpha \cos \beta, \sin \alpha \sin \beta, \cos \alpha)$ and $d\Omega_{\hat{s}} = \sin \alpha \, d\alpha \, d\beta$. The contour of integration $C^{\pm}$ in the complex α plane is indicated in Fig. 9.6 (C^{+}: $0 \to \pi/2 \to \pi/2 - i\infty$; C^{-}: $\pi/2 + i\infty \to \pi/2 \to \pi$).

9.4.3.2. Angular Spectrum Representation of a Scalar Multipole Propagator

Once we have the plane-wave expansion of a scalar multipole field in terms of the angular spectrum representation, we can translate the origin of the spherical representation to an arbitrary point in space by means of parallel displacement. This provides a general procedure for generating a multipole propagator which corresponds to a Green's function represented on the multipolar basis.

Let $\mathbf{r}_1$ be the position vector with respect to the origin O of the spherical representation, and consider a spatial displacement to an arbitrary point O' with a displacement vector $\overrightarrow{OO'} = \mathbf{R}$. Let $\mathbf{r}_2$ be the position vector with respect to new origin O', and consider the case of $|\mathbf{r}_2| < |\mathbf{R}|$. Then we consider the interaction propagator between O and O' in terms of the angular spectrum representation with respect to the assumed boundary planes at O and O' placed parallel to each other, as shown in Fig. 9.8.

The spatial transform is done simply by replacing the original coordinate vector $\mathbf{r}_1$ in each angular spectrum component with $\mathbf{R} = \mathbf{r}_1 - \mathbf{r}_2$,

$$i^{l_1} h_{l_1}^{(1)}(Kr_1) Y_{l_1}^{m_{l_1}}(\hat{\mathbf{r}}_1) = \frac{1}{2\pi} \int_C d\Omega_{\hat{s}} \, Y_{l_1}^{m_{l_1}}(\hat{s}) e^{iK\hat{s}\cdot\mathbf{R}} e^{iK\hat{s}\cdot\mathbf{r}_2} \qquad (9.4.23)$$

Then, substituting the plane wave $\exp(iK\hat{s}\cdot\mathbf{r}_2)$ by its spherical expansion given in Eq. (9.2.34),

$$e^{iK\hat{s}\cdot\mathbf{r}} = 4\pi \sum_{l,m_l} i^l j_l(Kr) \left[Y_l^{m_l}(\hat{s}) \right]^{*} Y_l^{m_l}(\hat{\mathbf{r}}) \qquad (9.4.24)$$

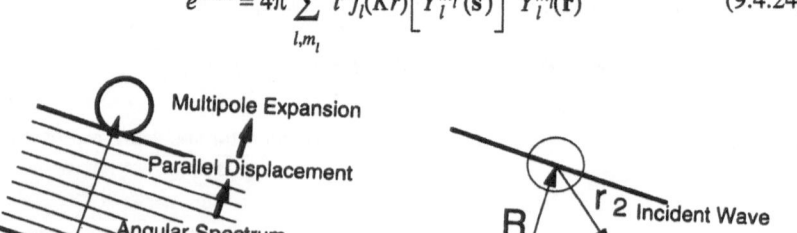

Figure 9.8.   Multipole propagator and its angular spectrum representation.

we get the multipole expansion of the scalar field with respect to the new origin O as

$$\underbrace{i^{l_1}h^{(1)}_{l_1}(Kr_1)Y^{m_{l_1}}_{l_1}(\hat{\mathbf{r}}_1)}_{\text{scattered field}} = 4\pi \sum_{l_2,m_{l_2}} \underbrace{i^{l_2}j_{l_2}(Kr_2)Y^{m_{l_2}}_{l_2}(\hat{\mathbf{r}}_2)}_{\text{incident field}}$$

$$\times \frac{1}{2\pi} \int_C d\Omega_{\hat{s}} \, [Y^{m_{l_2}}_{l_2}(\hat{s})]^* \, e^{iK\hat{s}\cdot\mathbf{R}} \, Y^{m_{l_1}}_{l_1}(\hat{s}) \tag{9.4.25}$$

This expression shows that for $|\mathbf{r}_2| < |\mathbf{R}|$ the outgoing wave or scattered field with respect to the origin O is expressed in terms of the incoming wave or incident field with respect to the new origin O'. The coefficients of the transform describe the contributions of scattered multipole fields at one point into incident multipole fields at another. Thus we obtain the angular spectrum representation of the multipole propagator:

$$G(\mathbf{r}, l_2, m_{l_2}; \mathbf{r}_1, l_1, m_{l_1}; \omega)$$

$$= \frac{1}{2\pi} \int_C d\Omega_{\hat{s}} \, [Y^{m_{l_2}}_{l_2}(\hat{s})]^* \, 4\pi e^{iK\hat{s}\cdot\mathbf{R}} Y^{m_{l_1}}_{l_1}(\hat{s}) \tag{9.4.26}$$

$$= \frac{1}{2\pi} \int_C d\Omega_{\hat{s}} \, \tilde{G}_{K;l_2,m_{l_2};l_1,m_{l_1}}(\hat{s}) e^{iK\hat{s}\cdot\mathbf{R}} \tag{9.4.27}$$

with

$$\tilde{G}_{K;l_2,m_{l_2};l_1,m_{l_1}}(\hat{s}) = 4\pi \, [Y^{m_{l_2}}_{l_2}(\hat{s})]^* \, Y^{m_{l_1}}_{l_1}(\hat{s}) \tag{9.4.28}$$

For example, let us examine this relation for the case of spherical symmetry, i.e., $l_1 = 0$, $m_{l_1} = 0$ with $Y^0_0(\hat{s}) = 1/\sqrt{4\pi}$. Let $\mathbf{R}' = -\mathbf{R}$ and perform the integration,

$$\frac{iK}{4\pi} h^{(1)}_0(Kr_1) = \frac{e^{iK|\mathbf{r}_2 - \mathbf{R}'|}}{4\pi|\mathbf{r}_2 - \mathbf{R}'|}$$

$$= \sqrt{4\pi} iK \sum_{l_2,m_{l_2}} i^{l_2} j_{l_2}(Kr_2) Y^{m_{l_2}}_{l_2}(\hat{\mathbf{r}}_2)$$

$$\times \frac{1}{2\pi} \int_C d\Omega_{\hat{s}} \frac{1}{\sqrt{4\pi}} [Y^{m_{l_2}}_{l_2}(\hat{s})]^* e^{-iK\hat{s}\cdot\mathbf{R}'}$$

$$= \sum_{l_2,m_{l_2}} iK i^{l_2} j_{l_2}(Kr_2) Y^{m_{l_2}}_{l_2}(\hat{\mathbf{r}}_2) \frac{1}{2\pi} \int_C d\Omega_{\hat{s}} \left[Y^{m_{l_2}}_{l_2}(\hat{s}) \right]^* e^{-iK\hat{s}\cdot\mathbf{R}'}$$

$$= \sum_{l_2, m_{l_2}} iKi^{l_2}j_{l_2}(Kr_2)Y_{l_2}^{m_{l_2}}(\hat{\mathbf{r}}_2)\,(-i)^{l_2}h_{l_2}^{(1)}(KR')\big[Y_{l'_2}^{m_{l'_2}}(\hat{\mathbf{R}}')\big]^*$$

$$= \sum_{l_2, m_{l_2}} iKj_{l_2}(Kr_2)h_{l_2}^{(1)}(KR')\big[Y_{l'_2}^{m_{l'_2}}(\hat{\mathbf{R}}')\big]^* Y_{l_2}^{m_{l_2}}(\hat{\mathbf{r}}_2)$$

$$= G(\mathbf{r}_2, \mathbf{R}', \omega)$$

The angular spectrum representation of the multipole propagator provides a very useful measure with which to evaluate the range and localization of the interaction especially in near-field problems.

It is instructive to rewrite the angular spectrum of the multipole propagator in terms of two successive spherical propagators as

$$\big[Y_{l_2}^{m_{l_2}}(\hat{\mathbf{s}})\big]^* e^{iK\hat{\mathbf{s}}\cdot\mathbf{R}}\, Y_{l_1}^{m_{l_1}}(\hat{\mathbf{s}}) = \big[Y_{l_2}^{m_{l_2}}(\hat{\mathbf{s}})e^{iK\hat{\mathbf{s}}\cdot\mathbf{r}_2}\big]^* e^{iK\hat{\mathbf{s}}\cdot\mathbf{r}_1}\, Y_{l_1}^{m_{l_1}}(\hat{\mathbf{s}}) \qquad (9.4.29)$$

Defining the propagator components by

$$\tilde{G}_{K;l,m_l}^{(\hat{s})}(\mathbf{r}) = \sqrt{4\pi}\, e^{iK\hat{\mathbf{s}}\cdot\mathbf{r}}\, Y_l^{m_l}(\hat{\mathbf{s}}) \qquad (9.4.30)$$

$$\big[\tilde{G}_{K;l,m_l}^{(\hat{s})}(\mathbf{r})\big]^\dagger = \big[\sqrt{4\pi}\, Y_l^{m_l}(\hat{\mathbf{s}})\, e^{iK\hat{\mathbf{s}}\cdot\mathbf{r}}\big]^* \qquad (9.4.31)$$

we get

$$G(\mathbf{r}_2, l_2, m_{l_2}; \mathbf{r}_1, l_1, m_{l_1}; \omega) = \frac{1}{2\pi}\int_C d\Omega_{\hat{\mathbf{s}}}\big[\tilde{G}_{K,l_2,m_{l_2}}^{(\hat{s})}(\mathbf{r}_2)\big]^\dagger \tilde{G}_{K,l_1,m_{l_1}}^{(\hat{s})}(\mathbf{r}_1) \qquad (9.4.32)$$

where the multipole propagator G is described in terms of a complex integral of two successive transforms described by angular spectrum components $\tilde{G}_{K,l,m_l}^{(\hat{s})}(\mathbf{r})$. This agrees with the geometry shown in Fig. 9.8.

9.4.4. Angular Spectrum Representation of Vector Multipole Field and Propagator

Here we study the slightly complicated case of vector multipole fields. We summarize this material here because this is rarely found in the usual textbooks.

9.4.4.1. Angular Spectrum Representation of a Vector Multipole Field

As shown in Section 9.2.2, the magnetic and electric $2J$ multipolar fields, which extend to spherical waves in the far field, are defined by

$$A_{K,J,M}^{(M,SC)}(\mathbf{r}) = i^J h_J^{(1)}(Kr) Y_{J,J,M}(\hat{\mathbf{r}}) \tag{9.4.33}$$

$$A_{K,J,M}^{(E,SC)}(\mathbf{r}) = i^{J+1} \sqrt{\frac{J}{2J+1}} \, h_{J+1}^{(1)}(Kr) Y_{J,J+1,M}(\hat{\mathbf{r}})$$

$$+ i^{J-1} \sqrt{\frac{J+1}{2J+1}} \, h_{J-1}^{(1)}(Kr) Y_{J,J-1,M}(\hat{\mathbf{r}}) \tag{9.4.34}$$

With respect to the following discussions, it is convenient to define vector spherical harmonics of magnetic and electric types as

$$Y_{J,M}^{(M)}(\hat{\mathbf{s}}) \equiv Y_{J,J,M}(\hat{\mathbf{s}}) \tag{9.4.35}$$

$$Y_{J,M}^{(E)}(\hat{\mathbf{s}}) \equiv \sqrt{\frac{J}{2J+1}} \, Y_{J,J+1,M}(\hat{\mathbf{s}}) + \sqrt{\frac{J+1}{2J+1}} \, Y_{J,J-1,M}(\hat{\mathbf{s}}) \tag{9.4.36}$$

where $\hat{\mathbf{s}}$ is the complex unit angular directional vector,

$$\hat{\mathbf{s}} \equiv (\sin\alpha\cos\beta, \sin\alpha\sin\beta, \cos\alpha) \tag{9.4.37}$$

We first show the results: The magnetic and electric $2J$ multipolar fields can be represented in terms of the angular spectrum by

$$\begin{cases} A_{K,J,M}^{(M,SC)}(\mathbf{r}) = \dfrac{1}{2\pi} \int_C d\Omega_{\hat{s}} \, Y_{J,M}^{(M)}(\hat{\mathbf{s}}) e^{iK\hat{s}\cdot\mathbf{r}} \\[4mm] A_{K,J,M}^{(E,SC)}(\mathbf{r}) = \dfrac{1}{2\pi} \int_C d\Omega_{\hat{s}} \, Y_{J,M}^{(E)}(\hat{\mathbf{s}}) e^{iK\hat{s}\cdot\mathbf{r}} \end{cases}$$

Now let us derive these formulas.

Let us first derive the angular spectrum representation of quadrupole fields, starting with the well-known formula for the electric dipole field.

$$A_{K,1,M}^{(E,SC)}(\mathbf{r}) = \frac{1}{2\pi} \int_C d\Omega_{\hat{s}} \, Y_{1,M}^{(E)}(\hat{\mathbf{s}}) \, e^{iK\hat{s}\cdot\mathbf{r}} \tag{9.4.38}$$

Taking the curl of this and using the relations

$$\nabla \times A_{K,J,M}^{(E,SC)}(\mathbf{r}) = -K A_{K,J,M}^{(M,SC)}(\mathbf{r}) \tag{9.4.39}$$

$$\nabla \times A_{K,J,M}^{(M,SC)}(\mathbf{r}) = -K A_{K,J,M}^{(E,SC)}(\mathbf{r}) \tag{9.4.40}$$

we get the expansion of the magnetic dipole field as

$$A_{K,1,M}^{(M,SC)}(\mathbf{r}) = \frac{1}{2\pi} \int_C d\Omega_{\hat{s}} \, [-i\hat{s}] \times \mathbf{Y}_{1,M}^{(E)}(\hat{s}) e^{iK\hat{s}\cdot\mathbf{r}} \tag{9.4.41}$$

With help of the formulas

$$\mathbf{Y}_{J,M}^{(M)}(\hat{s}) = -i\hat{s} \times \mathbf{Y}_{J,M}^{(E)}(\hat{s}) \tag{9.4.42}$$

$$\mathbf{Y}_{J,M}^{(E)}(\hat{s}) = i\hat{s} \times \mathbf{Y}_{J,M}^{(M)}(\hat{s}) \tag{9.4.43}$$

we obtain the angular spectrum representation of the magnetic dipole field

$$A_{K,1,M}^{(M,SC)}(\mathbf{r}) = \frac{1}{2\pi} \int_C d\Omega_{\hat{s}} \, \mathbf{Y}_{1,M}^{(M)}(\hat{s}) \, e^{iK\hat{s}\cdot\mathbf{r}} \tag{9.4.44}$$

As is shown in the previous section, the following relations hold:

$$h_0^{(1)}(\rho)\mathbf{Y}_{1,0,M}(\hat{\mathbf{r}}) = \frac{(-i)^0}{2\pi} \int_C d\Omega_{\hat{s}} \, \mathbf{Y}_{1,0,M}(\hat{s}) \, e^{iK\hat{s}\cdot\mathbf{r}} \tag{9.4.45}$$

$$h_1^{(1)}(\rho)\mathbf{Y}_{1,1,M}(\hat{\mathbf{r}}) = \frac{(-i)^1}{2\pi} \int_C d\Omega_{\hat{s}} \, \mathbf{Y}_{1,1,M}(\hat{s}) \, e^{iK\hat{s}\cdot\mathbf{r}} \tag{9.4.46}$$

$$h_2^{(1)}(\rho)\mathbf{Y}_{1,2,M}(\hat{\mathbf{r}}) = \frac{(-i)^2}{2\pi} \int_C d\Omega_{\hat{s}} \, \mathbf{Y}_{1,2,M}(\hat{s}) \, e^{iK\hat{s}\cdot\mathbf{r}} \tag{9.4.47}$$

Substituting these equations into the definition of vector spherical harmonics,

$$\mathbf{Y}_{J,l,M}(\hat{v}) \equiv \sum_{\mu} Y_l^{M-\mu}(\hat{v})\hat{e}_{\mu}\langle l, 1; M-\mu, \mu | J, M \rangle \tag{9.4.48}$$

and comparing both sides of these equations, we obtain

$$h_0^{(1)}(\rho)Y_0^M(\hat{\mathbf{r}}) = \frac{(-i)^0}{2\pi} \int_C d\Omega_{\hat{s}} \, Y_0^M(\hat{s}) \, e^{iK\hat{s}\cdot\mathbf{r}} \tag{9.4.49}$$

$$h_1^{(1)}(\rho)Y_1^M(\hat{\mathbf{r}}) = \frac{(-i)^1}{2\pi} \int_C d\Omega_{\hat{s}} \, Y_1^M(\hat{s}) \, e^{iK\hat{s}\cdot\mathbf{r}} \tag{9.4.50}$$

$$h_2^{(1)}(\rho)Y_2^M(\hat{\mathbf{r}}) = \frac{(-i)^2}{2\pi} \int_C d\Omega_{\hat{s}} \, Y_2^M(\hat{s}) \, e^{iK\hat{s}\cdot\mathbf{r}} \tag{9.4.51}$$

Here $M - \mu$ is replaced simply by M. Set M as $M - \mu$ and multiply by the polarization vector $\hat{\mathbf{e}}_\mu$; a vector field with total angular momentum $J = 2$ is synthesized as

$$h_1^{(1)}(\rho)\mathbf{Y}_{2,1,M}(\hat{\mathbf{r}}) = \frac{(-i)^1}{2\pi} \int_C d\Omega_{\hat{s}} \, \mathbf{Y}_{2,1,M}(\hat{\mathbf{s}}) \, e^{iK\hat{s}\cdot\mathbf{r}} \tag{9.4.52}$$

$$h_2^{(1)}(\rho)\mathbf{Y}_{2,2,M}(\hat{\mathbf{r}}) = \frac{(-i)^2}{2\pi} \int_C d\Omega_{\hat{s}} \, \mathbf{Y}_{2,2,M}(\hat{\mathbf{s}}) \, e^{iK\hat{s}\cdot\mathbf{r}} \tag{9.4.53}$$

Compared with the definition of the magnetic multipole, we get the angular spectrum representation of the magnetic quadrupole

$$\mathbf{A}_{K,2,M}^{(M,SC)}(\mathbf{r}) = \frac{1}{2\pi} \int_C d\Omega_{\hat{s}} \, \mathbf{Y}_{2,M}^{(M)}(\hat{\mathbf{s}}) \, e^{iK\hat{s}\cdot\mathbf{r}} \tag{9.4.54}$$

and taking the curl of this equation, we get the angular spectrum of the electric quadrupole field

$$\mathbf{A}_{K,2,M}^{(E,SC)}(\mathbf{r}) = \frac{1}{2\pi} \int_C d\Omega_{\hat{s}} \, \mathbf{Y}_{2,M}^{(E)}(\hat{\mathbf{s}}) \, e^{iK\hat{s}\cdot\mathbf{r}} \tag{9.4.55}$$

We can express this equation in the following forms.

$$h_3^{(1)}(\rho)\mathbf{Y}_{2,3,M}(\hat{\mathbf{r}}) = \frac{(-i)^3}{2\pi} \int_C d\Omega_{\hat{s}} \, \mathbf{Y}_{2,3,M}(\hat{\mathbf{s}}) \, e^{iK\hat{s}\cdot\mathbf{r}} \tag{9.4.56}$$

$$h_3^{(1)}(\rho)Y_3^M(\hat{\mathbf{r}}) = \frac{(-i)^3}{2\pi} \int_C d\Omega_{\hat{s}} \, Y_3^M(\hat{\mathbf{s}}) \, e^{iK\hat{s}\cdot\mathbf{r}} \tag{9.4.57}$$

In the same manner as in the derivation of the angular spectrum representation of quadrupole fields synthesized from those of dipole fields, we can derive a general formula for the case of $2J$ multipole fields.

Let us assume the following relations hold for $2(J - 1)$ multipole fields.

$$h_{J-2}^{(1)}(\rho)\mathbf{Y}_{J-1,J-2,M}(\hat{\mathbf{r}}) = \frac{(-i)^{J-2}}{2\pi} \int_C d\Omega_{\hat{s}} \, \mathbf{Y}_{J-1,J-2,M}(\hat{\mathbf{s}}) \, e^{iK\hat{s}\cdot\mathbf{r}} \tag{9.4.58}$$

$$h_{J-1}^{(1)}(\rho)\mathbf{Y}_{J-1,J-1,M}(\hat{\mathbf{r}}) = \frac{(-i)^{J-1}}{2\pi} \int_C d\Omega_{\hat{s}} \, \mathbf{Y}_{J-1,J-1,M}(\hat{\mathbf{s}}) \, e^{iK\hat{s}\cdot\mathbf{r}} \tag{9.4.59}$$

$$h_J^{(1)}(\rho)\mathbf{Y}_{J-1,J,M}(\hat{\mathbf{r}}) = \frac{(-i)^J}{2\pi}\int_C d\Omega_{\hat{s}}\,\mathbf{Y}_{J-1,J,M}(\hat{\mathbf{s}})\,e^{iK\hat{s}\cdot\mathbf{r}} \tag{9.4.60}$$

Using the definition of the vector spherical harmonics, we get

$$h_{J-1}^{(1)}(\rho)Y_{J-1}^M(\hat{\mathbf{r}}) = \frac{(-i)^{J-1}}{2\pi}\int_C d\Omega_{\hat{s}}\,Y_{J-1}^M(\hat{\mathbf{s}})\,e^{iK\hat{s}\cdot\mathbf{r}} \tag{9.4.61}$$

$$h_J^{(1)}(\rho)Y_J^M(\hat{\mathbf{r}}) = \frac{(-i)^J}{2\pi}\int_C d\Omega_{\hat{s}}\,Y_J^M(\hat{\mathbf{s}})\,e^{iK\hat{s}\cdot\mathbf{r}} \tag{9.4.62}$$

We set M as $M - \mu$ and multiply by the polarization vector $\hat{\mathbf{e}}_\mu$; a vector field with total angular momentum J is synthesized as

$$h_{J-1}^{(1)}(\rho)\mathbf{Y}_{J,J-1,M}(\hat{\mathbf{r}}) = \frac{(-i)^{J-1}}{2\pi}\int_C d\Omega_{\hat{s}}\,\mathbf{Y}_{J,J-1,M}(\hat{\mathbf{s}})\,e^{iK\hat{s}\cdot\mathbf{r}} \tag{9.4.63}$$

$$h_J^{(1)}(\rho)\mathbf{Y}_{J,J,M}(\hat{\mathbf{r}}) = \frac{(-i)^J}{2\pi}\int_C d\Omega_{\hat{s}}\,\mathbf{Y}_{J,J,M}(\hat{\mathbf{s}})\,e^{iK\hat{s}\cdot\mathbf{r}} \tag{9.4.64}$$

Following the same procedures as for the quadrupole case, we get the angular spectrum representation of $2J$ multipole fields,

$$\mathbf{A}_{K,J,M}^{(M,SC)}(\mathbf{r}) = \frac{1}{2\pi}\int_C d\Omega_{\hat{s}}\,\mathbf{Y}_{J,M}^{(M)}(\hat{\mathbf{s}})\,e^{iK\hat{s}\cdot\mathbf{r}} \tag{9.4.65}$$

$$\mathbf{A}_{K,J,M}^{(E,SC)}(\mathbf{r}) = \frac{1}{2\pi}\int_C d\Omega_{\hat{s}}\,\mathbf{Y}_{J,M}^{(E)}(\hat{\mathbf{s}})\,e^{iK\hat{s}\cdot\mathbf{r}} \tag{9.4.66}$$

and then

$$h_{J+1}^{(1)}(\rho)\mathbf{Y}_{J,J+1,M}(\hat{\mathbf{r}}) = \frac{(-i)^{J+1}}{2\pi}\int_C d\Omega_{\hat{s}}\,\mathbf{Y}_{J,J+1,M}(\hat{\mathbf{s}})\,e^{iK\hat{s}\cdot\mathbf{r}} \tag{9.4.67}$$

$$h_J^{(1)}(\rho)Y_J^M(\hat{\mathbf{r}}) = \frac{(-i)^J}{2\pi}\int_C d\Omega_{\hat{s}}\,Y_J^M(\hat{\mathbf{s}})\,e^{iK\hat{s}\cdot\mathbf{r}} \tag{9.4.68}$$

As we see, this holds for $J = 2$; the angular spectrum representation for multipole fields of arbitrary order is obtained above for both magnetic and electric multipoles.

9.4.4.2. Angular Spectrum Representation of a Vector Multipole Propagator

It is convenient to start with the angular spectrum representation of the $l = J - 1$ vector multipole field,

$$i^{J-1} h_{J-1}^{(1)}(Kr_1) \mathbf{Y}_{J,J-1,M}(\hat{\mathbf{r}}_1) = \frac{1}{2\pi} \int_C d\Omega_{\hat{s}} \mathbf{Y}_{J,J-1,M}(\hat{s}) e^{iK\hat{s}\cdot\mathbf{r}_1} \qquad (9.4.69)$$

where the contour of integration is the same as in the scalar case. We can obtain the vector multipole field by operating with $\nabla_1 \times$ on this as

$$\nabla_1 \times \left[i^{J-1} h_{J-1}^{(1)}(Kr_1) \mathbf{Y}_{J,J-1,M}(\hat{\mathbf{r}}_1) \right] = -K\sqrt{\frac{J+1}{2J+1}}\, \mathbf{A}_{K,J,M}^{(M,IN)}(\hat{\mathbf{r}}_1) \qquad (9.4.70)$$

$$\nabla_1 \times \mathbf{A}_{K,J,M}^{(M,IN)}(\hat{\mathbf{r}}_1) = -K\, \mathbf{A}_{K,J,M}^{(E,IN)}(\hat{\mathbf{r}}_1) \qquad (9.4.71)$$

Then a parallel displacement of the origin of the spherical representation from O to an arbitrary point O' is achieved simply by transferring the coordinate vector in each angular spectrum component by $\overrightarrow{OO'} = \mathbf{R}$. Let $\mathbf{r}_1$ and $\mathbf{r}_2$ be the coordinate vectors with respect to the origins O and O', respectively; therefore $\mathbf{R} = \mathbf{r}_1 - \mathbf{r}_2$, and we get the angular spectrum representation of vector multipole fields:

$$i^{J-1} h_{J-1}^{(1)}(Kr_1) \mathbf{Y}_{J,J-1,M}(\hat{\mathbf{r}}_1) = \frac{1}{2\pi} \int_C d\Omega_{\hat{s}} \mathbf{Y}_{J,J-1,M}(\hat{s}) e^{iK\hat{s}\cdot\mathbf{R}} e^{iK\hat{s}\cdot\mathbf{r}_2} \qquad (9.4.72)$$

Then inserting the identity operator

$$1 = \sum_{\mu}^{TE,TM,L} \hat{\mathbf{e}}_{\mu}(\hat{s}) \hat{\mathbf{e}}_{\mu}(\hat{s}) \cdot \qquad (9.4.73)$$

and using the vector spherical expansion of vector plane waves,

$$\hat{\mathbf{e}}_{\mu}(\hat{s})\, e^{iK\hat{s}\cdot\mathbf{r}} = 4\pi \sum_{\lambda}^{E,M,L} \sum_{J,M} \mathbf{A}_{K,J,M}^{(\lambda,IN)}(\mathbf{r}) \left[\mathbf{Y}_{J,M}^{(\lambda)}(\hat{s}) \right]^{\dagger} \cdot \hat{\mathbf{e}}_{\mu}(\hat{s}) \qquad (9.4.74)$$

we obtain

$$i^{J-1} h_{J-1}^{(1)}(Kr_1) \mathbf{Y}_{J,J-1,M}(\hat{\mathbf{r}}_1)$$

$$= 4\pi \sum_{\lambda'}^{E,M,L} \sum_{J',M'} \mathbf{A}_{K,J',M'}^{(\lambda',IN)}(\mathbf{r}_2)$$

$$\times \frac{1}{2\pi} \int_C d\Omega_{\hat{s}} \left[\mathbf{Y}^{(\lambda)}_{J',M'}(\hat{s}) \right]^\dagger \left\{ \sum_\mu^{TE,TM,L} \cdot \hat{\mathbf{e}}_\mu(\hat{s}) e^{iK\hat{s}\cdot\mathbf{R}} \hat{\mathbf{e}}_\mu(\hat{s}) \cdot \right\} \mathbf{Y}_{J,J-1,M}(\hat{s}) \quad (9.4.75)$$

Operating with $\nabla_1 \times$ twice on this, we can synthesize vector multipole fields $\mathbf{A}^{(M,SC)}_{K,J,M}(\mathbf{r})$ and $\mathbf{A}^{(E,SC)}_{K,J,M}(\mathbf{r})$ according to Eqs. (9.4.70) and (9.4.71). Then the longitudinal (L) mode is eliminated since $\hat{s} \times \mathbf{Y}^{(L)}_{J',M'}(\hat{s}) = 0$. Using Eqs. (9.2.81) and (9.4.42), which yield

$$-i\hat{s} \times \mathbf{Y}_{J,J-1,M}(\hat{s}) = -i\hat{s} \times \sqrt{\frac{J+1}{2J+1}} \; \mathbf{Y}^{(E)}_{J,M}(\hat{s}) = \sqrt{\frac{J+1}{2J+1}} \; \mathbf{Y}^{(M)}_{J,M}(\hat{s}) \quad (9.4.76)$$

we obtain

$$\mathbf{A}^{(M,SC)}_{K,J,M}(\mathbf{r}_1) = 4\pi \sum_{\lambda'} \sum_{J',M'}^{E,M} \mathbf{A}^{(\lambda',IN)}_{K,J',M'}(\mathbf{r}_2)$$

$$\times \frac{1}{2\pi} \int_C d\Omega_{\hat{s}} \left[\mathbf{Y}^{(\lambda)}_{J',M'}(\hat{s}) \right]^\dagger \left\{ \sum_\mu^{TE,TM} \cdot \hat{\mathbf{e}}_\mu(\hat{s}) e^{iK\hat{s}\cdot\mathbf{R}} \hat{\mathbf{e}}_\mu(\hat{s}) \cdot \right\} \mathbf{Y}^{(M)}_{J,M}(\hat{s}) \quad (9.4.77)$$

$$\mathbf{A}^{(E,SC)}_{K,J,M}(\mathbf{r}_1) = 4\pi \sum_{\lambda'} \sum_{J',M'}^{E,M} \mathbf{A}^{(\lambda',IN)}_{K,J',M'}(\mathbf{r}_2)$$

$$\times \frac{1}{2\pi} \int_C d\Omega_{\hat{s}} \left[\mathbf{Y}^{(\lambda)}_{J',M'}(\hat{s}) \right]^\dagger \left\{ \sum_\mu^{TE,TM} \cdot \hat{\mathbf{e}}_\mu(\hat{s}) e^{iK\hat{s}\cdot\mathbf{R}} \hat{\mathbf{e}}_\mu(\hat{s}) \cdot \right\} \mathbf{Y}^{(E)}_{J,M}(\hat{s}) \quad (9.4.78)$$

The above procedure is pictorially explained as follows, as one reads the equation from right to left:

Pick up the E or M multipole from the scattered field
$\Rightarrow$ Project the multipole onto TE and TM vector plane waves
$\Rightarrow$ Propagate TE and TM vector plane waves from O to O'
$\Rightarrow$ Project plane waves onto E and M incident multipole fields
This is shown diagrammatically in Fig. 9.9.

The result shows that for $|\mathbf{r}_2| < |\mathbf{R}|$ a scattered vector multipole field $\mathbf{A}^{(E,SC)}_{K,J,M}(\mathbf{r}_1)$ with respect to the origin O is rewritten as a sum of incident vector multipole fields $\mathbf{A}^{(\lambda',IN)}_{K,J',-M'}(\mathbf{r}_2)$ with respect to a new origin O'. This yields the vector multipole propagator $\mathbf{G}(\mathbf{r}_2, \lambda', J', M'; \mathbf{r}_1, \lambda, J, M; \omega)$ in terms of the angular spectrum representation as

$$\mathbf{G}(\mathbf{r}_2, \lambda', J', M'; \mathbf{r}_1, \lambda, J, M; \omega) = \frac{1}{2\pi} \int_C d\Omega_{\hat{s}} \; \widetilde{\mathbf{G}}^{(\lambda',\lambda)}_{K;J',M';J,M}(\hat{s}) \; e^{iK\hat{s}\cdot(\mathbf{r}_1-\mathbf{r}_2)} \quad (9.4.79)$$

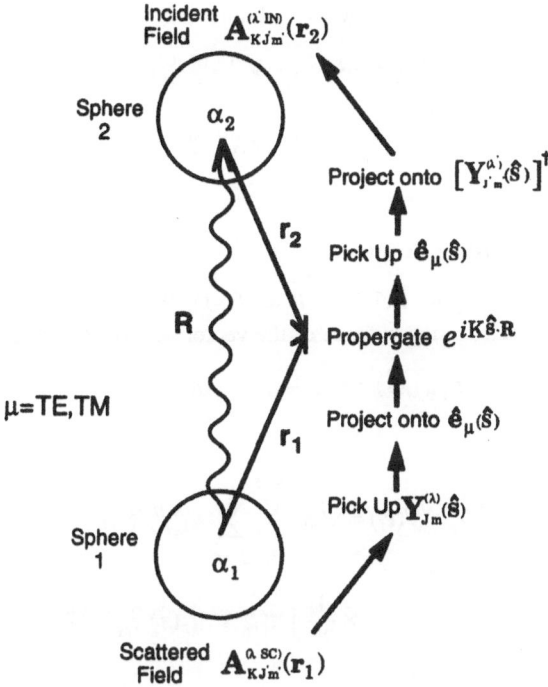

Figure 9.9. Fundamental processes involved in angular spectrum representation of the multipole propagator.

$$\tilde{G}^{(\lambda',\lambda)}_{K;J',M';J,M}(\hat{s}) = 4\pi \sum_{\mu}^{TE,TM} \left[Y^{(\lambda')}_{J',M'}(\hat{s})\right]^{\dagger} \cdot \hat{e}_{\mu}(\hat{s})\hat{e}_{\mu}(\hat{s}) \cdot Y^{(\lambda)}_{J,M}(\hat{s}) \qquad (9.4.80)$$

where λ' and λ stand for either of the E- or M-type vector multipole fields.

Similar to the case of scalar multipole propagators, it is useful to separate the outgoing and incoming parts of the vector multipole propagator as follows:

$$\mathcal{G}^{(\lambda,\hat{s})}_{K;J,M}(\mathbf{r}) = e^{iK\hat{s}\cdot\mathbf{r}}\hat{e}_{\mu}(\hat{s}) \cdot Y^{(\lambda)}_{J,M}(\hat{s}) \qquad (9.4.81)$$

$$\left[\mathcal{G}^{(\lambda,\hat{s})}_{K;J',M'}(\mathbf{r})\right]^{\dagger} = \left[Y^{(\lambda)}_{J',M'}(\hat{s})\right]^{\dagger} \cdot \hat{e}_{\mu}(\hat{s})\, e^{-iK\hat{s}\cdot\mathbf{r}} \qquad (9.4.82)$$

which yields

$$\mathbf{G}(\mathbf{r}_2, \lambda', J', M'; \mathbf{r}_1, \lambda, J, M; \omega) = \frac{1}{2\pi}\int_C d\Omega_{\hat{s}}\left[\mathcal{G}^{(\lambda,\hat{s})}_{K;J',M'}(\mathbf{r}_2)\right]^{\dagger}\mathcal{G}^{(\lambda,\hat{s})}_{K;J,M}(\mathbf{r}_1) \qquad (9.4.83)$$

By inverting the radial direction we obtain

$$A_{K,J,M}^{(E,IN)}(-\mathbf{R}) = (-1)^{J+1}A_{K,J,M}^{(E,SC)}(\mathbf{R}) \tag{9.4.84}$$

$$A_{K,J,M}^{(M,IN)}(-\mathbf{R}) = (-1)^{J}A_{K,J,M}^{(M,SC)}(\mathbf{R}) \tag{9.4.85}$$

9.4.4.3. Electric Dipole Propagator

For the electric dipole field, $(\lambda, l) = (E, 1)$, the transform becomes simple because of the angular independence of the vector spherical harmonic,

$$\mathbf{Y}_{1,0,M}(\hat{\mathbf{s}}) = \hat{\mathbf{e}}_M/\sqrt{4\pi} \qquad (M = 0, \pm1) \tag{9.4.86}$$

and we obtain

$$A_{K,1,M}^{(E,SC)}(\mathbf{r}_1) = \sqrt{6\pi} \sum_{\lambda'}^{E,M} \sum_{J',M'} A_{K,J',M'}^{(\lambda',IN)}(\mathbf{r}_2)$$

$$\times \frac{1}{2\pi} \int_C d\Omega_{\hat{s}} \, \mathbf{Y}_{J',M'}^{(\lambda')}(\hat{\mathbf{s}}) \cdot \hat{\mathbf{e}}_M e^{iK\hat{\mathbf{s}} \cdot \mathbf{R}} \tag{9.4.87}$$

$$= \sqrt{6\pi} \sum_{\lambda'}^{E,M} \sum_{J',M'} (-1)^{M'} \left[\hat{\mathbf{e}}_M \cdot A_{K,J',M'}^{(\lambda',SC)}(\mathbf{r}_2 - \mathbf{r}_1) \right]$$

$$\times A_{K,J',M'}^{(\lambda',IN)}(\mathbf{r}_2) \tag{9.4.88}$$

For $\lambda' = E$ and $J' = 1$ this gives the electric dipole propagator as

$$\mathbf{G}(\mathbf{r}_2, 1, M'; \mathbf{r}_1, 1, M; \omega) = \sqrt{6\pi}(-1)^{M'}\hat{\mathbf{e}}_M \cdot A_{K,1,-M'}^{(E,SC)}(\mathbf{r}_2 - \mathbf{r}_1) \tag{9.4.89}$$

$$= \mathbf{G}_{K;M',M}^{(ED)}(\mathbf{r}_2 - \mathbf{r}_1) \qquad (M, M' = 0, \pm1) \tag{9.4.90}$$

Here for further convenience we have defined the electric dipole propagator specified by the initial and final polarizations M and M'.

9.4.5. Angular Spectrum Representation of Cylindrical Field and Propagator

For the case of cylindrical modes we do not need the analytic continuation of the rotation angle to derive their angular spectrum representation.

The integral representation of the vector cylindrical wave in Eq. (9.3.18),

$$\mathbf{J}^{(\mu)}_{K,\alpha,M}(\mathbf{r}) = \frac{1}{2\pi} \int\limits_{-\pi}^{+\pi} d\beta \, \hat{\mathbf{e}}_{\mu}(\hat{s}) e^{iM\beta} e^{iK\hat{s}\cdot\mathbf{r}} \qquad (\mu = TE, TM) \qquad (9.4.91)$$

already gives the angular spectrum representation of the vector Bessel mode function. Then let us consider a spatial parallel displacement of the axis of the cylindrical representation to an arbitrary position. Let the original and new position vectors be $\mathbf{r}_1$ and $\mathbf{r}_2$, respectively, so that the displacement vector is $\mathbf{R} = \mathbf{r}_1 - \mathbf{r}_2$; then displace the coordinates as

$$\mathbf{J}^{(\mu)}_{K,\alpha,M}(\mathbf{r}_1) = \frac{1}{2\pi} \int\limits_{-\pi}^{+\pi} d\beta \, \hat{\mathbf{e}}_{\mu}(\hat{s}) e^{iM\beta} e^{iK\hat{s}\cdot\mathbf{R}} e^{iK\hat{s}\cdot\mathbf{r}_2} \qquad (\mu = TE, TM) \qquad (9.4.92)$$

Substituting the plane wave with respect to $\mathbf{r}_2$ using the well-known formula

$$e^{iK\hat{s}\cdot\mathbf{r}} = \sum_{M=-\infty}^{+\infty} e^{-iM\beta} J_{K,\alpha,M}(\mathbf{r}) \qquad (9.4.93)$$

we obtain the transformation between vector cylindrical modes

$$\mathbf{J}^{(\mu)}_{K,\alpha,M}(\mathbf{r}_1) = \sum_{M'=-\infty}^{+\infty} \frac{1}{2\pi} \int\limits_{-\pi}^{+\pi} d\beta \, \hat{\mathbf{e}}_{\mu}(\hat{s})$$

$$\times e^{i(M-M')\beta} e^{iK\hat{s}\cdot\mathbf{R}} \mathbf{J}^{(\mu)}_{K,\alpha,M'}(\mathbf{r}_2) \qquad (9.4.94)$$

$$= \sum_{M'=-\infty}^{+\infty} J_{K,\alpha,M-M'}(\mathbf{R}) \mathbf{J}^{(\mu)}_{K,\alpha,M'}(\mathbf{r}_2) \qquad (9.4.95)$$

In the above we do not need any evanescent wave since the transformation maintains the spatial translational symmetry in the axial direction.

9.4.6. Transformation between Spherical and Cylindrical Representations

The spherical-to-cylindrical transformation is also obtained by using the angular spectrum representation of vector spherical mode functions:

$$\mathbf{A}_{K,J,M}^{(\lambda,SC)}(\mathbf{r}) = \frac{1}{2\pi} \int_C d\Omega_{\hat{s}} \sum_{\mu}^{TE,TM} \hat{\mathbf{e}}_{\mu}(\hat{s}) \left[\hat{\mathbf{e}}_{\mu}(\hat{s}) \cdot \mathbf{Y}_{J,M}^{(\lambda)}(\hat{s}) \right] e^{iK\hat{s}\cdot\mathbf{r}} \qquad (\lambda = E, M) \quad (9.4.96)$$

For convenience let us use TE and TM rotation matrix elements defined by

$$\mathcal{D}_{J,M}^{(TE)}(\beta, \alpha) \equiv \left(-\frac{1}{\sqrt{2}} \right) \left[\mathcal{D}_{M,+1}^{(J)}(\hat{s}) - \mathcal{D}_{M,-1}^{(J)}(\hat{s}) \right] \qquad (9.4.97)$$

$$\mathcal{D}_{J,M}^{(TM)}(\beta, \alpha) \equiv \left(+\frac{i}{\sqrt{2}} \right) \left[\mathcal{D}_{M,+1}^{(J)}(\hat{s}) + \mathcal{D}_{M,-1}^{(J)}(\hat{s}) \right] \qquad (9.4.98)$$

$$\mathcal{D}_{J,M}^{(\mu)}(\beta, \alpha) \equiv d_{J,M}^{(\mu)}(\alpha) \, e^{-iM\beta} \qquad (9.4.99)$$

where

$$\mathcal{D}_{M,M'}^{(J)}(\hat{s}) \equiv \mathcal{D}_{M,M'}^{(J)}(\beta, \alpha, 0)$$

Using the rotational transformation of the polarization vectors

$$\hat{\mathbf{e}}_{\mu}(\hat{s}) = \sum_{M=-1}^{+1} \mathcal{D}_{1,M}^{(\mu)}(\beta, \alpha)\hat{\mathbf{e}}_M(\hat{z})$$

$$= \sum_{M=-1}^{+1} d_{1,M}^{(\mu)}(\alpha) \, e^{-iM\beta}\hat{\mathbf{e}}_M(\hat{z}) \qquad (\mu = TE, TM) \qquad (9.4.100)$$

we obtain the projections of the vector spherical harmonics

$$\hat{\mathbf{e}}_{TE}(\hat{s}) \cdot \mathbf{Y}_{J,M}^{(E)}(\hat{s}) = \sqrt{\frac{2J+1}{8\pi}} \left[\mathcal{D}_{J,M}^{(TM)}(\beta, \alpha) \right]^*$$

$$= \sqrt{\frac{2J+1}{8\pi}} \left[d_{J,M}^{(TM)}(\alpha) \right]^* e^{iM\beta} \qquad (9.4.101)$$

$$\hat{\mathbf{e}}_{TM}(\hat{s}) \cdot \mathbf{Y}_{J,M}^{(E)}(\hat{s}) = \sqrt{\frac{2J+1}{8\pi}} \left[\mathcal{D}_{J,M}^{(TE)}(\beta, \alpha) \right]^*$$

$$= \sqrt{\frac{2J+1}{8\pi}} \left[d_{J,M}^{(TE)}(\alpha) \right]^* e^{iM\beta} \qquad (9.4.102)$$

With the plane-wave representation of the vector Bessel mode function in Eq. (9.3.19),

$$\mathbf{J}_{K,\alpha,M}^{(\mu)} = \frac{1}{2\pi} \int_{-\pi}^{+\pi} d\beta\, \hat{\mathbf{e}}_{\mu}(\hat{\mathbf{s}}) e^{iM\beta} e^{iK\hat{\mathbf{s}}\cdot\mathbf{r}}$$

we get the spherical–cylindrical transformation as

$$A_{K,J,M}^{(\lambda,SC)}(\mathbf{r}) = \sqrt{\frac{2J+1}{8\pi}} \int_C \sin\alpha\, d\alpha$$

$$\times \left\{ \left[d_{1,M}^{(TM)}(\alpha) \right]^* \mathbf{J}_{K,\alpha,M}^{(TE)}(\mathbf{r}) + \left[d_{1,M}^{(TE)}(\alpha) \right]^* \mathbf{J}_{K,\alpha,M}^{(TM)}(\mathbf{r}) \right\} \qquad (9.4.103)$$

9.4.7. Summary: Representations of Electromagnetic Fields and Transformations between Mode Functions

Here we summarize the formulas for the transformations between the planar, spherical, and cylindrical representations of the electromagnetic mode functions. Figure 3.10 illustrates these relationships as listed in Section 9.4.7.3, items 1–6.

Note that the wavenumber notation $|K; \cdot \rangle$ is omitted in the following state representations.

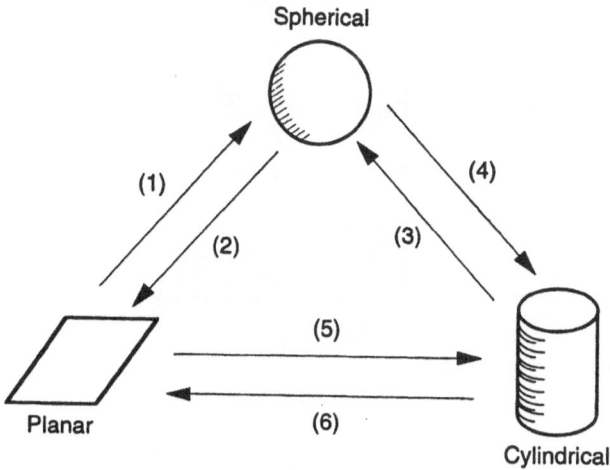

Figure 9.10. Transforms between planar, spherical, and cylindrical vector mode functions. See Section 9.4.7.3 for the respective relationships 1–6.

9.4.7.1. State of Polarization

First we define as follows the polarization states for plane waves with wavevector $\mathbf{K} = K\hat{\mathbf{s}}$ on the basis of the rotational transforms from the z-directed plane wave:

- Transverse magnetic, TM, polarization

$$|TM, \alpha, \beta\rangle = \hat{\mathbf{D}}(\hat{s})|\hat{\mathbf{x}}, 0, 0\rangle \tag{9.4.104}$$

- Transverse electric, TE, polarization

$$|TE, \alpha, \beta\rangle = \hat{\mathbf{D}}(\hat{s})|\hat{\mathbf{y}}, 0, 0\rangle \tag{9.4.105}$$

- Here the rotation operation is

$$\hat{\mathbf{D}}(\hat{s}) = \hat{\mathbf{D}}_z(\beta)\hat{\mathbf{D}}_y(\alpha) \tag{9.4.106}$$

The relations between the rotation angles and the polarizations are shown in Fig. 9.11.

9.4.7.2. State of Vector Field

The states of the planar, spherical, and cylindrical modes are specified as follows:

- Planar modes:

$$|TM, \alpha, \beta\rangle; \qquad \langle\mathbf{r}|TM, \alpha, \beta\rangle = \hat{\mathbf{e}}_{TM}(\hat{s})e^{iK\hat{s}\cdot\mathbf{r}} \tag{9.4.107}$$

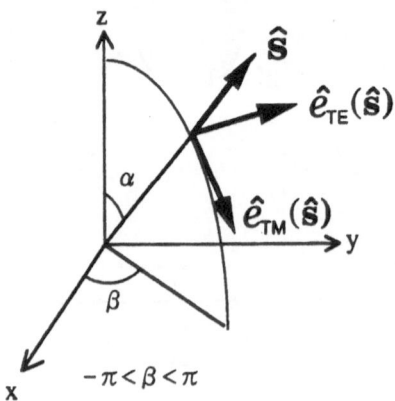

Figure 9.11. Angular coordinate reference system for rotational transforms between vector mode functions.

$$|TE, \alpha, \beta\rangle; \qquad \langle r|TE, \alpha, \beta\rangle = \hat{e}_{TE}(\hat{s})e^{iK\hat{s}\cdot r} \qquad (9.4.108)$$

where TM and TE stand for the transverse magnetic and electric polarizations, respectively.

- Spherical modes:

$$|M, J, M_J\rangle; \qquad \langle r|M, J, M_J\rangle = A_{K,J,M_J}^{(M,SC/IN)}(r) \qquad (9.4.109)$$

$$|E, J, M_J\rangle; \qquad \langle r|E, J, M_J\rangle = A_{K,J,M_J}^{(E,SC/IN)}(r) \qquad (9.4.110)$$

where M and E stand for the magnetic- and electric-type fields, and J and M_J the total angular momentum and its z-projection, respectively.

- Cylindrical modes:

$$|TM, \alpha, M_J\rangle; \qquad \langle r|TM, \alpha, M_J\rangle = J_{K,\alpha,M_J}^{(TM)}(r) \qquad (9.4.111)$$

$$|TE, \alpha, M_J\rangle; \qquad \langle r|TE, \alpha, M_J\rangle = J_{K,\alpha,M_J}^{(TE)}(r) \qquad (9.4.112)$$

where TM and TE stand for the transverse magnetic and electric fields, respectively, and M_J is the angular momentum along the cylinder axis z.

9.4.7.3. Transformations between Representations

The transformations between planar, spherical, and cylindrical vector mode functions are summarized as follows.

1. Planar to spherical modes:

$$A_{K,J,M_J}^{(\lambda,SC)}(r) = \frac{1}{2\pi} \sum_\mu \int_C^{TE,TM} d\Omega_{\hat{s}} \langle \mu, \alpha, \beta|\lambda, J, M_J\rangle \hat{e}_\mu(\hat{s})e^{iK\hat{s}\cdot r} \qquad (\lambda = E, M) \quad (9.4.113)$$

2. Spherical to planar modes:

$$\hat{e}_\mu(\hat{s})e^{iK\hat{s}\cdot r} = 4\pi \sum_\lambda \sum_{J=1}^{M,E\ \infty} \sum_{M_J=-J}^{+J} \langle \lambda, J, M_J|\mu, \alpha, \beta\rangle A_{K,J,M_J}^{(\lambda,IN)}(r) \qquad (\mu = TE, TM)$$

$$(9.4.114)$$

3. Cylindrical to spherical modes:

$$A_{K,J,M_J}^{(\lambda,SC)}(r) = \sum_\mu \int_C^{TE,TM} \sin\alpha\, d\alpha \langle \mu, \alpha, M_J|\lambda, J, M_J\rangle J_{K,\alpha,M_J}^{(\mu)}(r) \qquad (\lambda = E, M)$$

$$(9.4.115)$$

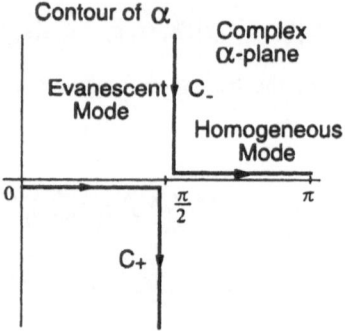

Figure 9.12.   Contour of integration in the complex α plane.

4.   Spherical to cylindrical modes:

$$J^{(\mu)}_{K,\alpha,M_J}(\mathbf{r}) = 4\pi \sum_{\lambda}^{M,E} \sum_{J=1}^{\infty} \langle \lambda, J, M_J | \mu, \alpha, M_J \rangle A^{(\lambda,IN)}_{K,J,M_J}(\mathbf{r}) \qquad (\mu = TE, TM)$$

(9.4.116)

5.   Planar to cylindrical modes:

$$\hat{\mathbf{e}}_\mu(\hat{\mathbf{s}}) e^{i K \hat{\mathbf{s}} \cdot \mathbf{r}} = \sum_{M_J = -\infty}^{+\infty} \langle \mu, \alpha, M_J | \mu, \alpha, \beta \rangle J^{\mu}_{K,\alpha,M_J}(\mathbf{r}) \qquad (\mu = TE, TM)$$

(9.4.117)

6.   Cylindrical to planar modes:

$$J^{(\mu)}_{K,\alpha,M_J}(\mathbf{r}) = \frac{1}{2\pi} \int_{-\pi}^{-\pi} d\beta \, \langle \mu, \alpha, \beta | \mu, \alpha, M_J \rangle \hat{\mathbf{e}}_\mu(\hat{\mathbf{s}}) e^{i K \hat{\mathbf{s}} \cdot \mathbf{r}} \qquad (\mu = TE, TM)$$

(9.4.118)

The contour of integration is indicated in Fig. 9.12.

9.4.7.4.   Coefficients of the Transformations

The coefficients of the transformations are given as follows:
• Coefficients of the transformations between planar and spherical modes:

$$\langle M, J, M_J | TE, \alpha, \beta \rangle = -i \sqrt{\frac{2J+1}{8\pi}} \, \mathcal{D}^{(TE)}_{J,M_J}(\beta, \alpha)$$

(9.4.119)

$$\langle E, J, M_J | TE, \alpha, \beta \rangle = \sqrt{\frac{2J+1}{8\pi}} \; \mathcal{D}_{J,M_J}^{(TM)}(\beta, \alpha) \qquad (9.4.120)$$

$$\langle M, J, M_J | TM, \alpha, \beta \rangle = -i \langle E, J, M_J | TE, \alpha, \beta \rangle \qquad (9.4.121)$$

$$\langle E, J, M_J | TM, \alpha, \beta \rangle = \langle M, J, M_J | TE, \alpha, \beta \rangle \qquad (9.4.122)$$

- Coefficients of transformations between spherical and cylindrical modes:

$$\langle M, J, M_J | TE, \alpha, M_J \rangle = -i \sqrt{\frac{2J+1}{8\pi}} \; d_{J,M_J}^{(TE)}(\alpha) \qquad (9.4.123)$$

$$\langle E, J, M_J | TE, \alpha, M_J \rangle = \sqrt{\frac{2J+1}{8\pi}} \; d_{J,M_J}^{(TE)}(\alpha) \qquad (9.4.124)$$

$$\langle M, J, M_J | TM, \alpha, M_J \rangle = -i \langle E, J, M_J | TE, \alpha, M_J \rangle \qquad (9.4.125)$$

$$\langle E, J, M_J | TM, \alpha, M_J \rangle = \langle M, J, M_J | TE, \alpha, M_J \rangle \qquad (9.4.126)$$

- Coefficients of transformations between planar and cylindrical modes:

$$\langle TM, \alpha, M_J | TM, \alpha, \beta \rangle = e^{-iM\beta} \qquad (9.4.127)$$

$$\langle TE, \alpha, M_J | TE, \alpha, \beta \rangle = e^{-iM\beta} \qquad (9.4.128)$$

The reverse transformations are obtained by using the fact that

$$\langle \xi | \chi \rangle = [\langle \chi | \xi \rangle]^* \qquad (9.4.129)$$

9.5. NEAR-FIELD INTERACTION OF DIELECTRIC SPHERES NEAR A PLANAR DIELECTRIC SURFACE

In this section, we study the electromagnetic interactions of dielectric spheres placed in a subwavelength vicinity of a planar dielectric surface. A system consisting of two subwavelength-size dielectric spheres simulates a basic sample–probe-tip NOM system. Illuminated by evanescent waves of Fresnel at total internal reflection, the sphere–sphere interaction produces via scattered and backscattered optical fields outgoing light waves extending to a photodetector. As discussed in Section 9.1, the NOM process also involves near- to far-field coupling and waveguiding schemes. In this section we concentrate on the near-field interaction

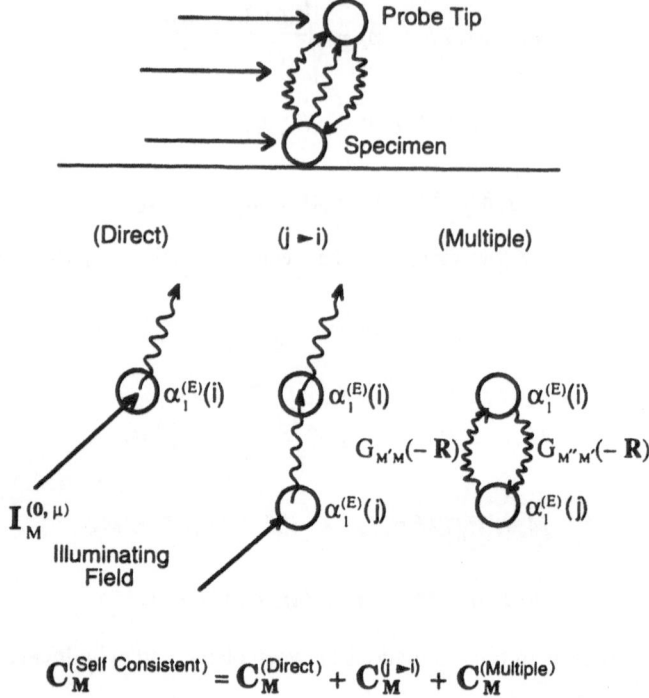

$$C_M^{(Self\ Consistent)} = C_M^{(Direct)} + C_M^{(j\,\blacktriangleright\,i)} + C_M^{(Multiple)}$$

Figure 9.13. Fundamental processes involved in the angular spectrum representation of the sphere–sphere interaction.

part. We demonstrate the localization of the near-field interaction and the origin of the high spatial frequency beyond the diffraction limit.

The theoretical description is based on the vector multipolar expansion and angular spectrum representation of the electromagnetic field in the near-field regime developed in the previous sections. We will examine the mathematical background and physical meaning of the general theoretical treatment of optical near-field phenomena.

9.5.1. Sample–Probe Interaction at a Dielectric Surface

Consider the simplified model of the NOM system shown in Fig. 9.13, in which a sample object consisting of a subwavelength-size dielectric sphere is put on a planar dielectric surface and is illuminated by an evanescent wave of Fresnel, and a probe tip in the form of a subwavelength-size dielectric sphere is scanned in a plane parallel to the surface above the sample.

Let us describe the electromagnetic interaction in the sample–probe system under the influence of a dielectric surface which acts both as a source of an illuminating evanescent wave and a reflector of the scattered field. This system involves two spherical objects placed at different positions and a planar boundary. The entire system has no specific symmetry for a general sample–probe position, so that it is not practical to try to find a unified basis set of mode functions which meets the entire boundary conditions of the system. Instead, we can trace the interaction processes one by one by using convenient mode functions which satisfy the boundary conditions of each subsystem and transform them into another convenient basis to evaluate the next step. In so doing we can extract the fundamental processes involved in the NOM process and understand their physical meanings.

The planar dielectric interface has a restricted translational symmetry in the direction parallel to the surface. When we expand the electromagnetic field associated with the dielectric interface in terms of plane waves in three dimensions, evanescent waves appear with complex propagation vectors. A useful mode description is provided by Carniglia and Mandel, which also serves as the basis of quantized evanescent electromagnetic waves.

Vector multipole functions serve as a convenient basis for the theoretical description of the scattered field from dielectric spheres. Provided that both incident and scattered electromagnetic fields are described in terms of vector multipoles, the scattering process by a sphere is described simply by the dielectric response function or polarizability of the sphere. Therefore, the vector multipole expansion of plane waves, including both homogeneous and evanescent waves, provides a useful basis for describing electromagnetic interactions in the optical near-field of a planar dielectric interface.

The interaction between dielectric spheres is described by the spatial displacement of vector multipole fields. Plane waves provide a good basis for describing the spatial displacement, and the angular spectrum representation can serve for translating the origin of the vector multipole expansion to another, arbitrary point. That is, the scattered field of the sample sphere is transformed into an incident field on the probing sphere via the angular spectrum representation. The inhomogeneous waves in the angular spectrum play dominant roles in the sample–probe interaction in the near-field regime.

The scattering process of incident vector multipole fields can be calculated according to Mie's theory, and the near-field intensity distribution and scattered intensity are evaluated in order to estimate the NOM signal. It is also very important to consider the polarization dependence of the signal collecting scheme such as an optical waveguide attached to the probe tip.

The scattered electromagnetic fields into the lower half-space both from the sample and probe spheres give rise to reflected waves, which in turn describe the multiple interaction between the sample and probe spheres as well as that between the spheres and the dielectric surface. The angular spectrum representation also indicates the dominant channel of the interaction process as well as the lateral localization since it is described in terms of the superposition of angular spectrum components near the dominant channel.

To summarize the theoretical treatment in the following, we first list the mathematical procedures:

1. Expansion of illuminating evanescent waves into vector multipoles.
2. Evaluation of the scattered field of the sample sphere in terms of vector multipole fields.

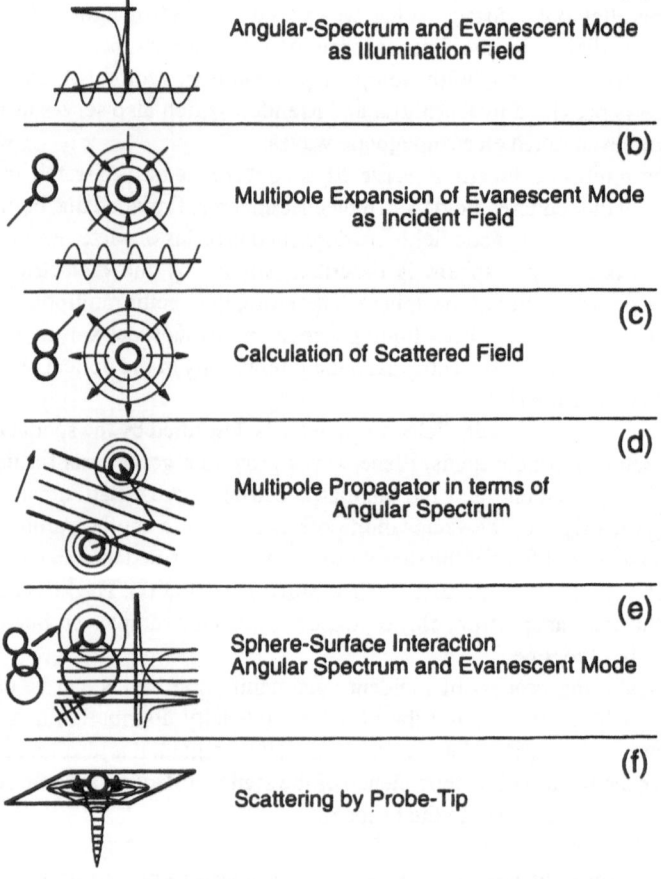

(a) Angular-Spectrum and Evanescent Mode as Illumination Field

(b) Multipole Expansion of Evanescent Mode as Incident Field

(c) Calculation of Scattered Field

(d) Multipole Propagator in terms of Angular Spectrum

(e) Sphere-Surface Interaction Angular Spectrum and Evanescent Mode

(f) Scattering by Probe-Tip

Figure 9.14. Summary of operations used in describing the fundamental processes involved in sample–probe-tip interaction.

3. Expansion of the scattered vector multipole fields of the sample sphere into the angular spectrum.

4. Spatial displacement of the field representation from sample to probe spheres in terms of the angular spectrum representation in the upper half-space of the sample sphere (the half-space where the probe sphere is placed).

5. Evaluation of the reflection of the scattered field by the dielectric surface in terms of the angular spectrum representation in the lower half-space of the sample sphere (the half-space where the dielectric surface is placed).

6. Expansion of the displaced plane waves in the angular spectrum representation into vector multipoles.

7. Evaluation of the scattered field of the probe sphere in terms of vector multipole fields.

These procedures are summarized in Fig. 9.14.

Before entering into a detailed description of the NOM process, let us first establish a convenient mathematical basis for expanding the electromagnetic field.

9.5.2. Mode Description of Evanescent Waves of Fresnel

First let us study appropriate mode functions for planar dielectric boundaries. Instead of starting from Fresnel's relation for incident, reflected, and transmitted electromagnetic waves, it is convenient to use mode functions formulated by Carniglia and Mandel as the basis for the quantization of evanescent waves [6].

Consider a planar boundary; to its left (L) $(z < 0)$ is a homogeneous dielectric medium of refractive index n, and to its right (R) $(z > 0)$ is a vacuum. Assume that the boundary surface lies in the x–y plane with $z < 0$ corresponding to the dielectric side (Fig. 9.15). Following Carniglia and Mandel, the L-mode and R-mode are defined, respectively, according to whether the wave is incident from the left (dielectric side, $z < 0$) or right (vacuum side, $z > 0$) of the boundary [6]. Hereafter we refer to these modes as CM modes for short.

Let the CM mode functions for the L and R modes be $\mathbf{F}_L(k, s, R)$ and $\mathbf{F}_R(K, s, R)$, respectively. Here s stands for electromagnetic polarization $s = TE$ or TM. (In the original paper they use $s = 1$ for the TE mode and $s = 2$ for the TM mode.)

Here we are interested only in the CM modes on the vacuum side $(z > 0)$:

$$\mathbf{F}_L(k, TE, \mathbf{R}) = \frac{1}{\sqrt{2}n} \hat{\varepsilon} \frac{2k_3}{k_3 + K_3} e^{i\mathbf{K}\cdot\mathbf{R}} \tag{9.5.1}$$

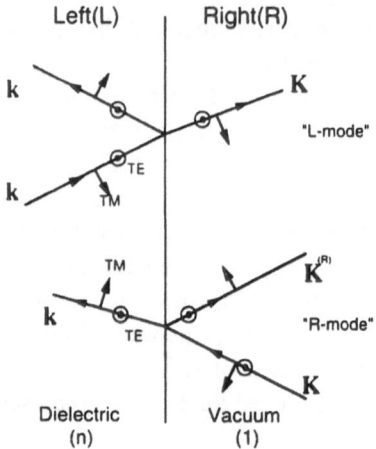

Figure 9.15.   Configurations of the evanescent modes as formulated by Carniglia and Mandel.

$$\mathbf{F}_L(k, TM, \mathbf{R}) = -\frac{1}{\sqrt{2}}\hat{\mathbf{c}}\times\hat{\varepsilon}\,\frac{2k_3}{k_3+n^2K_3}\,e^{i\mathbf{K}\cdot\mathbf{R}} \tag{9.5.2}$$

$$\mathbf{F}_R(K, TE, \mathbf{R}) = \frac{1}{\sqrt{2}}\hat{\varepsilon}e^{i\mathbf{K}\cdot\mathbf{R}} + \frac{1}{\sqrt{2}}\hat{\varepsilon}\,\frac{K_3-k_3}{K_3+k_3}\,e^{i\mathbf{K}^{(R)}\cdot\mathbf{R}} \tag{9.5.3}$$

$$\mathbf{F}_R(K, TM, \mathbf{R}) = -\frac{1}{\sqrt{2}}\hat{\mathbf{c}}\times\hat{\varepsilon}e^{i\mathbf{K}\cdot\mathbf{r}} - \frac{1}{\sqrt{2}}\hat{\mathbf{c}}^{(R)}\times\hat{\varepsilon}\,\frac{n^2K_3-k_3}{n^2K_3+k_3}\,e^{i\mathbf{K}^{(R)}\cdot\mathbf{R}} \tag{9.5.4}$$

where $\hat{\varepsilon}$ is the polarization vector of the TE mode, $\mathbf{k}$ and $\mathbf{K}$ are the wavevectors in the dielectric and vacuum, respectively, $\mathbf{K}^{(R)}$ is the wavevector of the reflected wave, in which the z component in $\mathbf{K}$ is reversed in sign, and k_3 and K_3 are the z components of the respective wavevectors.

Note that the L mode includes only the transmitted wave, whereas the R mode involves both the incident and reflected waves, as shown in Fig. 9.15.

9.5.3.   Multipolar Representation of Evanescent Modes

In describing the scattering of evanescent waves by dielectric spheres, it is convenient to represent the evanescent waves in terms of spherical waves. Evanescent waves provide a good basis for describing the illuminating field, the interaction between spheres, and the reflection of the scattered field from the dielectric substrate. Since there is no spherical symmetry at the planar dielectric–vacuum interface, we need to resort to analytic continuation with respect to the angular coordinate in the spherical representation of the evanescent waves.

Let us start with a simple geometrical understanding of the vector nature of evanescent waves with the help of quantum mechanical expressions for the states of the electromagnetic field.

9.5.3.1. Vector Nature of Evanescent Waves

As formulated in Section 9.3.1, a vector plane-wave state of an arbitrary wavevector $\mathbf{K} = K\hat{\mathbf{s}}$ is generated by rotational transform from the state $|\mathbf{K}; \hat{\mathbf{e}}_{\pm 1}\rangle$ as

$$|\mathbf{K}; \hat{\mathbf{e}}(\hat{\mathbf{s}})\rangle = \hat{\mathbf{D}}(\varphi, \theta, 0)|K\hat{\mathbf{z}}; \hat{\mathbf{e}}_{\pm 1}\rangle \tag{9.5.5}$$

and its multipole representation is given by

$$\langle \mathbf{r}|\mathbf{K}; \hat{\mathbf{e}}(\hat{\mathbf{s}})\rangle = \frac{1}{\sqrt{(2\pi)^3}} \sum_{J,M} \mathcal{D}_{M,\pm 1}^{(J)}(\varphi, \theta, 0) \sqrt{2\pi} \sqrt{2J+1} \left[\mathbf{A}_{K,J,M}^{(E,IN)}(\hat{\mathbf{r}}) - (\pm 1)\mathbf{A}_{K,J,M}^{(M,IN)}(\hat{\mathbf{r}}) \right] \tag{9.5.6}$$

That is, an arbitrary state of a vector plane wave corresponding to right/left circular polarization with respect to $\hat{\mathbf{s}}$ can be produced simply by a rotational transformation of the z axis in the direction of the wavevector $\mathbf{K}$. Here $\hat{\mathbf{e}}$ without the $(\hat{\mathbf{s}})$ stands for the polarization vector with respect to the absolute axis z, as $\hat{\mathbf{e}}_M = \hat{\mathbf{e}}_M(\hat{\mathbf{z}})$, whereas $\hat{\mathbf{e}}(\hat{\mathbf{s}})$ stands for the polarization vector with respect to an arbitrary directional unit vector $\hat{\mathbf{s}}$.

In describing evanescent waves, however, a set of linearly polarized states is usually employed. The x-polarized and y-polarized plane waves $|K\hat{\mathbf{z}}; \hat{\mathbf{x}}\rangle$ and $|K\hat{\mathbf{z}}; \hat{\mathbf{y}}\rangle$, respectively, are described in terms of circular polarizations defined by

$$\hat{\mathbf{x}} = -\frac{1}{\sqrt{2}} (\hat{\mathbf{e}}_{+1} - \hat{\mathbf{e}}_{-1}) \tag{9.5.7}$$

$$\hat{\mathbf{y}} = +\frac{i}{\sqrt{2}} (\hat{\mathbf{e}}_{+1} + \hat{\mathbf{e}}_{-1}) \tag{9.5.8}$$

Then the rotational transform produces the orthogonal set of φ and θ polarized states as

$$\langle \mathbf{r}|\mathbf{K}; \hat{\mathbf{e}}_{\varphi}(\hat{\mathbf{s}})\rangle = \langle \mathbf{r}|\hat{\mathbf{D}}(\varphi, \theta, 0)|K\hat{\mathbf{z}}; \hat{\mathbf{y}}\rangle \tag{9.5.9}$$

$$\langle \mathbf{r}|\mathbf{K}; \hat{\mathbf{e}}_{\theta}(\hat{\mathbf{s}})\rangle = \langle \mathbf{r}|\hat{\mathbf{D}}(\varphi, \theta, 0)|K\hat{\mathbf{z}}; \hat{\mathbf{x}}\rangle \tag{9.5.10}$$

The explicit forms of these are written, respectively, as follows:

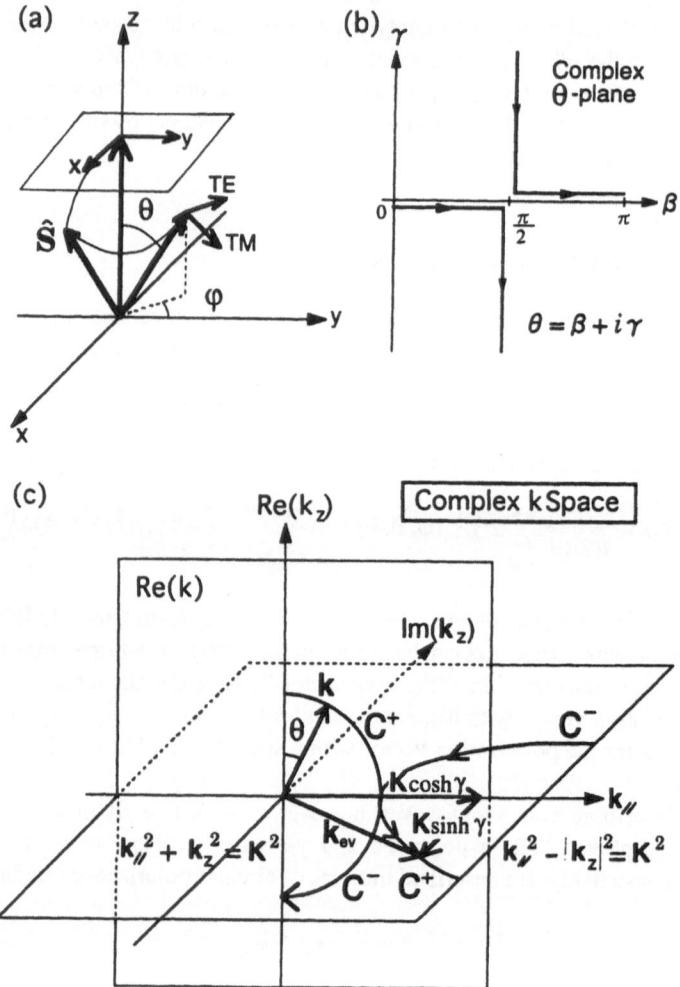

Figure 9.16. Evanescent waves described as the rotational transform of vector plane waves with complex rotation angle $\theta = \beta + i\gamma$. Contour of θ and corresponding dispersion relation in complex k space.

$$\langle \mathbf{r}|\mathbf{K}; \hat{\mathbf{e}}_\varphi(\hat{\mathbf{s}})\rangle = \frac{1}{\sqrt{(2\pi)^3}}\, \hat{\mathbf{e}}_\varphi(\hat{\mathbf{s}}) e^{i\mathbf{K}\cdot\mathbf{r}}$$

$$= +\frac{i}{\sqrt{2}}\Big[\langle \mathbf{r}|\hat{\mathbf{D}}(\varphi, \theta, 0)|K\hat{\mathbf{z}}; \hat{\mathbf{e}}_{+1}\rangle$$

$$+ \ \langle \mathbf{r}|\hat{\mathbf{D}}(\varphi, \theta, 0)|K\hat{\mathbf{z}}; \hat{\mathbf{e}}_{-1}\rangle \Big] \tag{9.5.11}$$

$$\langle \mathbf{r}|\mathbf{K}; \hat{\mathbf{e}}_{\theta}(\hat{\mathbf{s}})\rangle = \frac{1}{\sqrt{(2\pi)^3}} \hat{\mathbf{e}}_{\theta}(\hat{\mathbf{s}})e^{i\mathbf{K}\cdot\mathbf{r}}$$

$$= -\frac{1}{\sqrt{2}}\Big[\langle \mathbf{r}|\hat{\mathbf{D}}(\varphi, \theta, 0)|K\hat{\mathbf{z}}; \hat{\mathbf{e}}_{+1}\rangle$$

$$- \ \langle \mathbf{r}|\hat{\mathbf{D}}(\varphi, \theta, 0)|K\hat{\mathbf{z}}; \hat{\mathbf{e}}_{-1}\rangle \Big] \tag{9.5.12}$$

Provided that the rotation angle lies in the real space, $-\pi \le \varphi \le +\pi$ and $0 \le \theta \le \pi/2$, these φ and θ polarized states represent propagating (homogeneous) plane waves with linear polarizations orthogonal to each other, as shown in Fig. 9.16a. When the rotation angle θ is greater than $\pi/2$ and is analytically continued into the complex region as $\theta = \beta + i\gamma$, these φ and θ states represent, respectively, the TE and TM polarized evanescent waves (Figs. 9.16b and 9.16c). In the entire space of the rotation angle, plane waves are described as follows:

$$\theta = \beta + i\gamma \tag{9.5.13}$$

$$\text{homogeneous:} \quad 0 \le \beta \le \pi/2, \quad \gamma = 0 \tag{9.5.14}$$

$$\text{inhomogeneous:} \quad \beta = \pi/2, \quad -\ln\left(n^2 - 1 + \sqrt{n^2 - 1}\ \right) \le \gamma \le 0 \tag{9.5.15}$$

where n is the refractive index of the dielectric in the left half-space of the interface. The contour of θ and the entire behavior of the complex k are shown in Figs. 9.16b and 9.16c, respectively. These representations provide a simple geometrical understanding and convenient description of the vector nature of evanescent waves as well as the basis for expanding evanescent waves into vector multipoles. It should be noted that the rotational transform is a nonunitary operation when the angle θ lies in a complex space with nonzero imaginary part.

9.5.3.2. Multipole Representation of Evanescent Waves

Let us list the rotational transforms and multipole expansions of plane waves described in terms of the CM modes. Mode functions are written only for the outer half-space of the dielectrics, i.e., in the $z > 0$ half-space, which corresponds to our study of optical near-field processes for subwavelength-size dielectric spheres. These half-space mode functions are naturally continued to those on the dielectric side by Fresnel's relations or boundary conditions (actually, the CM modes are defined in the entire space).

- For the $L\text{-}TE$ mode (TE wave incident from the dielectric side):

$$\frac{1}{\sqrt{(2\pi)^3}} \hat{\varepsilon} e^{i\mathbf{K}\cdot\mathbf{r}} = \langle \mathbf{r} | \hat{\mathbf{D}}(\varphi, \theta, 0) | K\hat{\mathbf{z}}; \hat{\mathbf{y}} \rangle \tag{9.5.16}$$

- For the L-TM mode (TM wave incident from the dielectric side):

$$\frac{1}{\sqrt{(2\pi)^3}} \{ -\hat{\mathbf{c}} \times \hat{\varepsilon} \} e^{i\mathbf{K}\cdot\mathbf{r}} = \langle \mathbf{r} | \hat{\mathbf{D}}(\varphi, \theta, 0) | K\hat{\mathbf{z}}; \hat{\mathbf{x}} \rangle \tag{9.5.17}$$

- For the R-TE mode (TE wave incident from the vacuum side):

$$\frac{1}{\sqrt{(2\pi)^3}} \hat{\varepsilon} e^{i\mathbf{K}\cdot\mathbf{r}} = \langle \mathbf{r} | \hat{\mathbf{D}}(\varphi, \pi - \theta, 0) | K\hat{\mathbf{z}}; \hat{\mathbf{y}} \rangle \tag{9.5.18}$$

$$\frac{1}{\sqrt{(2\pi)^3}} \hat{\varepsilon} e^{i\mathbf{K}^{(R)}\cdot\mathbf{r}} = \langle \mathbf{r} | \hat{\mathbf{D}}(\varphi, \theta, 0) | K\hat{\mathbf{z}}; \hat{\mathbf{y}} \rangle \tag{9.5.19}$$

- For the R-TM mode (TM wave incident from the vacuum side):

$$\frac{1}{\sqrt{(2\pi)^3}} \{ -\hat{\mathbf{c}} \times \hat{\varepsilon} \} e^{i\mathbf{K}\cdot\mathbf{r}} = \langle \mathbf{r} | \hat{\mathbf{D}}(\varphi, \pi - \theta, 0) | K\hat{\mathbf{z}}; \hat{\mathbf{x}} \rangle \tag{9.5.20}$$

$$\frac{1}{\sqrt{(2\pi)^3}} \{ -\hat{\mathbf{c}}^{(R)} \times \hat{\varepsilon} \} e^{i\mathbf{K}^{(R)}\cdot\mathbf{r}} = \langle \mathbf{r} | \hat{\mathbf{D}}(\varphi, \theta, 0) | K\hat{\mathbf{z}}; \hat{\mathbf{x}} \rangle \tag{9.5.21}$$

Note that only the L modes involve the inhomogeneous component and give rise to evanescent waves.

Using the complex angle θ, we can express the wavenumbers K_3 and k_3 as

$$K_3 = -K \cos\theta$$

$$k_3 = nK \cos\theta_i \tag{9.5.22}$$

for the L modes and

$$K_3 = -K \cos\theta$$

$$k_3 = nK \cos\theta_i \tag{9.5.23}$$

for the R modes.

The explicit form of the CM modes is described in terms of the rotational matrix elements and the complex vector multipole potentials $\mathbf{A}_{K,J,M}^{(E,IN)}(r)$ and $\mathbf{A}_{K,J,M}^{(M,IN)}(r)$ as follows. Let us employ a new coordinate system; $\mathbf{R}$ is the center of spherical expansion, $\mathbf{r}$ is the local position vector with respect to $\mathbf{R}$, and $\mathbf{R}_e = \mathbf{R} + \mathbf{r}$ is the position vector with respect to the origin of the system. For further convenience let us utilize a set of composed rotation matrix elements $\mathcal{D}_{J,M}^{(TE)}(\varphi, \theta)$ and $\mathcal{D}_{J,M}^{(TM)}(\varphi, \theta)$ defined in Eqs. (9.4.97) and (9.4.98):

$$\mathcal{D}_{J,M}^{(TE)}(\varphi, \theta) \equiv -\frac{1}{\sqrt{2}}\left\{\mathcal{D}_{M,+1}^{(J)}(\varphi, \theta, 0) - \mathcal{D}_{M,-1}^{(J)}(\varphi, \theta, 0)\right\}$$

$$\mathcal{D}_{J,M}^{(TM)}(\varphi, \theta) \equiv +\frac{i}{\sqrt{2}}\left\{\mathcal{D}_{M,+1}^{(J)}(\varphi, \theta, 0) + \mathcal{D}_{M,-1}^{(J)}(\varphi, \theta, 0)\right\}$$

The multipole representations of the CM modes are given by

$$\mathbf{F}_L(k, TE, \mathbf{R}_e)$$

$$= \frac{\sqrt{2}\,\cos\theta_i}{n\cos\theta_i + \cos\beta\cosh\gamma - i\sin\beta\sinh\gamma}\,e^{Kr\sin\beta\sinh\gamma + iKr\cos\beta\cosh\gamma}$$

$$\times \sum_{J=1}^{\infty}\sum_{M}\sqrt{2\pi}\,\sqrt{2J+1}$$

$$\times\left[\mathcal{D}_{J,M}^{(TM)}(\varphi, \beta + i\gamma, 0)\mathbf{A}_{K,J,M}^{(E,IN)}(\mathbf{r}) - i\mathcal{D}_{J,M}^{(TE)}(\varphi, \beta + i\gamma, 0)\mathbf{A}_{K,J,M}^{(M,IN)}(\mathbf{r})\right] \quad (9.5.24)$$

$$\mathbf{F}_L(k, TM, \mathbf{R}_e)$$

$$= \frac{\sqrt{2}\,\cos\theta_i}{\cos\theta_i + n\cos\beta\cosh\gamma - in\sin\beta\sinh\gamma}\,e^{Kr\sin\beta\sinh\gamma + iKr\cos\beta\cosh\gamma}$$

$$\times \sum_{J=1}^{\infty}\sum_{M}\sqrt{2\pi}\,\sqrt{2J+1}$$

$$\times\left[\mathcal{D}_{J,M}^{(TE)}\varphi, \beta + i\gamma, 0)\mathbf{A}_{K,J,M}^{(E,IN)}(\mathbf{r}) - i\mathcal{D}_{J,M}^{(TM)}(\varphi, \beta + i\gamma, 0)\mathbf{A}_{K,J,M}^{(M,IN)}(\mathbf{r})\right] \quad (9.5.25)$$

$$\mathbf{F}_R(K, TE, \mathbf{R}_e)$$

$$= \frac{1}{\sqrt{2}}\,e^{-iKr\cos\beta}\sum_{J=1}^{\infty}\sum_{M}\sqrt{2\pi}\,\sqrt{2J+1}$$

$$\times\left[\mathcal{D}_{J,M}^{(TM)}(\varphi, \pi - \beta, 0)\mathbf{A}_{K,J,M}^{(E,IN)}(\mathbf{r}) + \mathcal{D}_{J,M}^{(TE)}(\varphi, \pi - \beta, 0)\mathbf{A}_{K,J,M}^{(M,IN)}(\mathbf{r})\right]$$

$$+\frac{1}{\sqrt{2}}\,\frac{\cos\beta - n\cos\theta_i}{\cos\beta + n\cos\theta_i}\,e^{+iKr\cos\beta}\sum_{J=1}^{\infty}\sum_{M}\sqrt{2\pi}\,\sqrt{2J+1}$$

$$\times\left[\mathcal{D}_{J,M}^{(TM)}(\varphi, \beta, 0)\mathbf{A}_{K,J,M}^{(E,IN)}(\mathbf{r}) - i\mathcal{D}_{J,M}^{(TE)}(\varphi, \beta, 0)\mathbf{A}_{K,J,M}^{(M,IN)}(\mathbf{r})\right] \quad (9.5.26)$$

$$\mathbf{F}_R(K, TM, \mathbf{R}_e)$$

$$= \frac{1}{\sqrt{2}} e^{-iKr\cos\beta} \sum_{J=1}^{\infty} \sum_M \sqrt{2\pi} \sqrt{2J+1}$$

$$\times \left[-i\mathcal{D}_{J,M}^{(TE)}(\varphi, \pi - \beta, 0)\mathbf{A}_{K,J,M}^{(E,IN)}(\mathbf{r}) + \mathcal{D}_{J,M}^{(TM)}(\varphi, \pi - \beta, 0)\mathbf{A}_{K,J,M}^{(M,IN)}(\mathbf{r}) \right]$$

$$+ \frac{1}{\sqrt{2}} \frac{n\cos\beta - \cos\theta_i}{n\cos\beta + \cos\theta_i} e^{+iKr\cos\beta} \sum_{J=1}^{\infty} \sum_M \sqrt{2\pi} \sqrt{2J+1}$$

$$\times \left[-\mathcal{D}_{J,M}^{(TE)}(\varphi, \beta, 0)\mathbf{A}_{K,J,M}^{(E,IN)}(\mathbf{r}) + \mathcal{D}_{J,M}^{(TM)}(\varphi, \beta, 0)\mathbf{A}_{K,J,M}^{(M,IN)}(\mathbf{r}) \right] \quad (9.5.27)$$

Here the R modes involve both the incident and reflected homogeneous waves.

Figure 9.17 shows the angular distribution of the dipole component involved in the spherical representation of the CM L mode as a function of the incident angle of the TM wave. One can see how the shape of the field patterns for TE incidence does not depend on the incident angle of either the transmitted or evanescent waves. In contrast, for TM incidence the field patterns depend on the angle of incidence,

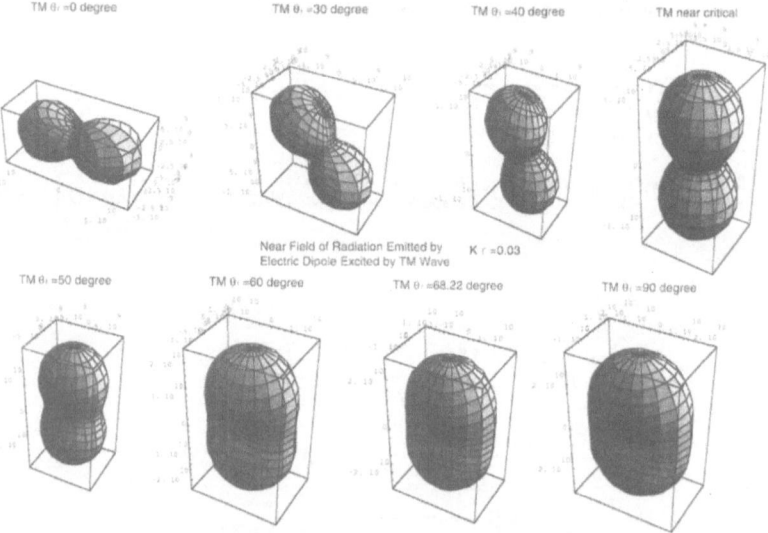

Figure 9.17. Near-field radial intensity distributions corresponding to the dipole components of the CM L mode. Only those for incident TM light are displayed. Those for TE incidence are not different in shape from that for $\theta_i = 0$.

and especially for the case of total internal reflection they exhibit quadrupole-like features.

9.5.4. Near-Field Interaction of Dielectric Spheres at a Planar Dielectric Surface

Based on the theoretical background in the previous sections, let us study a simplified model of the near-field optical microscope (NOM) in terms of a light scattering problem by two small dielectric spheres placed in the near-field of a planar dielectric surface. Consider that one of the spheres is fixed on the surface as a subwavelength-size specimen and the other is the probe tip scanned in a plane parallel to the planar surface. This corresponds to the constant-height operation of NOM in the collection mode. The model calculation is also applicable for illumination-mode NOM.

This problem includes the multiple scattering process between the dielectric spheres as well as the near-field interaction of the spheres with the substrate surface. We address the localization and penetration depth of the near-field interaction, which provide a measure of the resolution of the NOM image. We describe theoretically the interaction propagator and reflectivity tensor on the basis of the angular spectrum representation. For example, by analyzing the dominant spatial frequency and spectral width in the angular spectrum for the dipole propagator we can evaluate the origin of the high spatial frequency and localization relevant to the NOM process. It is shown in the near-field limit that the interaction is dominated by the angular spectrum component with a central frequency and spectral width approximately equal to the inverse of the sphere–sphere and sphere–surface distances.

We discuss in the following subsections (1) the scattering of an evanescent field by a dielectric sphere, (2) the interaction between spheres, and (3) the interaction of a sphere with a dielectric surface. This provides a physical understanding of the fundamental NOM process which cannot be obtained with an elaborate numerical approach alone based on a global theoretical treatment as usually employed in NOM theories.

9.5.4.1. Scattering of an Electric Dipole Field by a Planar Dielectric Surface in the Near-Field Regime

The sphere–surface configuration and coordinate system employed are shown in Fig. 9.18. Assume that d_1 and d_2 are the position vectors directed at the centers of the sample and probe spheres with radii a_1 and a_2 and bulk refractive indices n_1 and n_2, respectively. The origin of the coordinates is on the dielectric surface just

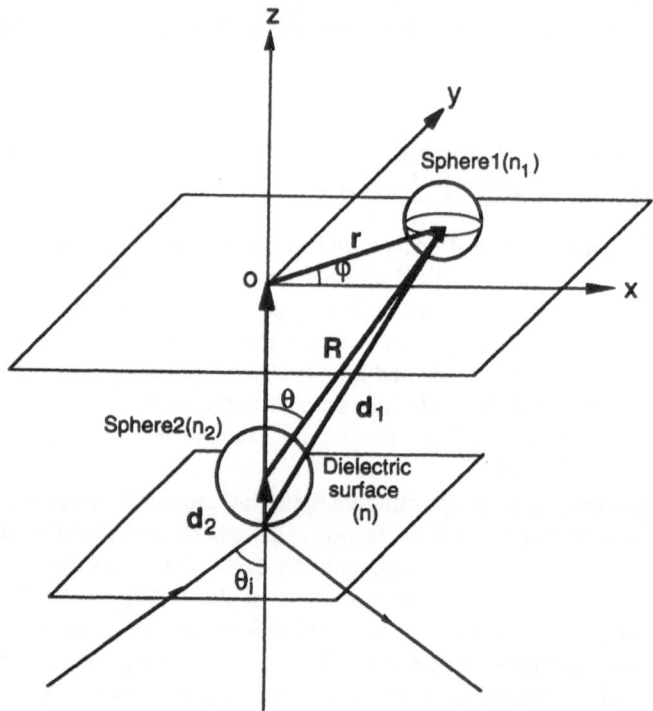

Figure 9.18. Coordinate system used for the evaluation of the sample–probe interaction of the NOM. Dielectric sphere 2 (sample) is placed on a planar dielectric surface and is illuminated by an evanescent wave. Another dielectric sphere 1 (probe) is scanned in a plane above sphere 2.

below the sample sphere. Let the specimen–probe-tip displacement be $\mathbf{R} = \mathbf{d}_1 - \mathbf{d}_2$ and consider the system as illuminated by an evanescent wave of Fresnel.

Let $\tilde{\mathbf{A}}_K^{(SC)}(\mathbf{r})$ be the vector potential corresponding to the scattered field from one of the dielectric spheres described in terms of a temporal Fourier basis with monochromatic frequency $\omega = Kc$. We address the near-field regime characterized by the size parameter $Ka \ll 1$, $K|\mathbf{d}| \ll 1$, $|K\mathbf{R}| \ll 1$, and assume a locally homogeneous dielectric response. In this case the electric dipole interaction dominates the scattering process, as discussed in the previous section, so that the scattered field is approximated by the superposition of complex vector electric dipole potentials $\mathbf{A}_{K,J=1,M}^{(E,SC)}(\mathbf{r})$ as

$$\tilde{\mathbf{A}}_K^{(SC)}(\mathbf{r}) \equiv \sum_{M=0,\pm 1} C_M \mathbf{A}_{K,1,M}^{(E,SC)}(\mathbf{r}) \tag{9.5.28}$$

where $A_{K,1,M}^{(E,SC)}(\mathbf{r})$ is defined in terms of vector spherical harmonics $\mathbf{Y}_{J,l,M}(\hat{\mathbf{r}})$ with directional unit vector $\hat{\mathbf{r}} = \mathbf{r}/r$ and spherical Hankel function $h_l^{(1)}(Kr)$:

$$A_{K,1,M}^{(E,SC)}(\mathbf{r}) = -\sqrt{\frac{1}{3}}\, h_2^{(1)}(Kr)\mathbf{Y}_{1,2,M}(\hat{\mathbf{r}}) + \sqrt{\frac{2}{3}}\, h_0^{(1)}(Kr)\mathbf{Y}_{1,0,M}(\hat{\mathbf{r}}) \quad (9.5.29)$$

On this basis, the electric dipole moment induced on the sphere is written as

$$\mathbf{p} \equiv \sum_{M=0,\pm1} C_M \hat{\mathbf{e}}_M \quad (9.5.30)$$

The dielectric response of sphere i of radius a_i with refractive index n_i to the incident electric dipole field is described by the complex coefficient $\alpha_{K,J=1}^{(E)}(i)$, the Mie coefficient [2], in the long-wavelength limit $Ka \ll 1$:

$$\alpha_{K,1}^{(E)}(i) \simeq i\,\frac{2}{3}\left(\frac{n_i^2 - 1}{n_i^2 + 2}\right)(Ka_i)^3 \quad (Ka_i \ll 1) \quad (9.5.31)$$

The complex response coefficient of the sphere results in a phase shift in the scattered field, so that inhomogeneous waves arise in its proximity.

We have to determine the amplitude C_M in a self-consistent manner accounting for the multiple scattering processes between the spheres as well as those between the spheres and the surface. Let us first study the sphere–sphere interaction in order to demonstrate its localization by evaluating the scattered intensity from the probing sphere scanned in a plane above the sample sphere. Next we study the sphere–surface interaction on the basis of the angular spectrum representation to see the importance of high-spatial-frequency components in the optical near-field regime.

9.5.4.2. Sphere–Sphere Interaction

It is convenient to describe the illuminating evanescent wave in terms of the multipolar representation of the CM modes developed in Section 9.5.3. Taking the origin of the spherical representation at the center of the sphere $\mathbf{d}_i$ and including only the electric dipole components, we can describe the incident evanescent field as follows:

$$I_{K,M}^{(0,TE)}(\mathbf{d}_i) = B_M^{(\mu)}\, e^{i\mathbf{k}\cdot\mathbf{d}}\, \mathcal{D}_{1,M}^{(TM)}(\varphi_{IN},\, \beta + i\gamma,\, 0) \quad (9.5.32)$$

$$I_{K,M}^{(0,TM)}(\mathbf{d}_i) = iB_M^{(\mu)}\, e^{i\mathbf{k}\cdot\mathbf{d}}\, \mathcal{D}_{1,M}^{(TE)}(\varphi_{IN},\, \beta + i\gamma,\, 0) \quad (9.5.33)$$

Note that the symbols TE and TM are exchanged on the two sides of these equations. Here $B_M^{(\mu)}$ and $\mathbf{k}\cdot\mathbf{d}_i$ are, respectively, the angular spectrum coefficient and phase of the $\mathbf{F}_L$ mode with $J = 1$ given in Eq. (9.5.24) or Eq. (9.5.25). Then the first-order scattered field amplitude from one of the spheres is given by

$$C_M^{(\text{Direct})} = \alpha_{K,1}^{(E)}(i) I_{K,M}^{(0,\mu)}(\mathbf{d}_i) \qquad (\mu = TE, TM) \tag{9.5.34}$$

Using the electric dipole propagator in Eq. (9.4.89),

$$\mathbf{G}_{K;M',M}^{(\text{ED})}(\mathbf{R}) = \sqrt{6\pi}(-1)^{M'}\hat{\mathbf{e}}_{M'} \cdot \mathbf{A}_{K,1,-M'}^{(E,SC)}(\mathbf{R}) \qquad (M, M' = 0, \pm 1) \tag{9.5.35}$$

we can describe the interaction between spheres i and j as follows.

First we consider two successive scatterings of the illuminating field first by sphere j and then by sphere i. The scattered amplitude is given by

$$C_M^{(j \rightarrow i)} = \alpha_{K,1}^{(E)}(i) \sum_{M'} G_{K;M',M}^{(\text{ED})}(\mathbf{R}) \alpha_{K,1}^{(E)}(j) I_{K,M'}^{(0,\mu)}(\mathbf{d}_j) \tag{9.5.36}$$

The right-hand side of this equation can be read from right to left in direct correspondence to the basic processes. The higher order process is a multiple scattering between two spheres, which is written in the form of the scattering of a self-consistent field as

$$C_M^{(\text{Multiple})} = \alpha(i) \sum_{M'} \mathbf{G}_{K;M',M}^{(\text{ED})}(\mathbf{R}) \alpha(j) \sum_{M''} \mathbf{G}_{K;M'',M'}^{(\text{ED})}(-\mathbf{R}) C_{M''}^{(\text{Self-Consistent})} \tag{9.5.37}$$

This can also be read from right to left.

Summing up the scattering processes described above, we have a self-consistent relation for the total scattered field amplitudes:

$$C_M^{(\text{Self-Consistent})} = C_M^{(\text{Direct})} + C_M^{(j \rightarrow i)} + C_M^{(\text{Multiple})} \tag{9.5.38}$$

where the explicit form is

$$C_M^{(i)} = \alpha_{K,1}^{(E)}(i) I_{K,M}^{(0,\mu)}(\mathbf{d}_i) + \alpha_{K,1}^{(E)}(i) \sum_{M'} \mathbf{G}_{K;M',M}^{(\text{ED})}(\mathbf{R}) \alpha_{K,1}^{(E)}(j) I_{K,M'}^{(0,\mu)}(\mathbf{d}_j)$$

$$+ \alpha(i) \sum_{M'} \mathbf{G}_{K;M',M}^{(\text{ED})}(\mathbf{R}) \alpha(j) \sum_{M''} \mathbf{G}_{K;M'',M'}^{(\text{ED})}(-\mathbf{R}) C_{M''}^{(i)} \tag{9.5.39}$$

These calculations are summarized in Fig. 9.19.

The total scattered amplitude is given by

$$I(\mathbf{r}) \propto \int d\Omega_{\hat{\mathbf{r}}} \left| \tilde{\mathbf{A}}^{(SC)}(\mathbf{r}) \right|^2 \tag{9.5.40}$$

$$= \sum_M \left| C_M^{(i)} \right|^2 \left\{ \frac{1}{3} \left| h_2^{(1)}(Kr) \right|^2 + \frac{2}{3} \left| h_0^{(1)}(Kr) \right|^2 \right\} \tag{9.5.41}$$

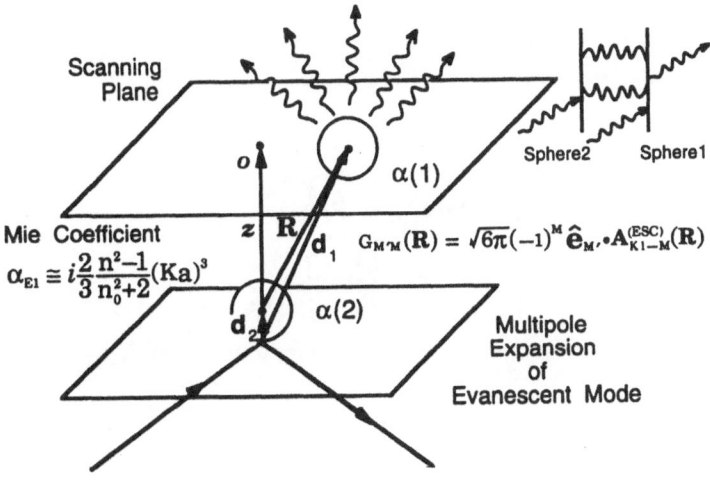

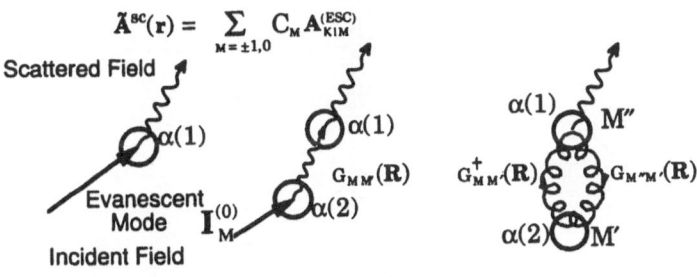

$$C_M^{(Direct)} = \alpha(1)I_M^{(0)}(\mathbf{d}_1) \qquad C_M^{(Sample\text{-}Probe)} = \alpha(1)\sum_{M'} G_{M'M}(\mathbf{R})\alpha(2)I_{M'}$$

$$C_M^{(Multiple)} = \alpha(1)\sum_{M'} G_{M'M}(\mathbf{R})\alpha(2)\sum_{M''} G_{M''M'}(\mathbf{R})C_{M''}^{(Selfconsistent)}$$

$$C_M^{(Selfconsistent)} = C_M^{(Direct)} + C_M^{(Sample\text{-}Probe)} + C_M^{(Multiple)} \Rightarrow \text{Selfconsistent Equation}$$

Figure 9.19. The basic procedures used in calculating the electric dipole interaction between two closely spaced dielectric spheres.

Since the Mie coefficient is considered to be a small parameter in the perturbation series, only the several lowest terms contribute to the total scattering intensity in the iterative calculation of the self-consistent coefficient.

The total scattered amplitude is shown in Figs. 9.19 and 9.20 as a function of the position of the scanned probe tip relative to the fixed sphere representing the

TE θ = 70 degree kz=0.02

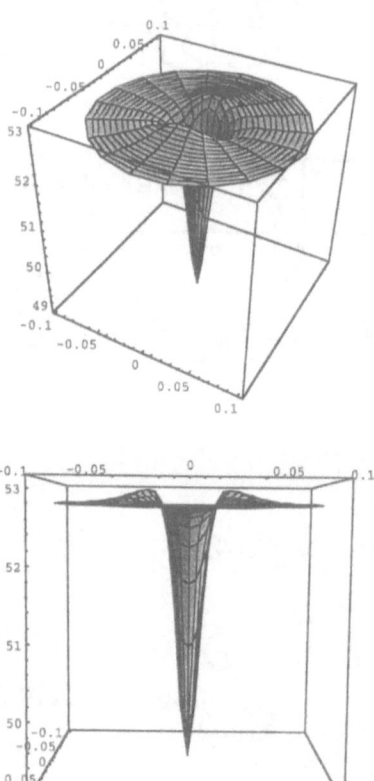

Figure 9.20. Total scattered intensity from a probing sphere scanned above the sample sphere on a plane parallel to the substrate. Sample–probe interaction in the near-field regime, $kz = 0.02$, and *TE* illumination are assumed. The center of the probing sphere sweeps the horizontal plane as indicated.

object for *TE* and *TM* illumination, respectively, for $a_1 = a_2 = a \leq z$, $Kz = 0.02$, and illumination by evanescent waves at an incident angle in the dielectric of 70 deg. Only terms in C_M up to the second order with respect to $(Ka)^3$ are significant in the near-field regime of $Ka \leq 0.02$.

One can see that the scattered amplitude or the interaction between sample and probe tip is highly localized in the range corresponding to the distance between two spheres. This is because the term related to $h_2^{(1)}(Kr)$ is significant only in the

TM $\theta = 70$ degree kz=0.02

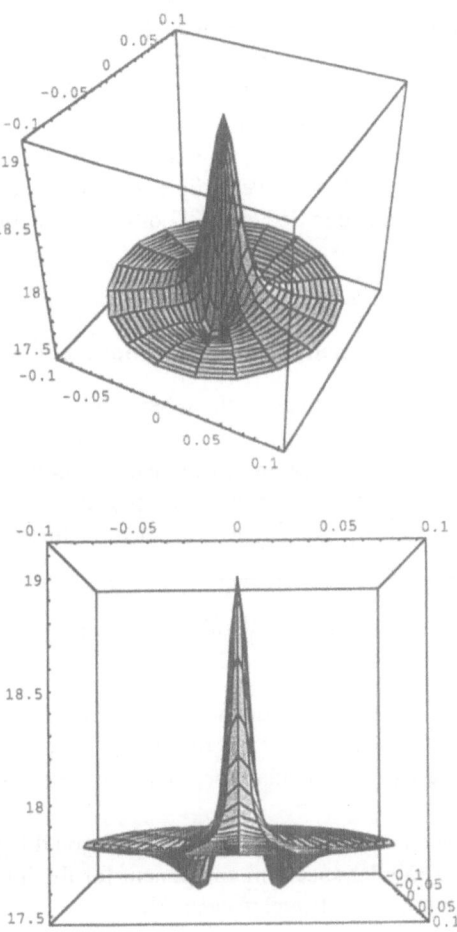

Figure 9.21.   Same as Fig. 9.20, for *TM* illumination.

near-field regime. This simulates the fundamental NOM process where a probe tip with a radius of curvature as small as the specimen is scanned on a plane as close as the size of the specimen. As the size of the probe tip increases, the center of the spherical representation moves far from the specimen, so that the higher spatial frequency components fall off rapidly. In contrast, as the size of the probe tip a decreases, the Mie coefficient decreases by the third power of Ka, so that the total intensity of the scattered field decreases very rapidly. This confirms our experience that a good NOM image is obtained when a probe tip as small as the specimen is scanned in a proximity characterized also by the size of the specimen. Compared with the scattering potential consideration in Section 8.3.2 of Chapter 8, the results calculated here are in good agreement with those estimated intuitively. This also confirms the potential of the model description of NOM in the regime of the near-field detection of a near-field event.

It should be emphasized that since we have just calculated the total scattered intensity by a probing sphere, no waveguiding effect has been considered in the above calculations. As studied in Section 8.3.2 of Chapter 8, the coupling process of a probe tip with an optical waveguide significantly alters the NOM image.

9.5.4.3. Origin of Localization in the Sphere–Sphere Interaction

As shown numerically in the above, the sphere–sphere interaction is localized within a spatial range corresponding to the sphere–sphere distance in the near-field regime. The origin of the localization can be understood clearly in terms of the angular spectrum representation of the dipole propagator.

The angular spectrum representation of the electric dipole propagator is given from Eq. (9.4.87) by

$$G^{(ED)}_{K;M',M}(\mathbf{r}_2 - \mathbf{r}_1) = \sqrt{6\pi}\, \frac{1}{2\pi} \int_C d\Omega_{\hat{s}}\, \mathbf{Y}^{(E)}_{1,M'}\!\cdot\!\hat{\mathbf{e}}_M e^{iK\hat{s}\cdot\mathbf{R}} \qquad (M, M' = 0, \pm 1) \qquad (9.5.42)$$

Since the inhomogeneous waves corresponding to the imaginary part of the contour $C^{\pm}$ are significant in the near-field regime of $Kz \ll 1$, it is meaningful to discuss the amplitude $\mathbf{Y}^{(E)}_{1,M'}\!\cdot\!\hat{\mathbf{e}}_M$ in the angular spectrum together with the rapid decay $\exp(iK\hat{s}\cdot\mathbf{R})$.

In Fig. 9.22, we show the magnitude of the integrand $|\mathbf{Y}^{(E)}_{1,M'}\!\cdot\!\hat{\mathbf{e}}_M \exp(iK\hat{s}\cdot\mathbf{R})|$ corresponding to the angular spectrum component for the imaginary part of the contour C^+ as a function of the lateral wavenumber,

$$K_{EV} = K\hat{s}\cdot(\hat{\mathbf{x}} + \hat{\mathbf{y}}) = K \cosh(\mathrm{Im}\{\alpha\})$$

for several values of the sphere–sphere distance $d = |\mathbf{R}\cdot\hat{\mathbf{z}}|$ in the near-field regime, $Kd \ll 1$. In Fig. 9.22 we use $\rho = Kd$ and $\xi = \cosh(\mathrm{Im}\{\alpha\})$. Note that $K = \omega/c$ is the wavenumber in vacuum, and the penetration depth is given by the inverse of the normal wavenumber,

$$|K_z| = K\hat{s}\cdot\hat{\mathbf{z}}$$

$$= |K \sinh(\mathrm{Im}\{\alpha\})|$$

$$= |\sqrt{K^2 - K^2_{EV}}|$$

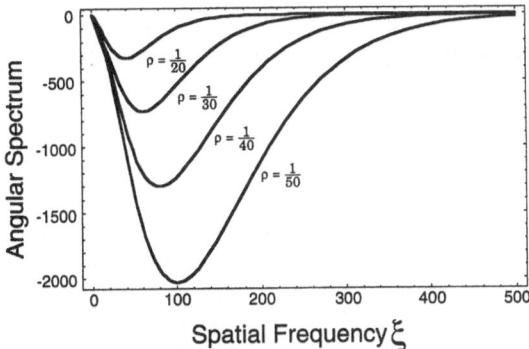

Figure 9.22. Angular spectrum of the electric dipole propagator for several different values of the propagation distance in the near-field regime, $Kr < 1$. Both the wavenumber $k_\parallel$ at the peak and the spectral width (FWHM) are nearly equal to the inverse of the propagation distance r^{-1}.

As one can see in Fig. 9.22, the angular spectrum takes its maximum at the lateral wavenumber $K_{EV} = K \cosh(\mathrm{Im}\{\alpha\})$ approximately equal to the inverse of the sphere–sphere distance, i.e., $2/d$, or $\xi = 2/\rho$. That is, the sphere–sphere dipole interaction is dominated by the angular spectrum component with lateral wavenumber approximated by the inverse of the distance of the interacting electric dipoles. The closer one approaches the spheres, the higher is the lateral wavenumber which dominates the interaction and also the higher is the strength of the interaction. One can understand why the amplitude diverges in the short-distance limit by recalling that the field intensity has a singularity around a point charge in the static limit of the Coulomb interaction.

One can also see in Fig. 9.22 that the spectral width becomes broader as the sphere–sphere distance decreases. The full-width at half-maximum (FWHM) is approximated also by the inverse of the sphere–sphere distance, i.e., $2/d$. It can be numerically shown that even if one cuts off the spectrum at the FWHM frequencies and reconstructs the electric dipole interaction by integrating over the dominant parts, one can reproduce the principal part of the electric dipole interaction, which shows the radial dependence with third power of the inverse distance $1/d$, corresponding mainly to the Hankel function $h_2^{(1)}$. Then the omitted part corresponds to the far-field component described by the Hankel function $h_0^{(1)}$.

In conclusion, the electromagnetic interaction between spheres in the near-field regime is dominated by the interaction channel via short-range terms with very large lateral wavenumber K_{EV} which is approximated by the inverse of the sphere–sphere distance, i.e., $2/d$. The broadening of the angular spectrum in turn implies that the lateral range of the corresponding interaction is also localized due to the interference of the evanescent waves spreading over a wide band. A numerical demonstration is left for the following section, where we deal with the sphere–surface interaction in terms of the angular spectrum and evanescent waves.

9.5.4.4. Sphere–Surface Interaction

The interaction process between the dielectric sphere and the surface can be described in a similar way as we derived the vector multipole propagator in Section 9.4.4. We have already prepared the basis for representing the scattered field from the sphere in its upper (probe side) and lower (surface side) half-spaces in terms of the angular spectrum, i.e., plane waves with both homogeneous and evanescent components. What we have to do next is insert the reflection of the scattered field at the dielectric surface, which adds up to the incident field on the sphere itself. Let us refer this term as the "reflected field." The multiple scattering and reflection processes provide the complete story of the sphere–surface interaction. Describing the plane waves in terms of TE and TM modes, the reflection at the surface is evaluated simply by referring to the R mode of the CM-mode description. This is equivalent to considering the Fresnel relation coming from the boundary conditions at the planar dielectric surface.

As we consider the sphere–surface interaction in the lower half-space of the sphere, we take the integration contour C^- (the z axis is taken in the direction normal to the surface). The position vector $\mathbf{d}$ is pointing to the center of the dielectric sphere with respect to the frame of reference (Fig. 9.18). The angular spectrum component of the vector multipole propagator is then written, following the definitions (9.4.81) and (9.4.82) in Section 9.4.4, by

$$\mathcal{G}^{(\lambda,\hat{s})}_{K;J,M}(\mathbf{r}) = e^{iK\hat{s}\cdot\mathbf{r}}\hat{\mathbf{e}}_{\mu}(\hat{s})\cdot\mathbf{Y}^{(\lambda)}_{J,M}(\hat{s})$$

$$\left[\mathcal{G}^{(\lambda',\hat{s}')}_{K;J',M'}(\mathbf{r})\right]^{\dagger} = \left[\mathbf{Y}^{(\lambda')}_{J',M'}(\hat{s}')\right]^{\dagger}\cdot\hat{\mathbf{e}}_{\mu}(\hat{s}')\,e^{iK\hat{s}'\cdot\mathbf{r}}$$

where λ and λ' represent E- or M-type vector multipoles, μ represents TE or TM polarizations, and $\hat{s}$ and $\hat{s}'$ are the directional unit vectors of wave propagation. Substituting $\mathbf{r} = -\mathbf{d}$ into $\mathcal{G}^{(\lambda,\hat{s})}_{K;J,M}(\mathbf{r})$, we obtain the incident wave onto the dielectric surface coming from the vector multipole component $\mathbf{Y}^{(\lambda)}_{J,M}(\hat{s})$ of the scattered field due to the sphere via a plane wave with polarization vector $\hat{\mathbf{e}}_{\mu}(\hat{s})$. Then the incident plane wave consisting of the R mode in the CM description produces a reflected field from the surface. Let $R^{(\mu)}_{K}$ be the reflection coefficient for the μ-polarized plane wave with wavenumber K. Then the reflected field is described by

$$\mathbf{A}^{(\lambda,REF)}_{K;J,M}(\mathbf{R}) = \frac{1}{2\pi}\int_{C^-} d\Omega_{\hat{s}}\sum_{\mu}^{TE,TM}\hat{\mathbf{e}}_{\mu}(\hat{s}^{(R)})e^{iK\hat{s}^{(R)}\cdot\mathbf{R}}\,R^{(\mu)}_{K}\,\mathcal{G}^{(\lambda,\hat{s})}_{K;J,M}(-\mathbf{d}) \qquad (9.5.43)$$

This can be read from right to left that the scattered field is propagated from the sphere to the surface, $\mathcal{G}_{K;J,M}^{(\lambda,\hat{s})}(-\mathbf{d})$ (e.g., $\mathbf{r} = -\mathbf{d}$), and reflected by the surface, $R_K^{(\mu)}$, and then propagated as a vector plane wave, $\hat{\mathbf{e}}_{\mu}(\hat{s}^{(R)}) \exp(iK\hat{s}^{(R)}\cdot\mathbf{R})$, to an arbitrary point $\mathbf{R}$ with respect to the surface coordinate with direction of propagation $\hat{s}^{(R)}$. Here $\hat{s}^{(R)}$ and $\hat{s}$ differ only in the sign of their z components since the former is the reflected field of the latter.

The reflected field from the surface then propagates back to the dielectric sphere as a μ-polarized plane wave and produces a vector multipole incident field on the sphere, $A_{K;J,M}^{(\lambda',IN)}(\mathbf{r})$. This process is described by projecting the reflected field $A_{K;J,M}^{(\lambda,REF)}(\mathbf{R})$ with $\mathbf{R} = \mathbf{d}$ onto the vector multipole field $[\mathbf{Y}_{J',M'}^{(\lambda')}(\hat{s})]^{\dagger}$ which is related to the incident field on the sphere. Then the round-trip interaction of the sphere via reflection from the surface is written by summing up the succeeding processes over polarizations $\mu = TE, TM$ and by integrating over the contour C^-. As a result we can define the round-trip interaction propagator $Q_{K;J,M,J',M'}^{(\lambda',\lambda)}(\mathbf{d})$:

$$Q_{K;J,M,J',M'}^{(\lambda',\lambda)}(\mathbf{d}) = \frac{1}{2\pi} \int_{C^-} d\Omega_{\hat{s}} \, 4\pi \sum_{\mu}^{TE,TM} \left[\mathcal{G}_{K;J',M'}^{(\lambda',\hat{s}^{(R)})}(\mathbf{d}) \right]^{\dagger} R_K^{(\mu)} \, \mathcal{G}_{K;J,M}^{(\lambda,\hat{s})}(-\mathbf{d}) \quad (9.5.44)$$

$$= \frac{1}{2\pi} \int_{C^-} d\Omega_{\hat{s}} \, 4\pi \sum_{\mu}^{TE,TM} \left[\mathbf{Y}_{J',M'}^{(\lambda')}(\hat{s}^{(R)}) \right]^{\dagger} \cdot \hat{\mathbf{e}}_{\mu}(\hat{s}^{(R)}) \, e^{+iK\hat{s}^{(R)}\cdot\mathbf{d}}$$

$$\times R_K^{(\mu)} \, e^{-iK\hat{s}\cdot\mathbf{d}} \, \hat{\mathbf{e}}_{\mu}(\hat{s}) \cdot \mathbf{Y}_{J,M}^{(\lambda)}(\hat{s}) \quad (9.5.45)$$

The procedure described above is explained as follows; as one reads the equation from right to left (see Fig. 9.23)

Pick up the $\lambda = E$ or M multipole field from the scattered field
$\Rightarrow$ Project the multipole field onto $\mu = TE$ and TM plane waves
$\Rightarrow$ Propagate as TE or TM plane wave from sphere to surface
$\Rightarrow$ Reflect by surface as R mode in CM description
$\Rightarrow$ Propagate as TE or TM plane wave from surface to sphere
$\Rightarrow$ Project plane waves onto the $\lambda' = E$ or M incident multipole field

Based on the description of the vector field and interactions in the previous section, we can easily understand the fundamental processes involved in the sphere–surface interaction. We can find the round-trip interaction propagator by arranging successive terms from right to left and summing up the polarizations and integrating over the contour C^-. The procedure is now almost intuitive.

Since we are interested in the near-field regime where the electric dipole interaction is predominant, we set $J = J' = 1$, $\lambda = \lambda' = E$, and $Q_{K;M,M'}^{(ED)}(\mathbf{d}) \equiv Q_{K;1,M,1,M'}^{(E,E)}(\mathbf{d})$. Let us calculate the total scattered field from the sphere in the form

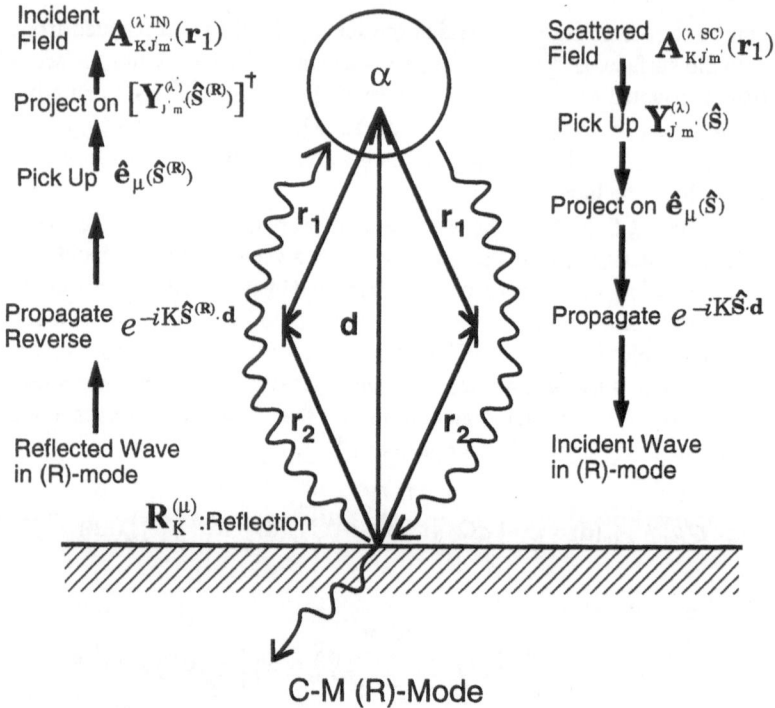

C-M (R)-Mode

Figure 9.23. Fundamental processes involved in the angular spectrum representation of the round-trip propagator involved in a dielectric sphere–surface interaction.

$$\tilde{\mathbf{A}}_K^{(SC)}(\mathbf{r}) \equiv \sum_{M=0,\pm1} C_{K,M} \mathbf{A}_{K,1,M}^{(E,SC)}(\mathbf{r})$$

in the lower (surface side) half-space of the sphere.

First, the direct scattering of an illuminating evanescent wave by the sphere gives rise to the "direct" scattering amplitude

$$C_{K,M}^{(\text{Direct})} = \alpha_{K,1}^{(E)} I_{K,M}^{(0,\mu)}(\mathbf{d}) \qquad (\mu = TE, TM) \tag{9.5.46}$$

The "multiple" interaction of the sphere with the dielectric surface contributes to the scattered field with a coefficient

$$C_{K,M}^{(\text{Multiple})} = \alpha_{K,1}^{(E)} \sum_{M'=0,\pm1} Q_{K;M,M'}^{(ED)}(\mathbf{d}) C_{K,M'}^{(\text{Self-Consistent})} \tag{9.5.47}$$

Then the overall interaction of a single sphere with the dielectric surface is written in the form of the self-consistent equation

$$C_{K,M}^{(\text{Self-Consistent})} = \alpha_{K,1}^{(E)} \left[I_{K,M}^{(0,\mu)}(\mathbf{d}) + \sum_{M'=0,\pm 1} Q_{K;M,M'}^{(ED)}(\mathbf{d}) C_{K,M'}^{(\text{Self-Consistent})} \right] \quad (9.5.48)$$

The explicit form of the round-trip electric dipole interaction propagator $Q_{K;M,M'}^{(ED)}(\mathbf{d})$ is given by integrating over β and using the new parameter $\xi = -\cos\alpha$,

$$Q_{K;M,M'}^{(ED)}(\mathbf{d}) = \frac{3}{4} \int_{C_1} d\xi \left(R_K^{(TE)} - R_K^{(TM)} \xi^2 \right) e^{+i2Kd\xi} \left(\delta_{M,+1}\delta_{M',+1} + \delta_{M,-1}\delta_{M',-1} \right)$$

$$+ \frac{3}{2} \int_{C_1} d\xi \left[\left(1 - \xi^2\right) R_K^{(TM)} \right] e^{+i2Kd\xi} \delta_{M,0}\delta_{M',0} \quad (9.5.49)$$

where the reflection coefficients for orthogonal polarizations $R_K^{(TE)}$ and $R_K^{(TM)}$ are calculated from the R-mode coefficients or equivalently Fresnel's relation as

$$R_K^{(TE)} \equiv \frac{n^2 - 1 + 2\xi^2 - 2\xi\sqrt{n^2 - 1 + \xi^2}}{1 - n^2} \quad (9.5.50)$$

$$R_K^{(TM)} \equiv \frac{n^2 - 1 + (n^4 + 1)\xi^2 - 2n^2\xi\sqrt{n^2 - 1 + \xi^2}}{(n^2 - 1)\left[(n^2 + 1)\xi^2 - 1\right]} \quad (9.5.51)$$

where n is the dielectric constant of the surface. The contour of integration C_1 corresponds to C^-.

9.5.4.5. Penetration Depth and Localization in the Sphere–Surface Interaction

The angular spectrum representation of the sphere–surface interaction provides a very instructive way of understanding the fundamental processes dominating the interaction, i.e., the penetration depth and localization.

The sphere–surface interaction involves the following three different types of interaction channels when described in terms of the angular spectrum (see Fig. 9.24):

(a) Both *incident* and *reflected* waves are *homogeneous*, as is the *transmitted* wave, so that the direction of transmission is within the angle of total internal reflection, i.e., smaller than the critical angle θ_c: $\pi/2 \leq \text{Re}\{\alpha\} \leq \pi$, $\text{Im}\{\alpha\} = 0$.

(b) Both *incident* and *reflected* waves are *inhomogeneous* (evanescent) with wavenumber in the range of the propagating waves in the dielectric

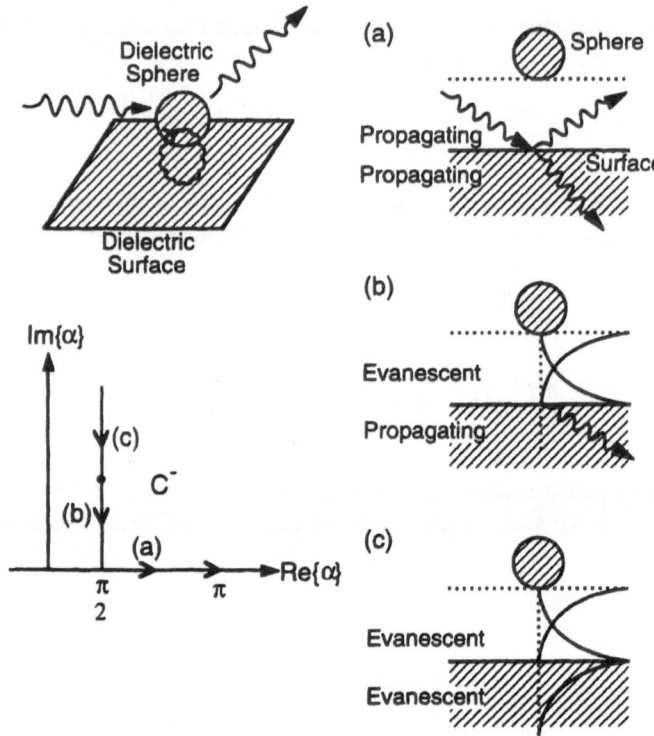

Figure 9.24. Three processes involved in dielectric sphere–surface interaction: (a) Homogeneous incident, reflected, and transmitted waves corresponding to region (a) of the contour of α. (b) Evanescent incident and reflected waves and homogeneous transmitted wave. (c) Evanescent incident, reflected, and transmitted waves. The contribution from process (c) is dominant in the near-field limit.

medium, so that the *transmitted* wave remains *homogeneous* and directed at an angle larger than the critical angle θ_c: $\text{Re}\{\alpha\} = \pi/2$, $0 < \text{Im}\{\alpha\} \leq \text{arccosh}(n)$.

(c) Both *incident* and *reflected* waves are *inhomogeneous* (evanescent) with wavenumber exceeding any propagating wave in the dielectric medium, so that the *transmitted* wave is also *inhomogeneous* (evanescent) and is localized within a thin layer inside the dielectric. In other words, the transmitted plus reflected field is localized on the planar dielectric surface and behaves like a surface wave: $\text{Re}\{\alpha\} = \pi/2$, $\text{arccosh}(n) < \text{Im}\{\alpha\} \leq +\infty$.

We will see that in the near-field limit the third channel via fully evanescent waves dominates the reflection or round-trip interaction of the sphere via the planar dielectric surface.

Let us evaluate the strength of the angular spectrum components involved in the round-trip interaction propagator, which corresponds to the integrand in $Q_{K;M,M'}^{(ED)}(\mathbf{d})$, as a function of normalized sphere–surface distance Kd. A numerical result is shown in Fig. 9.25 for the near-field case of $Kd = 0.1$. One can see that the angular spectrum takes its maximum at the spatial frequency component $|\xi| = 1/Kd \gg 1$ with peak amplitude proportional to $(Kd)^2$. That is, for the near-zone case of $Kd \ll 1$, the dominant channel of the sphere–surface interaction corresponds to type (c) above, as described in Fig. 9.24c. This explains why a short penetration depth and high localization are achieved in the near-field regime of the sphere–surface interaction.

The integral can then be approximated by taking the contour corresponding to the evanescent part around the spatial frequency $K_{EV} = 1/d$. By keeping the higher order terms with respect to ξ, we obtain the sphere–surface interaction propagator for the near-zone case as

$$Q_{K;M,M'}^{(ED)}(\mathbf{d}) \cong \frac{1}{2}\left(\frac{n^2-1}{n^2+1}\right)\left[-i\frac{3}{(2Kd)^3}\right]\varepsilon_M\delta_{M,M'} \tag{9.5.52}$$

$$\varepsilon_M = \begin{cases} 2, & M = 0 \\ 1, & M = \pm 1 \end{cases}$$

Thus the approximate results are in good agreement with those obtained by numerical integration of Eq. (9.5.49) for $Kd \ll 1$.

Note that the approximate form of $Q_{K;M,M'}^{(ED)}(\mathbf{d})$ includes the asymptotic form of the spherical Hankel function $h_2^{(1)}(2Kd) \sim -3i/(2Kd)^3$ and the screening factor $(n^2-1)/(n^2+1)$ corresponding to an image dipole field induced by the dielectric sphere. This implies that the sphere–surface interaction in the near-field limit describes the interaction between the induced dipole on the dielectric sphere with its image. This confirms in part the validity of the procedures used in deriving $Q_{K;M,M'}^{(ED)}(\mathbf{d})$.

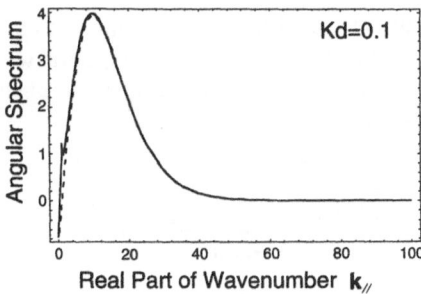

Figure 9.25. Angular spectrum of the round-trip propagator in the near-field regime.

With this near-field approximation ($Kd \ll 1$), the scattered amplitude C_M is described as a perturbation series with respect to $\alpha_{K,1}^{(E)} Q_{K;M,M'}^{(ED)}(\mathbf{d})$, which is given with the Mie coefficient for $Ka \ll 1$ defined in Eq. (9.5.31) by

$$\alpha_{K,1}^{(E)} Q_{K;M,M'}^{(ED)}(\mathbf{d}) \cong \frac{(n^2 - 1)(n_1^2 - 1)}{(n^2 + 1)(n_1^2 + 2)} \left(\frac{Ka}{2Kd} \right)^3 \varepsilon_M \delta_{M,M'} \qquad (9.5.53)$$

For example, consider the case of a dielectric sphere and a surface with refractive indices $n_1 = n = 1.5$,

$$\alpha_{K,1}^{(E)} Q_{K;M,M'}^{(ED)}(\mathbf{d}) \sim \frac{1}{80} \varepsilon_M \delta_{M,M'} \qquad (9.5.54)$$

Substituting the approximate form into the self-consistent equation (9.5.49), we get

$$C_{K,M} = \frac{\alpha_{K,1}^{(E)} I_{K,M}^{(0,\mu)}(\mathbf{d})}{1 - \overline{Q}(Kd)\varepsilon_M} \qquad (9.5.55)$$

where

$$\overline{Q}(\rho_0) = \frac{(n^2 - 1)(n_1^2 - 1)}{(n^2 + 1)(n_1^2 + 2)} \left(\frac{Ka}{2Kd} \right)^3 \qquad (9.5.56)$$

In conclusion, we obtain the scattered field from the sphere with renormalized sphere–surface interaction as

$$\tilde{\mathbf{A}}_K^{(SC)}(\mathbf{r}) = \sum_{M=0,\pm 1} \frac{\alpha_{K,1}^{(E)} I_{K,M}^{(0,\mu)}(\mathbf{d})}{1 - \overline{Q}(Kd)\varepsilon_M} \mathbf{A}_{K,1,M}^{(E,SC)}(\mathbf{r}) \qquad (9.5.57)$$

Here we can examine the lateral intensity distribution of the reflected field as a measure of localization relevant to the sphere–surface interaction in the near-field regime. We have seen that the round-trip interaction is dominated by the evanescent waves with lateral spatial frequency around $K_{EV} = 1/d$. Then the spectral width is considered as a measure of lateral localization, similar to the case of the sphere–sphere interaction. Since Fig. 9.25 implies that the spectral width is also of the order of $1/d$, we can expect that the lateral localization is also in the range of the sphere–surface distance d.

The total reflected field can be written in the form

$$\tilde{\mathbf{A}}_K^{(REF)}(\mathbf{R}) \equiv \sum_{M=0,\pm 1} C_{K,M} \mathbf{A}_{K,1,M}^{(E,REF)}(\mathbf{R}) \qquad (9.5.58)$$

by using the same coefficient as for the total scattered field and the basis of the reflected field in the electric dipole case given by

$$\mathbf{A}_{K;1,M}^{(E,REF)}(\mathbf{R}) = \frac{1}{2\pi} \int_{C^-} d\Omega_{\hat{\mathbf{s}}} \sum_{\mu}^{TE,TM} \hat{\mathbf{e}}_\mu(\hat{\mathbf{s}}^{(R)}) e^{iK\hat{\mathbf{s}}^{(R)}\cdot\mathbf{R}} R_K^{(\mu)} \, \mathcal{G}_{K;1,M}^{(E,\hat{\mathbf{s}})}(-\mathbf{d}) \tag{9.5.59}$$

$$= \frac{1}{2\pi} \int_{C^-} d\Omega_{\hat{\mathbf{s}}} \sum_{\mu}^{TE,TM} \hat{\mathbf{e}}_\mu(\hat{\mathbf{s}}^{(R)}) \, e^{iK\hat{\mathbf{s}}^{(R)}\cdot\mathbf{R}}$$

$$\times R_K^{(\mu)} \, e^{-iK\hat{\mathbf{s}}\cdot\mathbf{d}} \hat{\mathbf{e}}_\mu(\hat{\mathbf{s}}) \cdot \mathbf{Y}_{1,M}^{(E)}(\hat{\mathbf{s}}) \tag{9.5.60}$$

$$= \int_{C^-} \sin\alpha\, d\alpha \sum_{\mu}^{TE,TM} \frac{1}{2\pi} \int_{-\pi}^{+\pi} d\beta\, \hat{\mathbf{e}}_\mu(\hat{\mathbf{s}}^{(R)}) e^{iK\hat{\mathbf{s}}^{(R)}\cdot(\mathbf{R}+\mathbf{d})}$$

$$\times R_K^{(\mu)} \, \hat{\mathbf{e}}_\mu(\hat{\mathbf{s}}) \cdot \mathbf{Y}_{1,M}^{(E)}(\hat{\mathbf{s}}) \tag{9.5.61}$$

where $-iK\hat{\mathbf{s}}\cdot\mathbf{d} = iK\hat{\mathbf{s}}^{(R)}\cdot\mathbf{d}$ since $\mathbf{d} = d\hat{\mathbf{z}}$ and $\{\hat{\mathbf{s}}^{(R)}\}_z = -\{\hat{\mathbf{s}}\}_z$.

As discussed in the previous section, since the sphere–surface system exhibits local cylindrical symmetry, the field distribution relevant to this system can be expressed in a convenient form using the vector Bessel mode functions. Utilizing the transformation formulas summarized in Section 9.4.7, we can rewrite the M component of the reflected field using a procedure similar to that for the spherical-to-cylindrical mode transformation described in Section 9.4.6.

We can utilize the following formulas: the cylindrical representation of a vector plane wave

$$\hat{\mathbf{e}}_\mu(\hat{\mathbf{s}}) e^{iK\hat{\mathbf{s}}\cdot\mathbf{r}} = \sum_{M'} e^{-iM'\beta} \mathbf{J}_{K,\alpha,M'}^{(\mu)}(\mathbf{r})$$

or its reverse

$$\mathbf{J}_{K,\alpha,M}^{(\mu)} = \frac{1}{2\pi} \int_{-\pi}^{+\pi} d\beta\, \hat{\mathbf{e}}_\mu(\hat{\mathbf{s}}) e^{iM\beta} e^{iK\hat{\mathbf{s}}\cdot\mathbf{r}}$$

and the spherical-to-planar transformation coefficients $\langle \mu, \alpha, \beta | E, J, M \rangle$,

$$\hat{\mathbf{e}}_{TE}(\hat{\mathbf{s}}) \cdot \mathbf{Y}_{J,M}^{(E)}(\hat{\mathbf{s}}) = \sqrt{\frac{2J+1}{8\pi}} \left[\mathcal{D}_{J,M}^{(TM)}(\beta, \alpha) \right]^*$$

$$= \sqrt{\frac{2J+1}{8\pi}} \left[d_{J,M}^{(TM)}(\alpha) \right]^* e^{iM\beta} \tag{9.5.62}$$

$$\hat{e}_{TM}(\hat{s}) \cdot Y_{J,M}^{(E)}(\hat{s}) = \sqrt{\frac{2J+1}{8\pi}} \left[\mathcal{D}_{J,M}^{(TE)}(\beta, \alpha) \right]^*$$

$$= \sqrt{\frac{2J+1}{8\pi}} \left[d_{J,M}^{(TE)}(\alpha) \right]^* e^{iM\beta} \qquad (9.5.63)$$

with explicit expression of the rotation matrix elements in terms of the reduced matrix,

$$d_{J,M}^{(TE)}(\alpha) = \left(-\frac{1}{\sqrt{2}} \right) \left[d_{M,+1}^{(J)}(\alpha) - d_{M,-1}^{(J)}(\alpha) \right]$$

$$d_{J,M}^{(TM)}(\alpha) = \left(+\frac{i}{\sqrt{2}} \right) \left[d_{M,+1}^{(J)}(\alpha) + d_{M,-1}^{(J)}(\alpha) \right]$$

It might be helpful to show the explicit form of the reduced rotation matrix for $J = 1$, with rows and columns in the order of $(M, M' = +1, 0, -1)$

$$d_{M,M'}^{(1)}(\alpha) = \begin{bmatrix} \dfrac{1+\cos\alpha}{2} & \dfrac{-\sin\alpha}{\sqrt{2}} & \dfrac{1-\cos\alpha}{2} \\[2mm] \dfrac{\sin\alpha}{\sqrt{2}} & \cos\alpha & \dfrac{-\sin\alpha}{\sqrt{2}} \\[2mm] \dfrac{1-\cos\alpha}{2} & \dfrac{\sin\alpha}{\sqrt{2}} & \dfrac{1+\cos\alpha}{2} \end{bmatrix} \qquad (9.5.64)$$

Then a straightforward calculation yields

$$A_{K;1,M}^{(E,REF)}(\mathbf{R}) = \sum_{\mu}^{TE,TM} \int_{C^-} \sin\alpha \, d\alpha \, R_K^{(\mu)} \langle \mu, \alpha, \beta | E, 1, M \rangle J_{K,\pi-\alpha,M}^{(\mu)}(\mathbf{R}+\mathbf{d})$$

$$= -\frac{i}{4} \sqrt{\frac{3}{\pi}} \int_{C^-} \sin\alpha \, d\alpha \, \{ R_K^{(TE)}(\delta_{M,+1} + \delta_{M,-1}) J_{K,\pi-\alpha,M}^{(\mu)}(\mathbf{R}+\mathbf{d})$$

$$+ \cos\alpha \, R_K^{(TE)} \left(\delta_{M,+1} + \frac{\sqrt{2}\sin\alpha}{\cos\alpha} \delta_{M,0} - \delta_{M,-1} \right) J_{K,\pi-\alpha,M}^{(\mu)}(\mathbf{R}+\mathbf{d}) \}$$

$$(9.5.65)$$

This integral is a well-known type (sometimes referred to as the Sommerfeld integral with a certain transform) and can be analytically described in terms of the hypergeometric function $F(\alpha, \beta, \gamma, z)$ [12].

Let us calculate the intensity distribution of a reflected near-field from a planar dielectric surface. For $K(d + Z) < 1$, the reflected field can be written

$$A_{K;1,M}^{(E,REF)}(\mathbf{R}) = -i \frac{\sqrt{3}}{8\pi} \left(\frac{n^2-1}{n^2+1}\right) \left[\left\{ \frac{1}{\rho_1^3} F\left(\frac{3}{2}, 2, 1; -\frac{\rho^2}{\rho_1^2}\right) \hat{\mathbf{e}}_{+1} \right. \right.$$

$$+ \frac{3}{2} \frac{\rho}{\rho_1^4} F\left(2, \frac{5}{2}, 2; -\frac{\rho^2}{\rho_1^2}\right) e^{+i\Theta} \hat{\mathbf{e}}_0$$

$$+ \left. \frac{3}{2} \frac{\rho^2}{\rho_1^5} F\left(\frac{5}{2}, 3, 3; -\frac{\rho^2}{\rho_1^2}\right) e^{+2i\Theta} \hat{\mathbf{e}}_{-1} \right\} \delta_{M,+1}$$

$$+ \left\{ -\frac{3}{2} \frac{\rho}{\rho_1^4} F\left(2, \frac{5}{2}, 2; -\frac{\rho^2}{\rho_1^2}\right) e^{-i\Theta} \hat{\mathbf{e}}_{+1} \right.$$

$$+ 2 \frac{1}{\rho_1^3} F\left(\frac{3}{2}, 2, 1; -\frac{\rho^2}{\rho_1^2}\right) \hat{\mathbf{e}}_0$$

$$+ \left. \frac{3}{2} \frac{\rho}{\rho_1^4} F\left(2, \frac{5}{2}, 2; -\frac{\rho^2}{\rho_1^2}\right) e^{+i\Theta} \hat{\mathbf{e}}_{-1} \right\} \delta_{M,0}$$

$$+ \left\{ +\frac{3}{2} \frac{\rho^2}{\rho_1^5} F\left(\frac{5}{2}, 3, 3; -\frac{\rho^2}{\rho_1^2}\right) e^{-i2\Theta} \hat{\mathbf{e}}_{+1} \right.$$

$$- \frac{3}{2} \frac{\rho}{\rho_1^4} F\left(2, \frac{5}{2}, 2; -\frac{\rho^2}{\rho_1^2}\right) e^{-i\Theta} \hat{\mathbf{e}}_0$$

$$+ \left. \left. \frac{1}{\rho_1^3} F\left(\frac{3}{2}, 2, 1; -\frac{\rho^2}{\rho_1^2}\right) \hat{\mathbf{e}}_{-1} \right\} \delta_{M,-1} \right] \tag{9.5.66}$$

Here cylindrical coordinates $\mathbf{R} = (R, \Theta, d+Z)$ are employed, and normalized radial and axial variables are used as $\rho = KR$ and $\rho_1 = K(d+Z)$.

By using the asymptotic form of the hypergeometric function for $KR \sim 0$, we write the reflected field in the near-field regime as

$$A_{K;1,M}^{(E,REF)}(\mathbf{R}) \overset{KR\to 0}{\sim} -i \sqrt{\frac{3}{8\pi}} \left(\frac{n^2-1}{n^2+1}\right) \frac{1}{(Kr)^3} \varepsilon_M \hat{\mathbf{e}}_M \tag{9.5.67}$$

which shows that the reflected field in the near-field limit is that due to the screened image dipole. The asymptotic form for $\rho \sim \infty$ is in turn given by

$$A_{K;1,M}^{(E,REF)}(\mathbf{R}) \overset{KR\to\infty}{\sim} i \frac{1}{2} \sqrt{\frac{3}{8\pi}} \left(\frac{n^2-1}{n^2+1}\right) \frac{1}{K(d+Z)^3} \left[\left\{ +\hat{\mathbf{e}}_{+1} - 3e^{+2i\Theta} \hat{\mathbf{e}}_{-1} \right\} \delta_{M,+1} \right.$$

$$+ \left. \hat{\mathbf{e}}_0 \delta_{M,0} + \left\{ -3e^{-2i\Theta} \hat{\mathbf{e}}_{+1} + \hat{\mathbf{e}}_{-1} \right\} \delta_{M,-1} \right] \tag{9.5.68}$$

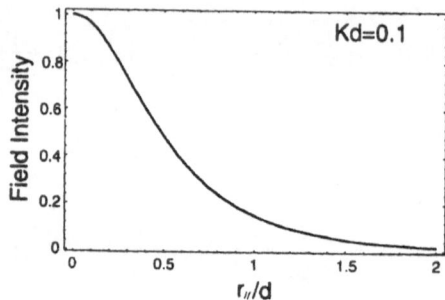

Figure 9.26. Intensity distribution of the reflected field on a sphere along the direction parallel to the surface in the near-field regime, $K\eta_\| \ll 1$.

which implies that the lateral intensity falls off rapidly as the third power of the height parameter $K(d + Z)$. This shows that the sphere–surface interaction is localized in the range of the sphere–surface distance $d + Z$. This is clearly shown in Figs. 9.26 and 9.27, which show the intensity distribution calculated for $K(d + Z) = 0.1$ and $\Theta = 0$.

In conclusion, by using the angular spectrum representation of the round-trip interaction propagator between the dielectric sphere and the surface, we have studied the character and localization of the reflected field in the near-field regime for the small-scatterer case. When the sphere–surface distance is in the subwave-

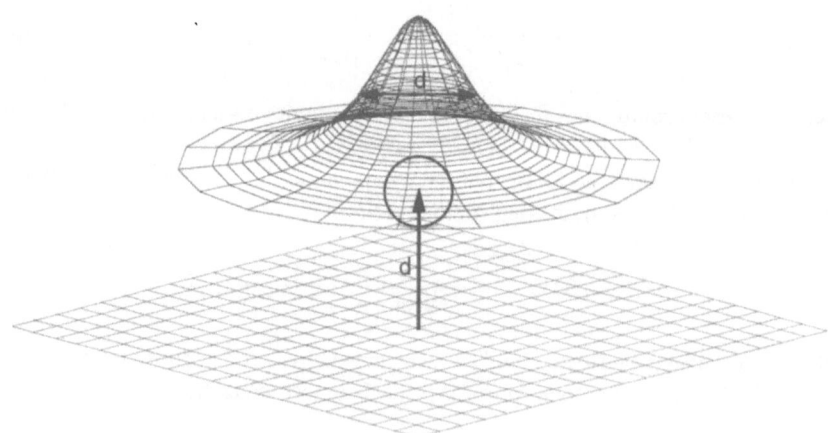

Figure 9.27. Three-dimensional display of Fig. 9.26. The dimension of localization of the reflected field corresponds to the distance d between the sphere center and the surface, which corresponds to the spectral width of the angular spectrum of the round-trip propagator.

length region $(Kd \ll 1)$, the spatial frequency $1/d$ is dominant and the lateral localization is also in the range of the sphere–surface distance d.

For the case of $Kd < 1$ and both n and $n_1 \sim 1.5$, multiple reflection can be neglected and the scattered field into the lower half-space (dielectric side) of the sphere is approximated by

$$\tilde{A}_K^{(SC)}(\mathbf{r}) \cong \alpha_{K,1}^{(E)} \sum_{M=0,\pm 1} I_{K,M}^{(0)} A_{K,1,M}^{(E,SC)}(\mathbf{r}) \tag{9.5.69}$$

which corresponds to the direct scattered field by the sphere. Also, in this situation, the reflected field by a planar dielectric surface is approximated by

$$\tilde{A}_K^{(REF)}(\mathbf{R}) \cong \alpha_{K,1}^{(E)} \sum_{M=0,\pm 1} I_{K,M}^{(0)} A_{K;1,M}^{(E,REF)}(\mathbf{R}) \tag{9.5.70}$$

The scattered and reflected fields are of the same order of magnitude since the incident field with very large lateral wavenumber $K_{EV} \sim 1/d$ produces all inhomogeneous transmitted and reflected waves, which can be considered as a kind of surface mode. It should be emphasized that such a confined field can be observed only by means of near-field detection.

9.6. REFERENCES

1. J. D. Jackson, *Classical Electrodynamics*, 2nd ed., Wiley, New York, 1975.
2. M. Born and E. Wolf, *Principles of Optics*, 3rd ed., Pergamon Press, Oxford, 1965.
3. A. Sommerfeld, *Partial Differential Equations in Physics*, Academic Press, New York, 1949.
4. P. M. Morse and H. Feshbach, *Methods of Theoretical Physics, Part 1*, McGraw-Hill, New York, 1953.
5. P. M. Morse and H. Feshbach, *Methods of Theoretical Physics, Part 2*, McGraw-Hill, New York, 1953.
6. C. K. Carniglia and L. Mandel, Quantization of evanescent electromagnetic waves, *Phys. Rev. D* 3: 280–296 (1971).
7. E. Wolf and M. Niet-Vesperinas, Analyticity of the angular spectrum amplitude of scattered fields and some of its consequence, *J. Opt. Soc. Am. A* 2: 886–890 (1985).
8. T. Inoue and H. Hori, Representations and transforms of vector fields as the basis of near-field optics, *Opt. Rev.* 3: 458–462 (1996).
9. T. Inoue and H. Hori, Theoretical treatment of electric and magnetic multipole radiation near a planar dielectric surface based on angular spectrum representation of vector field, *Opt. Rev.* 5: 295–302 (1998).
10. M. E. Rose, *Elementary Theory of Angular Momentum*, Wiley, New York, 1957.
11. W. J. Thompson, *Angular Momentum*, Wiley, New York, 1994.
12. L. D. Landau and E. M. Lifshitz, *Quantum Mechanics (Non-relativistic Theory)*, 3rd ed., Pergamon Press, Oxford, 1977.
13. V. B. Berestetskii, E. M. Lifshitz, and L. P. Pitaevskii, *Quantum Electrodynamics*, 2nd ed., Pergamon Press, Oxford, 1982.

14. C. Cohen-Tannoudji, J. Dupont-Roc, and G. Grynberg, *Photon and Atoms*, Wiley, New York, 1989.
15. A. Messiah, *Quantum Mechanics*, North-Holland, Amsterdam, 1961.

INDEX